Oilfield Chemistry

PE0298

Caili Dai · Fulin Zhao

Oilfield Chemistry

中国石油大学出版社
CHINA UNIVERSITY OF PETROLEUM PRESS

Springer

Caili Dai
School of Petroleum Engineering
China University of Petroleum (East China)
Qingdao, Shandong, China

Fulin Zhao
School of Petroleum Engineering
China University of Petroleum (East China)
Qingdao, Shandong, China

ISBN 978-981-13-2949-4 ISBN 978-981-13-2950-0 (eBook)
https://doi.org/10.1007/978-981-13-2950-0

Jointly published with China University of Petroleum Press, Qingdao, China

The print edition is not for sale in China Mainland. Customers from China Mainland please order the print book from: China University of Petroleum Press, Qingdao, Shandong, China.

ISBN of the China Mainland edition: 978-7-5636-6254-8

Library of Congress Control Number: 2018958357

This Springer imprint is published by the registered company Springer Nature Singapore Pte Ltd.
The registered company address is: 152 Beach Road, #21-01/04 Gateway East, Singapore 189721, Singapore

Foreword

The complex chemistry of the oilfield gives rise to numerous problems in the commercial exploitation of oil and gas fields. Addressing these problems is the domain of oilfield chemistry, beginning with the study of the chemical phenomena that give rise to these problems.

Because oilfield chemistry addresses chemical problems that arise during oil and gas production, it is of necessity an interdisciplinary field, spanning both chemistry and petroleum engineering. The oilfield chemist must understand both the process of oil and gas production and the severe limitations imposed by the sheer scope of oil and gas operations.

In the early years of oilfield exploitation, when many easily produced fields were newly developed, there was little need to be concerned with the production barriers created by poor understanding of the chemical phenomena involved. Today, however, as production continues to shift to more and more difficult fields, there is a critical need for understanding and controlling the chemistry of oilfield production. This reality makes it a high priority for the petroleum engineer to understand the chemical agents and chemical engineering technologies required for the commercial exploitation of the oilfields being produced today, and for the oilfield chemist to understand the production problems that require attention.

Oilfield chemistry is the youngest subdiscipline within petroleum science. It may be considered to consist of drilling chemistry, production chemistry, and transportation chemistry. These areas constitute the research targets of oilfield chemistry.

This English edition of "Oilfield Chemistry" is based on the 2010 Chinese second edition of "Oilfield Chemistry," edited by Fulin Zhao. It supplements that edition with new theories, new technologies, and new chemical agents from the intervening years. The book is divided into three parts, corresponding to drilling chemistry, production chemistry, and transportation chemistry. The chemical nature and principles of problems encountered during these operations are analyzed, the chemical systems used to address these problems are explained, and the mechanisms and synergistic effects of various chemical approaches are illustrated.

The Chinese first edition of "Oilfield Chemistry" is based on "Oil Production Chemistry," which was published in 1990 and received first prize for national excellence in teaching materials in 1991. After 10 years of effort and advances was the first edition published in July 2000. It has since been adopted as part of the curriculum by most of the petroleum-related programs in China. It has played an important role in broadening and deepening students' understanding of oil and gas chemistry as well as their ability to address practical problems in the oilfield using chemical methods. It was another 10 years before the second edition of "Oilfield Chemistry" was published in 2010. In recognition of the 20 years of dedication and academic accomplishments of Prof. Fulin Zhao and his team, as well as the enormous efforts of the publishers and editors, the Chinese second edition of "Oilfield Chemistry" was awarded TOP 1 recognition for highly cited books in the discipline of petroleum and natural gas engineering in China.

The author of this English edition of "Oilfield Chemistry," Prof. Caili Dai, obtained her doctorate under the supervision of Prof. Fulin Zhao. She also participated in the compilation of the Chinese edition of "Oilfield Chemistry." This English edition of "Oilfield Chemistry" has also been composed in consultation with Prof. Zhao. Its development was supported by the National Publishing Fund and will soon be published to meet the needs of large numbers of students and scholars worldwide.

It is my pleasure and honor to extend my hearty congratulations on the publication of this book, along with the sincere hope that it will expand the impact of the research achievements of China's oilfield chemistry enterprise on the international stage, provide English textbooks for international students in China, and promote the development of the scientific and engineering infrastructure of the "Belt and Road" countries in the field of oilfield chemistry.

Norman, OK, USA
July 2018

Jeffrey H. Harwell
Asahi Glass Chair of Chemical Engineering
George Lynn Cross Research Professor
The University of Oklahoma

Preface

Oilfield chemistry is the science of studying the chemical problems encountered in drilling, oil production, and oil gathering and transportation.

Oilfield chemistry is the youngest member in petroleum science.

Oilfield chemistry consists of three parts: drilling chemistry, oil production chemistry, and oil gathering and transportation chemistry, which constitute the research object of oilfield chemistry.

Although drilling, oil production, and oil gathering and transportation are different processes, they are connected to each other. Therefore, although the three components of oilfield chemistry have their own respective development directions, they are closely interrelated.

Drilling chemistry mainly studies the property of drilling fluid and cement slurry and its control and adjustment.

Oil production chemistry mainly studies various chemical methods for enhanced oil recovery and also for profile control, water shutoff, heavy oil viscosity reduction, acid performance adjustment, sand control in water and oil wells, and paraffin control and removal in oil wells.

Oil gathering and transportation chemistry mainly studies the issues of corrosion and corrosion control of buried pipelines, demulsification of emulsified crude oil and defoaming of foaming crude oil, pour point depression transportation of crude oil and drag reduction transportation of crude oil, and natural gas treatment and oilfield wastewater treatment.

When addressing different problems encountered in the three processes of oilfield chemistry, many of the applied chemical agents are the same. Surfactants and polymers are the most commonly used chemical agents.

The research content of oilfield chemistry mainly includes three parts.

1. Chemical nature of the problems encountered during processes of drilling, oil production, and oil gathering and transportation.
2. Chemical agents applied for solving the existing problems.
3. Function mechanisms and synergistic effects of various chemical agents.

Oilfield chemistry is studied by the following sequence.

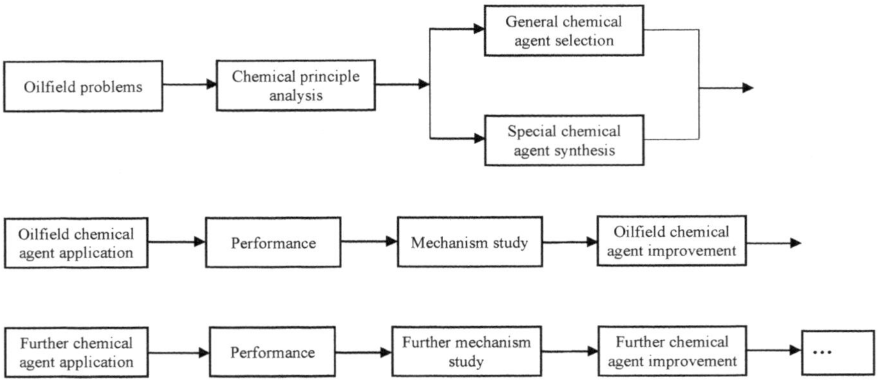

It can be seen from the above diagram that the study of oilfield chemistry is carried out according to the rules of practice–theory–practice, making oilfield chemistry develop continuously.

Oilfield chemistry is closely related to other disciplines.

1. Enhanced oil recovery is an important task of oilfield chemistry. To enhance oil recovery, the reservoir must be understood sufficiently. Oilfield geology is an integral part of geology, and it mainly studies substance composition, structure and distribution of oil, gas and water in reservoirs. Therefore, oilfield geology is an important means to understand reservoirs, and it has a close connection with oilfield chemistry.
2. Oilfield chemistry is an interdisciplinary subject between chemistry and drilling engineering, oilfield production engineering (including reservoir engineering and oil production engineering), and gathering and transportation engineering. The problems to be solved by oilfield chemistry are all raised by the above engineering science, and thus oilfield chemistry is closely linked with engineering science.
3. Since chemistry is an important means for understanding reservoirs and solving existing problems in oilfields, basic chemistry (inorganic chemistry, organic chemistry, analytical chemistry, physical chemistry, surface chemistry, colloid chemistry, etc.) naturally becomes the foundation of oilfield chemistry.
4. Oilfield chemistry solves existing problems in oilfields through oilfield chemicals. Oilfield chemicals are usually used in various solvents (fluids). The dissolution of oilfield chemicals, followed by adsorption at the interface and distribution in the phase, has significant impacts on the properties of the applied system. These effects can be studied by means of fluid mechanics and seepage mechanics. Therefore, oilfield chemistry is closely related to fluid mechanics and seepage mechanics.

Taking into consideration the significance of knowledge of oilfield chemistry in cultivating petroleum engineering students, the Oilfield Chemistry course was established and the textbook was prepared for this course.

Since the publication in July 2000, "Oilfield Chemistry" (Chinese edition) has been adopted by most petroleum-related majors in China. It has played an important role in broadening and deepening the chemistry foundation of students and training students to use chemical methods to solve practical problems in the oilfield, which has achieved expected teaching effects.

As the development of related disciplines in recent years has put forward higher requirements on oilfield chemistry, the achievements from the development of oilfield chemistry science and technology should also be reflected in the textbooks in a timely manner. Therefore, "Oilfield Chemistry" (Chinese edition) has been approved as a revision topic of the "11th Five-Year" national textbook construction plan. It was revised from 2007 to 2009 to become the current Chinese second edition.

"Oilfield Chemistry" (English edition) is based on the "Oilfield Chemistry" (Chinese second edition). Based on the principles of integrity, fundamentality, prospectiveness, associativity, and regularity, some structures of the textbook have been adjusted, and some new mechanisms of the oilfield chemical agents and new agents emerged in recent years have been added. Basic concepts, formulas, and laws of oilfield chemistry involved have also been reviewed, which has significantly improved the textbook quality.

The authors would like to thank Prof. Zhengsong Qiu, Prof. Kaihe Lv, Prof. Zexia Fan, Prof. Yefei Wang, Prof. Xiutai Zhao, Prof. Chuanxian Li, Prof. Dexin Liu for their valuable comments during the compilation of this book.

It is sincerely hoped that readers can put forward valuable opinions on the problems existing in this textbook, so that they can be revised in the next edition.

Qingdao, China
June 2018

Caili Dai
Fulin Zhao

Contents

Part I
Drilling Chemistry

Drilling chemistry is a part of oilfield chemistry.

Drilling chemistry is an edge discipline between drilling engineering and chemistry.

Drilling chemistry studies how to use chemical methods to solve problems encountered during the process of well drilling and cementing.

As the problems encountered during the process of well drilling and cementing mainly come from drilling fluid and cement slurry, drilling chemistry can be classified into two categories, drilling fluid chemistry and cement slurry chemistry.

Drilling fluid chemistry is a study on drilling fluid composition, performance, and their control and adjustment, aiming at achieving drilling process with high quality, high security, and time and economical efficiency. In achieving the control and adjustment of the performance of drilling fluid, chemical method is the most important one. Drilling fluid treatment agents are significant components of drilling fluid chemistry; therefore, their types, structures, properties, and function mechanisms must be clearly understood.

Cement slurry chemistry is a study on cement slurry composition, performance, and their control and adjustment, aiming at plugging complex and leakage formation and protecting production formation and casing. Therefore, the types, structures, properties, and function mechanisms of cement slurry admixtures become the main components of cement slurry chemistry.

Clay is an important raw material in preparing drilling fluid, and its major component is clay mineral. The structure and basic characteristics of clay minerals are closely associated with the performance of the drilling fluid and its control and adjustment, so the structure and basic characteristics of clay minerals should be understood. Thus, in Part I, three chapters are given including clay minerals, drilling fluid chemistry, and cement slurry chemistry.

Chapter 1
Clay Minerals

Clay mineral is the major component of clay. Its crystal structure and basic characteristics have a direct impact on the performance of drilling fluid.

1.1 Basic Structure of Clay Minerals

Although clay minerals differ vastly in types and structures, they all possess the same basic tectonic units. These basic tectonic units form basic tectonic sheets, then basic tectonic sheets form basic structure layers, and finally, these basic structure layers compose various kinds of clay minerals (Velde 1992; van Olphen 1977; Grim 1953).

1.1.1 The Basic Tectonic Units and Basic Tectonic Sheets

Clay minerals have two basic tectonic units: silicon-oxygen tetrahedron and alumina octahedron. These two basic tectonic units form two basic tectonic sheets (also known as the chips).

1.1.1.1 Silicon-Oxygen Tetrahedron and Silicon-Oxygen Tetrahedron Sheet

Silicon-oxygen tetrahedron is composed of a silicon atom and four much bigger oxygen atoms located equidistantly to this silicon atom (Fig. 1.1). It can be observed from Fig. 1.1 that in silicon-oxygen tetrahedron, three oxygen atoms lie in the same plane, named as bottom oxygen atoms, while the one located on the top is referred as the top oxygen atom.

C. Dai and F. Zhao, *Oilfield Chemistry*,
https://doi.org/10.1007/978-981-13-2950-0_1

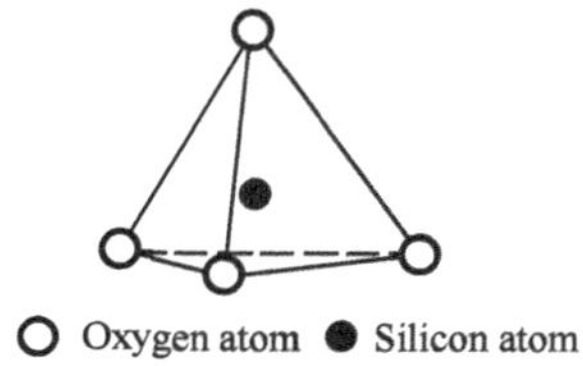

Fig. 1.1 Silicon-oxygen tetrahedron

Silicon-oxygen tetrahedron sheet is formed by multiple silicon-oxygen tetrahedrons sharing bottom oxygen atoms (Fig. 1.2). Therefore, each silicon-oxygen tetrahedron sheet has bottom oxygen atom plane and top oxygen atom plane. Obviously, the bottom oxygen plane contains more oxygen atoms than the top oxygen plane does. Silicon-oxygen tetrahedron sheet can extend infinitely in the plane, forming a structure with continuous hexagonal grids (Fig. 1.3). The inscribed circle diameter of this hexagonal grid is around 0.288 nm, while the thickness of silicon-oxygen tetrahedron sheet is about 0.5 nm.

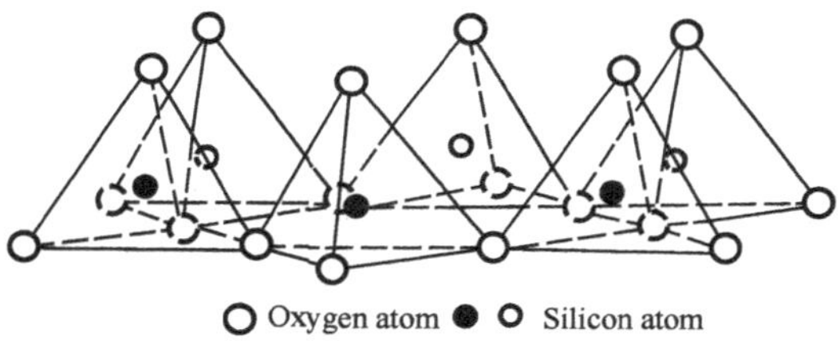

Fig. 1.2 Silicon-oxygen tetrahedron sheet

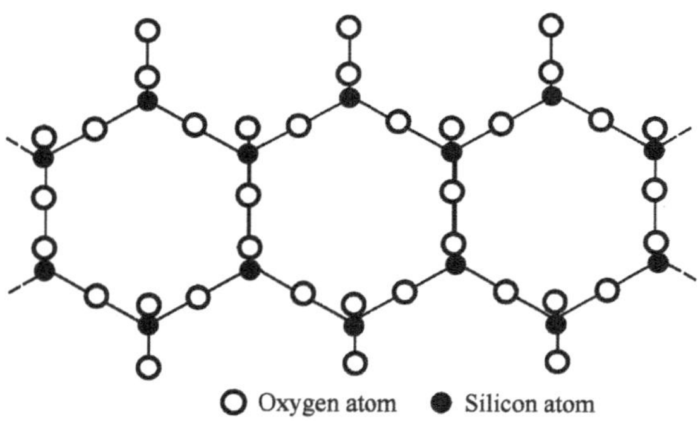

Fig. 1.3 Hexagonal grid structure of silicon-oxygen tetrahedron sheet

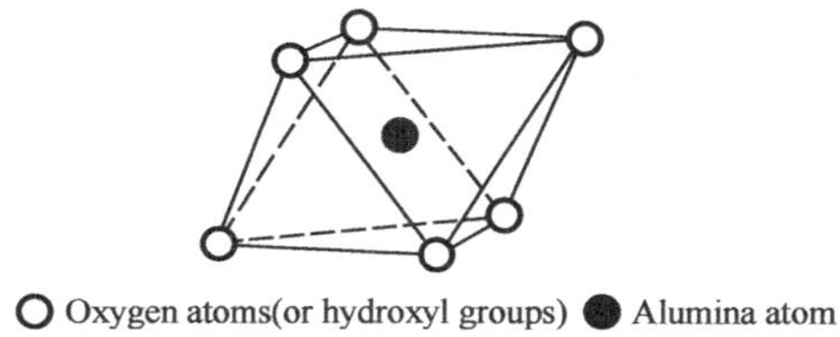

Fig. 1.4 Alumina octahedron

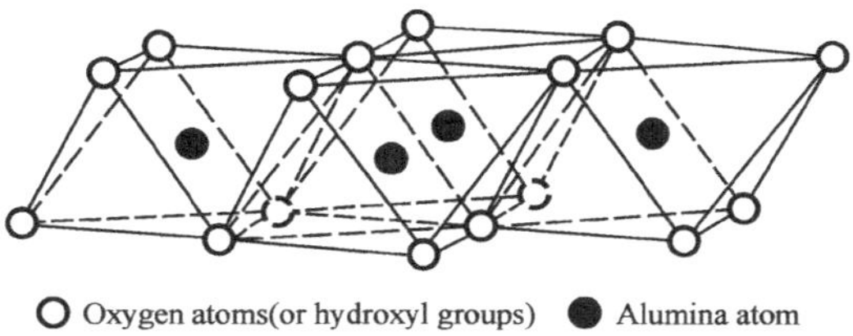

Fig. 1.5 Alumina octahedron sheet

1.1.1.2 Alumina Octahedron and Alumina Octahedron Sheet

Alumina octahedron is formed through coordination between an aluminum atom and six oxygen atoms (or hydroxyl groups) (Fig. 1.4).

Similarly, alumina octahedron sheet is constructed by sharing oxygen atoms. There are two parallel planes of oxygen atoms (or hydroxyl groups) in the alumina octahedron sheet; all the oxygen atoms (or hydroxyl groups) in the alumina octahedron sheet distribute in the two planes (Fig. 1.5).

1.1.2 The Basic Structure Layer

The basic structure layer (also called crystal layer) of clay mineral is constructed by silicon-oxygen tetrahedron sheet and alumina octahedron sheet bonding with each other at different ratios.

1.1.2.1 1:1 Basic Structure Layer

This kind of basic structure layer is constructed by a silicon-oxygen tetrahedron sheet bonding with an alumina octahedron sheet. It is the simplest crystal structure of layered-structured silicon aluminum acid salt clay mineral.

In the 1:1 basic structure layer, the top oxygen atoms of silicon-oxygen tetrahedron sheet form a part of alumina octahedron sheet, replacing part of the hydroxyl groups in alumina octahedron sheet. Therefore, there are five layers of atoms in the 1:1 basic structure, which are one layer of silicon atoms, one layer of aluminum atoms, and three layers of oxygen atoms (or hydroxyl groups). The crystal structure of kaolinite is composed of this basic structure layer.

1.1.2.2 2:1 Basic Structure Layer

This kind of basic structure is constructed by two silicon-oxygen tetrahedron sheets sandwiching one alumina octahedron sheet. The top oxygen atoms of two silicon-oxygen tetrahedron sheets substitute, respectively, the two oxygen atoms in the alumina octahedron sheet, or part of the hydroxyl groups in the hydroxyl plane. Therefore, there are seven layers of atoms in the 2:1 basic structure, which are one layer of aluminum atoms, two layers of silicon atoms, and four layers of oxygen atoms (or hydroxyl groups). The crystal structure of montmorillonite is constructed by this kind of basic structure layer.

1.1.3 Concepts Derived from Repeating and Stacking of the Basic Structure Layers

Clay minerals are constructed by stacking of the two kinds of basic structure layers mentioned above. When the two basic structure layers stack repeatedly, the space between two adjacent structure layers is called interlayer space. The basic structure layer together with the interlayer space is called the unit structure of clay minerals. The substances existing in the interlayer are called interlayer matter. If the matter is water, then this kind of water is called interlayer water; if cations exist in the interlayer space, these cations are called interlayer cations. The vertical distance between corresponding crystal planes of adjacent basic structural layers is called lattice spacing. These concepts are all derived from repeating and stacking of the basic structure layers.

1.2 Clay Minerals

Clay minerals include kaolinite, montmorillonite, illite, chlorite, palygorskite, sepiolite, etc. And the first three are the most common ones.

1.2.1 Kaolinite

The basic structure layer of kaolinite is constructed by an oxygen silicon tetrahedron sheet bonding with an alumina octahedron sheet, so kaolinite belongs to type 1:1 clay mineral. The kaolinite clay mineral crystal is constructed by the infinite extension of basic structure layer along the layer level (i.e., the axis *a* and axis *b* in a rectangular coordinate system) and the repetitive stacking of basic structure layer along the direction perpendicular to the layer level (i.e., the axis *c* in a rectangular coordinate system), and its lattice spacing is around 0.72 nm (Fig. 1.6).

In the structure of kaolinite, one side of the crystal layer is composed of oxygen atoms only while the other side consists nothing but hydroxyl groups. Crystal layers are joined tightly through hydrogen bonds and intermolecular forces, and thus, it is difficult for water to enter the crystal structure.

Lattice substitution is rare in kaolinite. Lattice substitution refers to the replacement of silicon atoms in oxygen silicon tetrahedron and aluminum atoms in alumina octahedron by other atoms (usually metal atoms with lower valence state), such as the replacement of silicon atoms by aluminum atoms and the replacement of aluminum atoms by magnesium atoms. Lattice substitution results in the unbalance of crystal electrovalence. To balance the electrovalence, certain amounts of cations

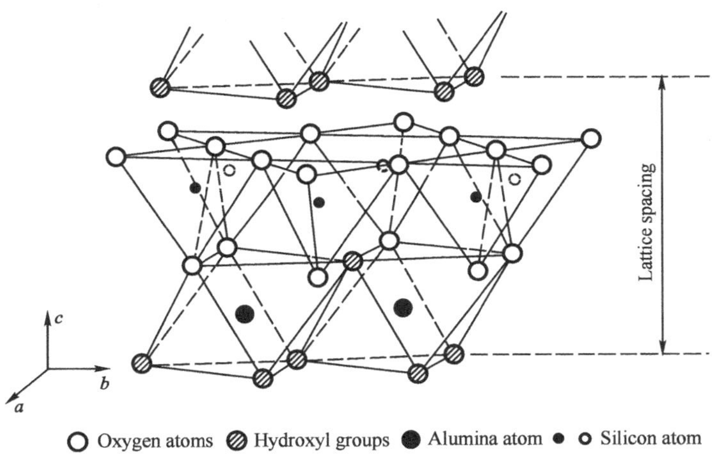

Fig. 1.6 Crystal structure of montmorillonite

are needed in crystal surface. These cations bonded for valence compensation are exchangeable, so they are called exchangeable cations (or compensating cations). Because lattice substitution occurs rarely in kaolinite, very few exchangeable cations exist in its crystal surface.

Kaolinite is a kind of non-expansive clay mineral, which can be understood from two aspects: the existence of hydrogen bonds between crystal layers and the rarity of exchangeable cations on crystal surfaces.

1.2.2 Montmorillonite

The basic structure layer of montmorillonite is constructed by two oxygen silicon tetrahedron sheets and an alumina octahedron sheet, so montmorillonite belongs to type 2:1 clay mineral. In this basic structure layer, all the top oxygen atoms in oxygen silicon tetrahedron point toward the alumina octahedron. Silicon-oxygen tetrahedron sheets and alumina octahedron sheets bond with each other by sharing oxygen atoms. The basic structure layer extends infinitely along axis *a* and *b* and stacks repeatedly along axis *c* to construct montmorillonite clay mineral crystal (Fig. 1.7).

Montmorillonite is a kind of expansive clay mineral. On one hand, in the structure of montmorillonite, the two sides of crystal layers are composed purely of oxygen atoms. The interactions between crystal layers are intermolecular forces (no hydrogen bonds exist), so the crystal layers are loosely bonded and water can get into the structure easily. On the other hand, a considerable amount of lattice substitution exists, combining a large number of exchangeable cations in the crystal surface. When water gets into the crystal layer, these exchangeable cations dissociate in the water to form diffusible electrical double layer. Thus, the surfaces of crystal layers are negatively charged, resulting in repulsion between layers and the clay expansion. Due to the above-mentioned characteristics, the lattice spacing of montmorillonite is variable, ranging usually between 0.96 and 4.00 nm.

The lattice substitution of montmorillonite occurs mainly in the alumina octahedron sheet, by replacing the aluminum atoms in alumina octahedron with iron atoms or magnesium atoms. The silicon atoms in silicon-oxygen tetrahedron are rarely replaced. After lattice substitution, various exchangeable cations can be bonded on the crystal surface. When the exchangeable cations are mainly sodium ions, this kind of montmorillonite is called sodium montmorillonite; while when the exchangeable cations are mainly calcium ions, this kind of montmorillonite is called calcium montmorillonite. Similarly, there are hydrogen montmorillonite and lithium montmorillonite, etc.

The main component of bentonite is montmorillonite, and it is an important drilling fluid material. When the main component of bentonite is sodium montmorillonite, it is called sodium soil; while when the main component of bentonite is calcium montmorillonite, it is called the calcium soil. They can be used as suspending agents and tackifiers for drilling fluid.

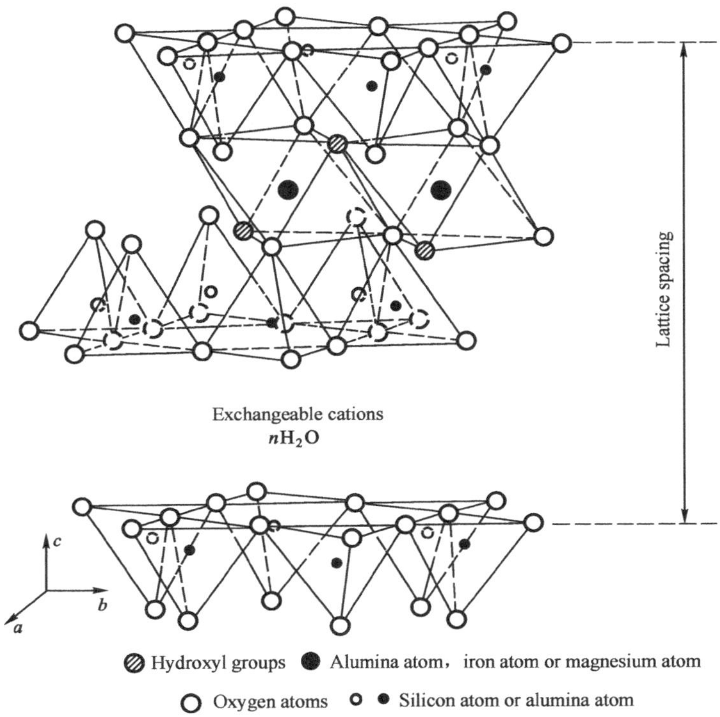

Fig. 1.7 Crystal structure of montmorillonite

1.2.3 Illite

Similar as montmorillonite, the basic structure layer of illite is constructed by two oxygen silicon tetrahedron sheets and one alumina octahedron sheet. Illite belongs to type 2:1 clay mineral (Fig. 1.8).

The structural difference between illite and montmorillonite is that for illite the lattice substitution mainly takes place in the silicon-oxygen tetrahedron sheets (about 1/6 of the silicon atoms are replaced by aluminum atoms), and exchangeable cations for the electrovalence compensation are mainly sodium ions. As the diameter of potassium ion (0.266 nm) is close to that of the inscribed circle in the six-party grid structure of the silicon-oxygen tetrahedron sheet (0.288 nm), it is easy for the potassium ions to enter the six-party grid structure and get closer to the aluminum atoms after lattice substitution, resulting in large electrostatic attraction. Due to this electrostatic attraction, it is difficult for potassium ions to release from the six-party grid structure of the silicon-oxygen tetrahedron sheet. So crystal layers are tightly

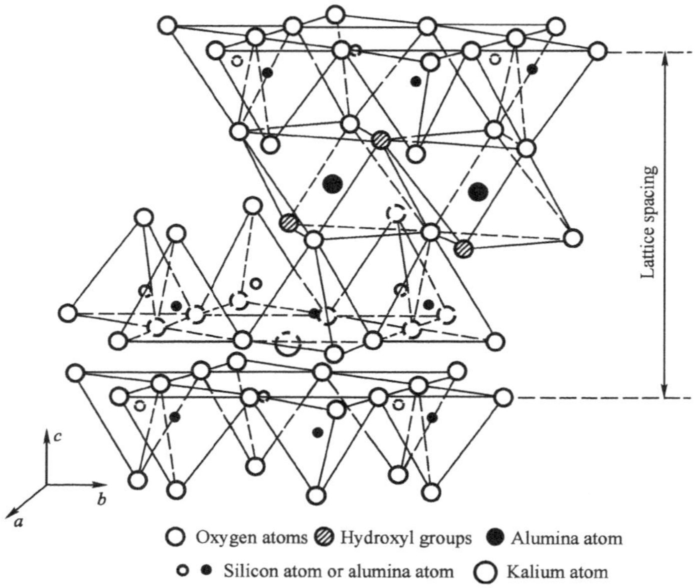

Fig. 1.8 Crystal structure of illite

bonded and water cannot enter easily, therefore illite is a kind of expansive clay mineral. The lattice spacing of illite is stable, generally around 1.0 nm.

1.2.4 Chlorite

The basic structure layer of chlorite consists of a sheet layer similar to illite (type 2:1) and a sheet layer of brucite (Fig. 1.9) (Zhao and Zhang 1990).

Chlorite also belongs to type 2:1 clay mineral. What differs it from other clay minerals is that its interlayer space is filled by brucite sheets. The brucite sheet is octahedral sheet, in which the silicon atoms are replaced by aluminum atoms, resulting in its overall electropositivity. It can replace exchangeable cations to compensate the unbalanced electrovalence generated by aluminum atoms substituting silicon atoms in a type 2:1 structure.

It can be observed in Fig. 1.9 that hydrogen bonds exist between crystal layers of chlorite. On account of the electrostatic attraction between brucite and crystal layer, the crystal layers of chlorite are tightly bonded and water cannot enter the interlayer easily. So chlorite is also a kind of non-expansive clay mineral.

The lattice spacing of chlorite is also quite stable, generally around 1.42 nm.

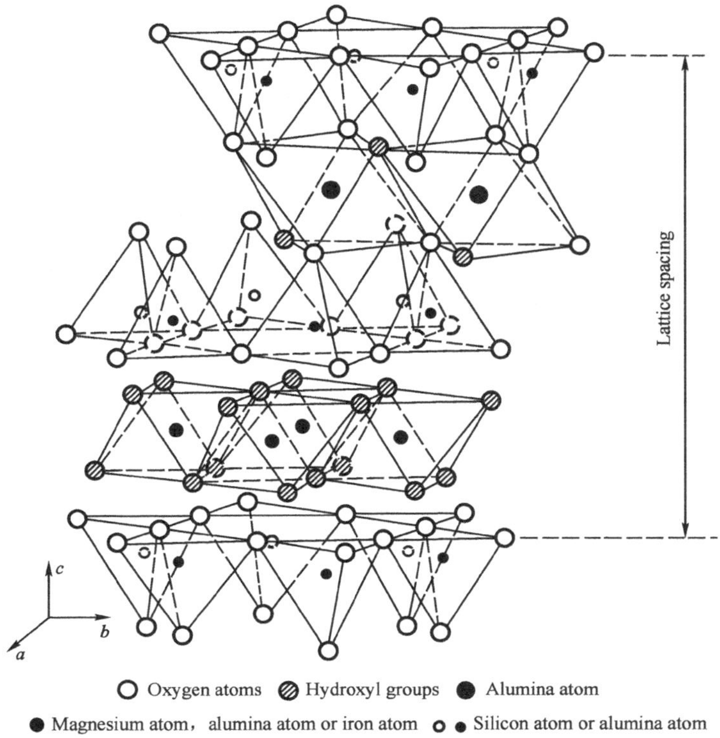

Fig. 1.9 Crystal structure of chlorite

1.2.5 Palygorskite and Sepiolite

Palygorskite and sepiolite are both chain-layered clay minerals, which are different from the aforementioned layered clay minerals (such as kaolinite, montmorillonite, illite) (Yang et al. 1994; Ren 1992).

The crystal structure of palygorskite and sepiolite can be understood as type 2:1 layered structure. If the silica tetrahedron of type 2:1 layered crystal structure is rotated 180° at regular intervals and stretched along axis *a*, the structure of palygorskite and sepiolite could be formed. This structure combines the characteristics of both chain-like clay minerals and layered clay minerals, and thus, they are called chain-layered clay minerals.

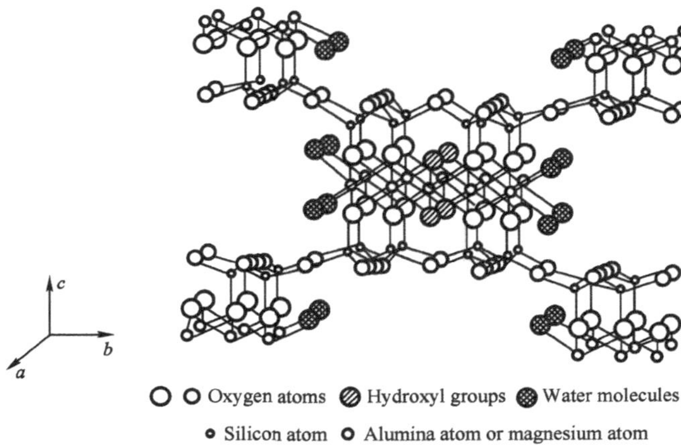

Fig. 1.10 Crystal structure of palygorskite

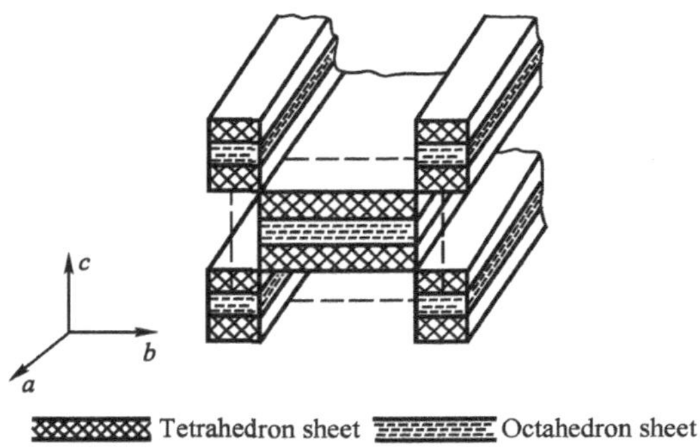

Fig. 1.11 Three-dimensional schematic of palygorskite crystal structure

The crystal structure of palygorskite is shown in Fig. 1.10. It is constructed by a type 2:1 layered sheet rotating 180° for every four silica tetrahedrons and extending along axis *a*. The three-dimensional schematic of palygorskite crystal structure is illustrated in Fig. 1.11, in which the dashed frame indicates the crystal structure demonstrated in Fig. 1.10. It can be also observed in Fig. 1.11 that due to the rotating and extending of silica tetrahedrons, a channel parallel to axis *a* is formed, accommodating water molecules and exchangeable cations.

Figure 1.12 shows the crystal structure of sepiolite. It is constructed by a type 2:1 layered sheet rotating 180° for every six silica tetrahedrons and extending along axis *a*. When rotating 180°, sepiolite needs to cross more silica tetrahedrons than palygorskite does; thus in its structure, sepiolite has wider channel parallel to axis *a*.

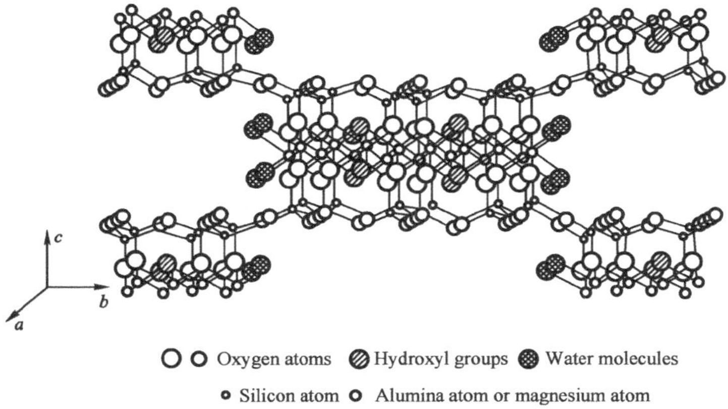

Fig. 1.12 Crystal structure of sepiolite

Lattice substitution occurs rarely in palygorskite and sepiolite. For these two kinds of clays, their crystals are slender and easily assembled to bundles, with fiber-like appearance. Usually they have good water solubility and can entangle with each other to generate structural viscosity; therefore, they can act as tackifier. On the other hand, this tackifying effect is salinity-insensitive, so these two kinds of clays possess good salt resistance. Clays containing palygorskite and sepiolite as major components are called salt-resistant clays.

1.2.6 Mixed Crystal-Layered Clay Mineral

Mixed crystal-layered clay minerals are also called mixed layer minerals. They are constructed by different kinds of clay mineral crystal layers stacking with each other (orderly or disorderly). Some common examples are the mineral mixed by illite and montmorillonite (referred as I/S mixed layer) and the mineral mixed by chlorite and montmorillonite (referred as C/S mixed layer). Generally, mixed layer minerals are more prone to expand and disperse in water as compared with single component clay minerals.

1.3 Properties of Clay Minerals

The following properties of clay minerals are closely related to the performances of drilling fluid (Swartzen-Allen and Matijevic 1974).

1.3.1 Electrical Property

The electrical property of the surface of clay minerals refers to the charge symbol and capacity when the surface of clay minerals and water contact with each other.

There are two sources for the electrical properties of clay minerals.

1.3.1.1 The Dissociation of Exchangeable Cations

Certain amounts of exchangeable cations exist on the surface of clay minerals. When clay minerals contact with water, these exchangeable cations will dissociate from the surfaces of clay minerals and align around mineral surfaces by diffusion to form diffused electric double layer, rendering mineral surfaces electronegative. Therefore, larger number of exchangeable cations on mineral surfaces would result in higher electrical capacity and stronger electronegativity after dissociation of the surfaces.

The electrical capacity of the clay mineral surfaces can be expressed by cation exchange capacity (CEC).

Cation exchange capacity is the total number of cations got exchanged from 1 kg clay minerals in condition of pH 7 (demonstrated as the molar amount of univalent cations). The unit is $mmol \cdot kg^{-1}$.

Different kinds of clay minerals have different cation exchange capacities. Table 1.1 shows cation exchange capacities of various clay minerals (Gray et al. 1980). It can be observed from Table 1.1 that montmorillonite has the largest CEC while kaolinite possesses the minimum CEC.

Generally speaking, the larger number of lattice substitution, the higher CEC. However, there are some exceptions. For example, illite has the largest number of lattice substitution among clay minerals. However, the positions where lattice substitutions occur locate in silica tetrahedron sheet, which are close to surfaces of crystal layers. On the other hand, the exchangeable cations are potassium ions, which are not easily exchanged by other cations. Therefore, the cation exchange capacity of illite is lower than that of montmorillonite.

Table 1.1 Cation exchange capacity of clay minerals

Clay minerals	Cation exchange capacity (mmol/kg)
Kaolinite	30–150
Montmorillonite	800–1,500
Illite	200–400
Chlorite	100–400
Palygorskite	100–200
Sepiolite	200–450

1.3.1.2 The Reaction Between Surface Hydroxyl Groups and H^+ or OH^-

There are two kinds of hydroxyl groups on the surfaces of clay minerals. One kind is the hydroxyl groups existed on the crystal layer surface of clay minerals (Fig. 1.6), and the other kind is the surface hydroxyl groups generated on the edge of clay minerals during bond breaking. The latter surface hydroxyl groups are generated through the following procedures:

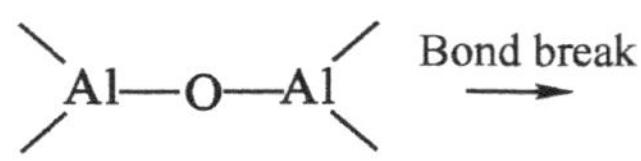

(Alumina octahedral)

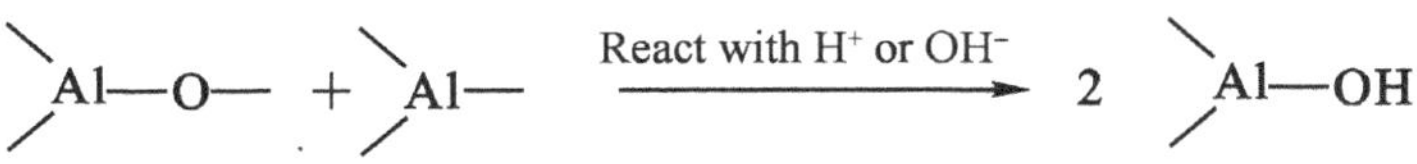

(Surface hydroxyl groups)

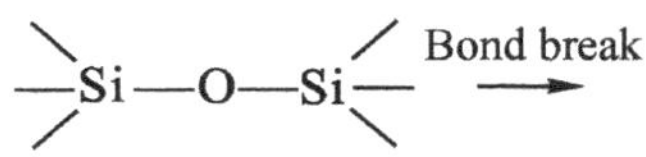

(Silica tetrahedron)

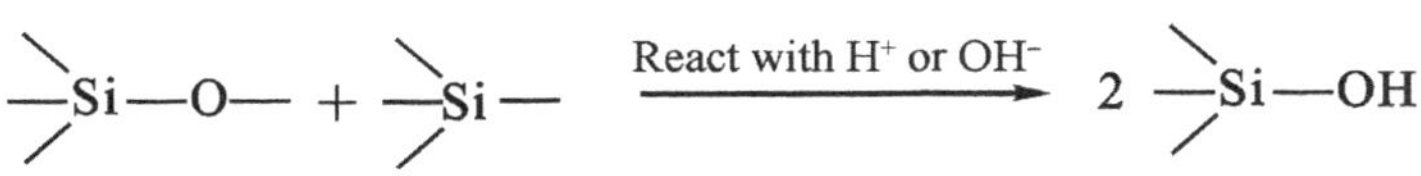

(Surface hydroxyl groups)

Under acidic or alkaline conditions, these surface hydroxyls can react with H^+ or OH^- to make the surfaces of clay minerals bear different electrical properties (electropositive or electronegative). Under acidic conditions, hydroxyls on the surface of clay minerals can react with H^+, which makes the surface of clay minerals electropositive.

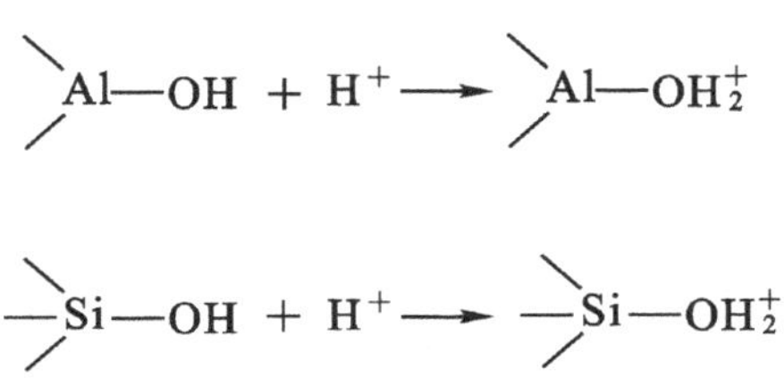

Table 1.2 Relationship between the cation exchange capacity of kaolinite and the size of particles

Particle size (μm)	Cation exchange capacity (mmol/kg)
0.05–0.1	95
0.1–0.25	54
0.25–0.5	39
0.5–1	38
2–4	36
5–10	26
10–20	24

Under alkaline conditions, hydroxyls on the surface of clay minerals can react with OH^-, rendering the surface of clay minerals electronegative.

$$>\mathrm{Al{-}OH} + \mathrm{OH^-} \longrightarrow >\mathrm{Al{-}O^-} + \mathrm{H_2O}$$

$$\equiv\mathrm{Si{-}OH} + \mathrm{OH^-} \longrightarrow \equiv\mathrm{Si{-}O^-} + \mathrm{H_2O}$$

Under specific acidic or alkaline conditions, the electricity generated by surface hydroxyls on broken bonds is concerned with the dispersity of clay minerals. Table 1.2 shows the relationship between the cation exchange capacity of kaolinite under alkaline condition and the size of particles. It can be concluded from Table 1.2 that the higher the dispersity degree of kaolinite, the larger its cation exchange capacity. That's because higher dispersity degree will result in larger number of broken bonds on the edge of kaolinite, which will produce more surface hydroxyls and therefore higher cation exchange capacity.

The algebraic sum of the above-mentioned two sources of electrical properties decides the final electricity of clay minerals. Generally, the surface of clay minerals is electronegative and its cation exchange capacity is within range shown in Table 1.1.

1.3.2 Adsorptivity

Adsorptivity is the property of substances accumulating on the surface of clay minerals. When studying the surface adsorptivity of clay minerals, clay minerals are adsorbents while the substances accumulating on them are adsorbates. Adsorption is the accumulation process of adsorbates on the surface of adsorbents.

The surface adsorption of clay minerals can be divided into two types: the physical adsorption and chemical adsorption.

1.3.2.1 Physical Adsorption

Physical adsorption means the adsorption between the adsorbent and adsorbate through intermolecular forces. The adsorption generated by hydrogen bonds also belongs to physical adsorption. The adsorptions of nonionic surfactant (e.g., polyoxyethylene alkyl alcohol ether and polyoxyethylene alkyl phenol ether) and nonionic polymer (e.g., polyvinyl alcohol and polyacrylamide) on surface of clay minerals are achieved through both intermolecular forces and hydrogen bonds, so they both belong to physical adsorption.

1.3.2.2 Chemical Adsorption

Chemical adsorption means the adsorption between the adsorbent and adsorbate through chemical bonds. Cationic surfactants (e.g., dodecyl trimethyl ammonium chloride) and cationic polymers (e.g., polydiallyl dimethyl ammonium chloride) can dissociate in water to generate cationic surfactants and polymers. These cations can form ionic bonds with electronegative surface of clay minerals to achieve adsorption. So they all belong to chemical adsorption.

1.3.3 Expansibility

Expansibility refers to the properties of volume increase of clay minerals when contacting with water. Expansibilities differ with different kinds of clay minerals. According to the crystal structure, clay minerals can be divided into expansive clay minerals and non-expansive clay minerals (Gorodnov 1992; Wang 1995).

Montmorillonite belongs to expansive clay minerals. Its expansibility derives from the presence of large number of exchangeable cations. When montmorillonite contacts water, water will enter its crystal layer, resulting in the dissociation of exchangeable cations and the establishment of diffused double layer on crystal surface. So the surface of montmorillonite is electronegative (Su 2002). The electrostatic repulsion among negatively charged crystal layers leads to the increase of lattice spacing, resulting in the expansibility of montmorillonite.

Kaolinite, illite, and chlorite, all belong to non-expansive clay minerals, and their weak expansibilities are attributed to different reasons. The poor expansibility of kaolinite is attributed to the small number of lattice substitution and the existence of hydrogen bonds between crystal layers; the reason for illite is that the lattice substitution mainly happens in silica tetrahedron sheets and the exchangeable cations among crystal layers are K^+ which will strengthen the connection. For chlorite, the weak expansibility comes from the existence of hydrogen bonds among crystal layers and the compensation of unbalanced electrovalence due to lattice spacing by replacing exchangeable cations with brucite.

1.3.4 Cohesion

Cohesion means the property of clay mineral particles (sheets precisely) cross-linking with each other in water under specific conditions.

Here, the specific condition means the certain concentration of electrolytes (such as NaCl and $CaCl_2$). With the increased concentration of electrolytes, the diffused double layers on the surface of clay minerals are compressed, resulting in reduced electricity on the edge and the surface. When the concentration of electrolytes surpasses certain value, it would induce cross-linking among the clay mineral particles.

After the occurrence of cross-linking, if the clay mineral cross-links can connect with each other and spread all over water, then spatial structures will be generated. Otherwise, these cross-links will sink.

There are three cross-linking ways for clay mineral particles, which are edge-to-edge cross-linking (Fig. 1.13a), edge-to-surface cross-linking (Fig. 1.13b), and surface-to-surface cross-linking (Fig. 1.13c) (Lummus and Azar 1986; Fan et al. 1996).

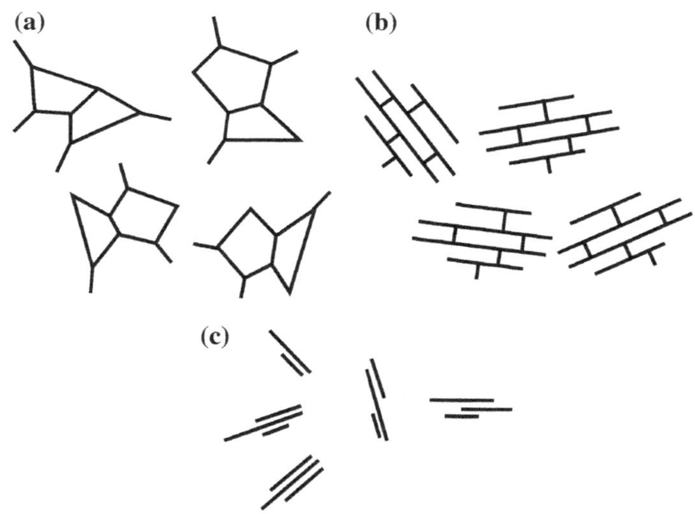

Fig. 1.13 Cross-linking ways of clay mineral particles. **a** Edge-to-edge cross-linking; **b** edge-to-surface cross-linking; and **c** surface-to-surface cross-linking

Formation factors can influence the properties of drilling fluids through cohesive characters of clay minerals, while the treating agents can adjust the properties of drilling fluids through cohesive characters of clay minerals. If the Ca^{2+} in the formation invades the drilling fluids (called calcium invasion), edge-to-edge cross-linking and edge-to-surface cross-linking would be induced among the clay mineral sheets, which will generate spatial structure and lead to enhanced viscosity. With further Ca^{2+} invasion, surface-to-surface cross-linking would be induced among clay mineral particles, causing the decrease of viscosity. By adsorption on mineral surfaces, viscosity reducer (also called dispersant) can increase the electronegativity of the surface of clay minerals and increase the thickness of hydration layers, which would lead to the redispersion of clay mineral sheets and recover the usability of drilling fluids.

References

Fan S, Yan J, Zhou D (1996) The technology of drilling and completion fluids and hydrocarbon reservoir protection. Petroleum University Publishing House, Dongying, p 52

Grim RE (1953) Clay mineralogy. McGraw-Hill, New York. Chinese edition: Grim RE (1960) Clay mineralogy (trans: Xu J). Geological Publishing House, Beijing, pp 48–80

Gray GR, Darley HCH, Rogers WF (1980) Composition and properties of oil well drilling fluids. Gulf Publishing Company, Houston, p 154

Gorodnov B (1992) Prevention for complicated conditions in process of drill and physiochemical methods. Li R, Zhou D, interpret. Petroleum Industry Press, Beijing, p 4

Lummus JL, Azar JJ (1986) Drilling fluids optimization: a practical field approach. Penn Well Publishing Company, Tulsa Oklahoma, pp 94–108

Ren L (1992) Clay minerals and clay rock. Geological Publishing House, Beijing, pp 123–134

Su C, Fu J, Guo B (2002) Research on the changing tendency of zeta potential of the clay minerals and drilling fluid. Drill Fluid Complet Fluid 19(6):1–4

Swartzen-Allen SL, Matijevic E (1974) Surface and colloid chemistry of clays. Chem Rev 74(3):385–400

Velde B (1992) Introduction to clay minerals. Chapman & Hall, London, pp 42–54

van Olphen H (1977) An introduction to clay colloid chemistry, 2nd edn. John Wiley & Sons, New York, pp 57–82

Wang J (1995) The structure and lamellar inflation of clay minerals. J Anqing Teach Coll (Nat Sci Ed) 1(4):14–16

Yang Y, Zhang N, Su Z et al (1994) Chinese clay minerals. Geological Publishing House, Beijing, pp 173–196

Zhao X, Zhang Y (1990) Clay minerals and clay mineral analysis. Ocean Press, Beijing, p 29

Chapter 2
Drilling Fluid Chemistry

Drilling fluid (formerly called mud) refers to the circulating working fluid used in the drilling process. It can be liquid, gas, or foam.

2.1 Functions and Compositions of Drilling Fluid

2.1.1 Functions of Drilling Fluid

The drilling fluid plays an important role in the drilling process. Figure 2.1 shows the circulation process of liquid drilling fluid in the drilling process (Chilingarian 1981). It can be seen that driven by the drilling fluid pump, drilling fluid in the drilling fluid pool passes through the ground pipeline, column, and water hose to get into the drilling pipe before it is sprayed to the well bottom through the drilling bit. Carrying the cuttings drilled down by the drilling bit, the drilling fluid then returns back to the ground from the annulus between drilling pipe and formation (or casing). After the removal of cuttings through solid-control equipment such as the vibrating screen, the drilling fluid returns to the drilling fluid pool for recycling.

The drilling fluid has the following functions.

2.1.1.1 Washing the Borehole Bottom

Drilling fluid can form a high-speed liquid flow in the drill bit port and be sprayed into the bottom of the borehole. The high-speed drilling fluid can wash the cuttings which are held down in the borehole bottom due to the pressure difference between drilling fluid and formation. Therefore, the drilling fluid can perform the function of washing the borehole bottom.

C. Dai and F. Zhao, *Oilfield Chemistry*,
https://doi.org/10.1007/978-981-13-2950-0_2

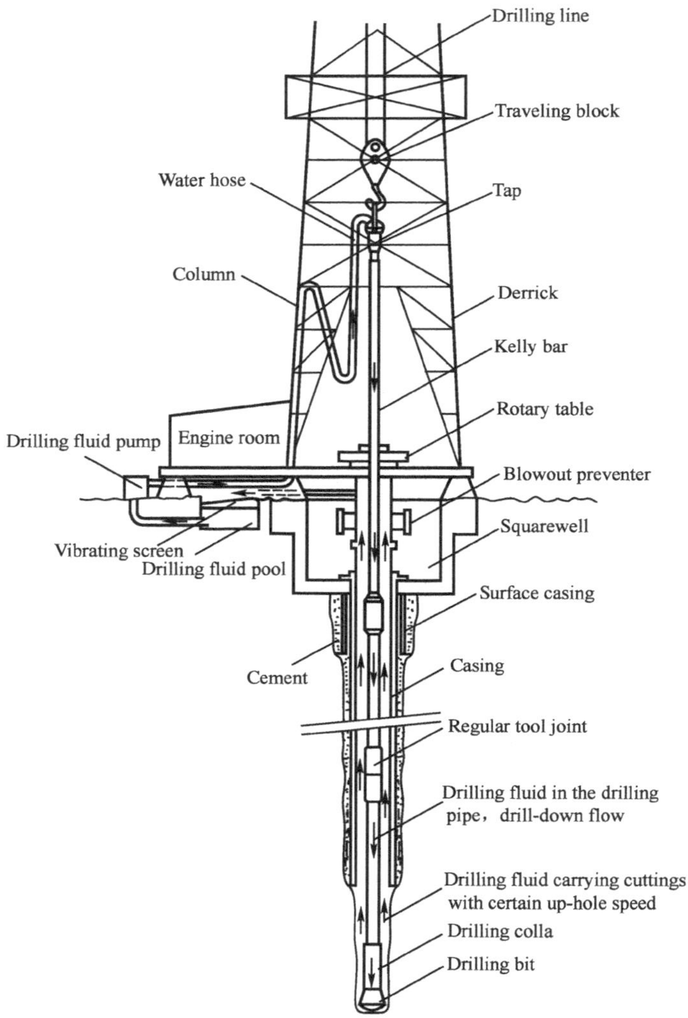

Fig. 2.1 Circulation process of liquid drilling fluid in the drilling process

2.1.1.2 Carrying Cuttings

When the up-hole speed of drilling fluid is greater than the settling speed of the cuttings, the drilling fluid can carry the cuttings out of the wells. That means at certain up-hole speed, the drilling fluid has the function of carrying cuttings.

2.1.1.3 Balancing Formation Pressure

The liquid pressure of the drilling fluid must be balanced with the formation pressure to prevent the well kick, well blowout, and leakage of drilling fluid into the formation. By adjusting the density of drilling fluid, the liquid pressure of the drilling fluid can be controlled to equilibrate with the formation pressure.

2.1.1.4 Cooling and Lubricating Bit

The drilling fluid can cool the drilling tools by bringing the heat generated from the friction between drilling tools (bit and drill string) and the formation to the ground. Meanwhile, drilling fluid can effectively reduce the friction between drilling tools and formation. Therefore, it can also perform the function of lubrication.

2.1.1.5 Stabilizing the Borehole Wall

Treating agents can be added into the drilling fluid so that drilling fluid bears the capabilities of inhibiting shale[1] expansion and dispersion. Thin and tough filter cakes can also be formed to stabilize the borehole wall. In addition, the hydraulic pressure produced by the drilling fluid in the borehole can provide an effective mechanical support to the borehole wall, which can help with the borehole wall stability.

2.1.1.6 Suspending Cuttings and Solid Density-Adjusting Materials

When the circulation stops, the drilling fluid is in a stationary state where the bentonite particles can link with each other to form structure and suspend the cuttings. If solid density-adjusting material (e.g., barite) is added to the drilling fluid, it can also be suspended by the drilling fluid when the circulation stops. The capability of suspending cuttings and solid density-adjusting materials allows drilling fluid to be easily restarted after circulation stops.

---

[1]The shale mentioned here refers to the rock with high clay content. Problems including clay expansion and/or dispersion and collapse of borehole wall can be induced by the contact of shale with water.

2.1.1.7 Obtaining Stratigraphic Information

Lots of stratigraphic information such as oil and gas shows and formation physical properties can be obtained through the cuttings carried out by the drilling fluid.

2.1.1.8 Transferring Power

Drilling fluid can form high-pressure jet at the bit port to transfer the power of drilling fluid pump to the borehole bottom, improving the rock breaking ability and drilling speed. If a turbodrill is used for drilling, drilling fluid can pass the power of the drilling fluid pump to the turbine when passing through the turbine blade in a high speed, which will drive the drill bit to break the rock.

2.1.2 Compositions of Drilling Fluid

Drilling fluid generally consists of dispersing medium, dispersion agent, and drilling fluid additives.

The dispersion medium in the drilling fluid can be water or oil.

For the dispersion agent in the drilling fluid, if it is suspension, it is clay and/or solid density-adjusting material; if it is emulsion, it is oil or water; if it is foam, it is gas.

The drilling fluid additives are chemical agents which are added to the drilling fluid to adjust the function of the drilling fluid. For the drilling fluid additives, if classified by element compositions, they can be divided into inorganic drilling fluid additives (including inorganic acid, alkali, salt, and oxide) and organic drilling fluid additives (such as surfactants and polymers); if classified by applications, they can be divided into the following 15 categories (Gulf Publishing Company 1998; Li et al. 1990), which are drilling fluid pH controlling agent, drilling fluid calcium removal agent, drilling fluid foaming agent, drilling fluid emulsifier, drilling fluid viscosity reducer, drilling fluid tackifier, drilling fluid filtrate reducer, drilling fluid flocculants, shale inhibitor (also known as anti-sloughing agent), drilling fluid corrosion inhibitor, drilling fluid lubricant, pipe-freeing agent, temperature stabilizer, density-adjusting material, and plugging material. The last two categories are expressed as materials because they are used in large quantities (usually over 5%).[2]

---

[2]Unless otherwise stated, the matter content in drilling fluid refers to the mass of the matter in certain volume of solvent or dispersion medium. Here, over 5% indicates the mass of additive in 100 mL water exceeds 5 g.

2.2 The Density of Drilling Fluid and Its Adjustment

The mass of unit volume drilling fluid is defined as drilling fluid density. It is adjusted accordingly to balance formation pressure and the stratigraphic tectonic stress. Drilling fluid with appropriate density can prevent well kick, well blowout or severe leakage of drilling fluid, and also prevent the collapse of borehole wall.

Ways of adjusting density of drilling fluid include the reduction and the increase of the density of drilling fluid.

The density of drilling fluid can be decreased by adding water, oil, or gas, as their densities are lower than that of the drilling fluid (Zhang et al. 2005). The density of drilling fluid can also be reduced by removing the useless solids in drilling fluid through mechanical or (and) chemical flocculation. While by adding high-density materials, the density of drilling fluid can be enhanced.

High-density materials include two kinds. One kind is the powder of high-density insoluble mineral or ore (Table 2.1) (Sawdon and Bradbury 2006). These powders can improve the density of drilling fluid by suspending in the spatial structure formed by clay mineral particles. Since barite has abundant sources and low cost, it has become the mostly used high-density material.

The other kind of high-density material is water-soluble salt (Table 2.2). It can improve drilling fluid density by dissolving in drilling fluid.

When improving drilling fluid density by adding water-soluble salt, corrosion inhibitor should be added to prevent the salt-induced corrosion of drilling tools. Meanwhile, the precipitation temperature of salt from drilling fluid should be paid attention to. Figure 2.2 shows the changing curves of ice or salt separating-out temperatures of aqueous solution with densities of salt water (Bleakley 1980). It can be observed from this figure that this curve (e.g., NaCl curve) is divided into two parts.

Table 2.1 Insoluble mineral or ore with high density

Name	Major constituent	Density (g/cm^3)
Chalk	$CaCO_3$	2.7–2.9
Baryte	$BaSO_4$	4.2–4.6
Siderite	$FeCO_3$	3.6–4.0
Ilmenite	$TiO_2 \cdot Fe_2O_3$	4.7–5.0
Magnetite	Fe_2O_3	4.9–5.2
Pyrite	FeS_2	4.9–5.2

Table 2.2 Water-soluble salt with high density

Water-soluble salt	Density of salt (g/cm^3)	Density of liquor (g/cm^3)
KCl	1.398	1.16 (20 °C)
NaCl	2.17	1.20 (20 °C)
$CaCl_2$	2.15	1.40 (60 °C)
$CaBr_2$	2.29	1.80 (10 °C)
$ZnBr_2$	4.22	2.30 (40 °C)

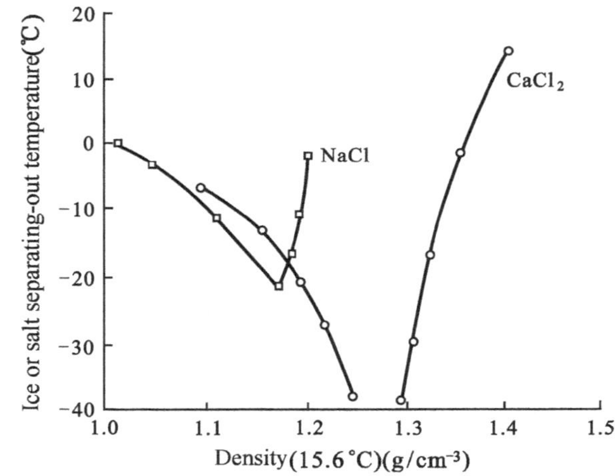

Fig. 2.2 Changing curves of ice or salt separating-out temperatures of aqueous solution with densities of salt water

At lower densities (the left part of curve), ice is precipitated when the temperature drops to a certain degree and this temperature is called the freezing point of salt solution, which decreases with the brine density; At higher densities (the right part of curve), salt is precipitated when the temperature decreases to a certain degree which is called salting temperature, which increases abruptly with the salt solution density. Therefore, when using water-soluble salts as high-density materials, the operating temperature of drilling fluid should be higher than its salting temperature at specific drilling fluid density. In order to prevent the salt-out effect on the performance of drilling fluid, salt crystallization inhibitors could be added in the drilling fluid.

The useful salt crystallization inhibitors are mostly amino carboxylic acid salts, and some examples are given below.

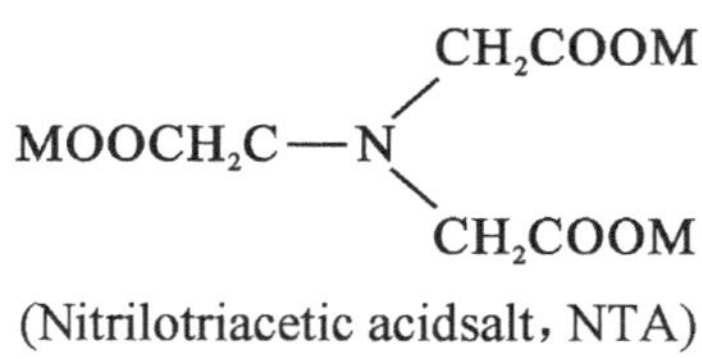

(Nitrilotriacetic acidsalt, NTA)

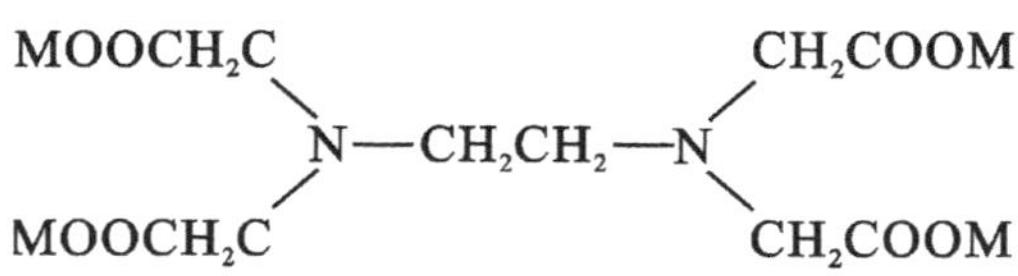

(Ethylenediaminetetraacetic acid salt, EDTA)

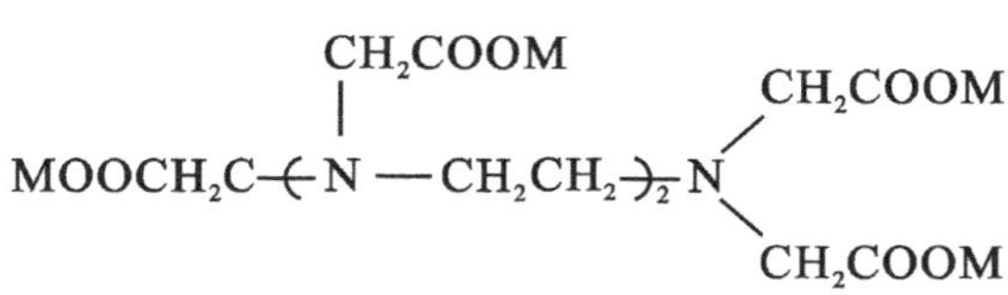

(Diethylenetriaminepentaacetic acid salt, DTPA)

M in formula could be K, Na, or NH_4. After dissolving in the drilling fluid, amino carboxylic acid salt can turn into corresponding salt by ion exchange (e.g., when the high-density material is calcium salt, the amino carboxylic acid salt will turn into calcium salt). The transformed salt can selectively adsorb on the surface of the newly precipitated salt crystals and lead to their structural distortion, which will prevent further salt-out effect on the crystal surface to achieve the salt control.

2.3 The Acid–Base Property of Drilling Fluid and Its Control

The acid–base property of drilling fluid is closely related to the clay dispersion degree in the drilling fluid, existential state of Ca^{2+}, Mg^{2+} and drilling fluid treating agents, and the rheological properties of drilling fluid. It also has direct impact on the corrosion of drilling tools.

The acid–base property of the drilling fluid can be evaluated by pH.

The pH of drilling fluid is usually controlled in the range of weak alkaline (pH 8–11). As in this range, clays can be dispersed properly and the treating agents can be dissolved sufficiently in the drilling fluid, thus the concentration increase of Ca^{2+} and Mg^{2+} in the drilling fluid could be inhibited to certain degree. And also in this pH range, the drilling fluid has low corrosiveness to the drilling tools.

Since beside OH^-, the inorganic ions employed to maintain the alkalinity of drilling fluid also include $CO_3{}^{2-}$ and $HCO_3{}^-$, which cannot be reflected in the pH value. Therefore, alkalinity is also used to express the acid–base property of drilling fluid.

Alkalinity is defined as the volume (with a unit of mL) of standard sulfuric acid (with a concentration of 0.01 mol/L) used to neutralize 1 mL sample to reach the indicator discoloration.

It can be concluded from the definition of alkalinity that the greater the alkalinity, the more basic the drilling fluid.

The sample used to measure the alkalinity should be the filtrate of drilling fluid, in order to prevent the effect of the non-neutralizing reaction between sulfuric acid and the components in the drilling fluid.

The alkalinity of the drilling fluid can indicate the sources of ions generating the basicity, because when using sulfuric acid with standard concentration to titrate samples these ions participate in reaction successively with the pH decrease. When the pH value reaches 8.3, the following reactions are almost complete.

$$OH^- + H^+ \rightarrow H_2O$$
$$CO_3{}^{2-} + H^+ \rightarrow HCO_3{}^-$$

However, at this pH the $HCO_3{}^-$ in the fluid does not participate in the reaction. When the pH value reaches 4.3, the reaction between $HCO_3{}^-$ and H^+ is mostly complete.

$$HCO_3{}^- + H^+ \rightarrow CO_2\uparrow + H_2O$$

Phenolphthalein and methyl orange can be used as neutralization indicators in the determination of alkalinity, as phenolphthalein changes color near the pH of 8.3 while methyl orange changes color at pH of around 4.3. The alkalinity measured by acid–base neutralization with phenolphthalein as indicator is called phenolphthalein alkalinity (denoted as P_f). The alkalinity measured by acid–base neutralization with methyl orange as indicator is called methyl orange alkalinity (denoted as M_f). Based on the relationship of phenolphthalein alkalinity and methyl orange alkalinity, we can distinguish the source of basicity of drilling fluid. If $P_f = 0$, it indicates the basicity of drilling fluid derives from HCO_3^-. If $P_f = M_f$, it indicates the basicity of drilling fluid derives from OH^-. If $P_f = 1/2M_f$, it indicates the basicity of drilling fluid comes from CO_3^{2-}.

The alkalinity of the drilling fluid is better kept at the range of 1.3–1.5 mL, and the M_f/P_f is better controlled within 3.

The acid–base property of the drilling fluid can be controlled using pH control agents (or basicity control agents). As the drilling fluid is generally used in weak alkaline conditions, the pH control agents are all alkaline chemical agents.

Some common pH control agents are shown below.

Sodium Hydroxide

Sodium hydroxide can dissociate in water to give OH^-.

$$NaOH \rightarrow Na^+ + OH^-$$

Therefore, the sodium hydroxide has strong pH control ability. The Na^+ generated from sodium hydroxide's dissociation in solution can change the calcium soil in drilling fluid to sodium soil, thus improving the stability of the drilling fluid. However, it can also induce the swelling and dispersion of the shale in the well wall, thus adverse to the well wall stability.

Potassium Hydroxide

Sodium hydroxide can also dissociate in water to give OH^-.

$$KOH \rightarrow K^+ + OH^-$$

Therefore, the pH control ability of potassium hydroxide is same with that of sodium hydroxide. However, unlike sodium hydroxide, the K^+ generated from potassium hydroxide's dissociation in water has the effect to suppress swelling and dispersion of the shale in well wall, which can improve the stability of well wall.

Sodium Carbonate

Sodium carbonate dissociates in water to generate carbonate, which hydrolyzes to produce OH^- to adjust the pH of drilling fluid.

$$Na_2CO_3 \rightarrow 2Na^+ + CO_3^{2-}$$
$$CO_3^{2-} + H_2O \rightarrow HCO_3^- + OH^-$$
$$HCO_3^- + H_2O \rightarrow H_2CO_3 + OH^-$$

Besides controlling pH of the drilling fluid, sodium carbonate can also reduce the concentration of Ca^{2+} and Mg^{2+} in drilling fluid. Therefore, sodium carbonate can be used as calcium or magnesium removal agents in the drilling fluid.

$$Ca^{2+} + CO_3^{2-} \rightarrow CaCO_3 \downarrow$$
$$Mg^{2+} + 2OH^- \rightarrow Mg(OH)_2 \downarrow$$
$$Mg^{2+} + CO_3^{2-} \rightarrow MgCO_3 \downarrow$$

Sodium carbonate can generate the same function as the sodium hydroxide does for calcium soil in drilling fluid and shale in the well wall. Thus, they can produce the same effect.

Sodium Bicarbonate

Sodium bicarbonate dissociates in water to generate bicarbonate, which hydrolyzes to produce OH^- to adjust the pH of drilling fluid.

$$NaHCO_3 \rightarrow Na^+ + HCO_3^-$$
$$HCO_3^- + H_2O \rightarrow H_2CO_3 + OH^-$$

The way sodium bicarbonate controls the acid–base property of drilling fluid is similar to that of sodium carbonate. Since sodium bicarbonate is acidic salt, it can control the pH of drilling fluid to reach a lower value (as low as 8.3).

2.4 The Filtration Property of Drilling Fluid and Its Control

2.4.1 *The Filtration Property of Drilling Fluid*

At certain temperature and pressure, the drilling fluid can filter into the formations. The filtration of the drilling fluid will result in the formation of a filter cake on the infiltration surface.

The filtration property of drilling fluid characterizes the difficulty of drilling fluid filtering into the formations. It can be evaluated by the amount of filtration loss of drilling fluid, which refers to the volume of filtrate obtained when the drilling fluid flows through a certain area of infiltration surface at certain temperature, pressure difference, and period of time (unit expressed in mL).

In general, the drilling fluid with lower filtration loss can form high-quality filter cake on the infiltration surface, which is thin and tough, has dense structure, good scouring resistance, and low friction coefficient.

The filtration loss of drilling fluid can be classified according to different standards. If classified according to the fluidity, it can be divided into static filtration loss and dynamic filtration loss. If classified according to the test temperature and pressure difference, it can be divided into conventional filtration loss (measured under the temperature of 24 ± 3 °C, the pressure difference of 0.69 MPa, the

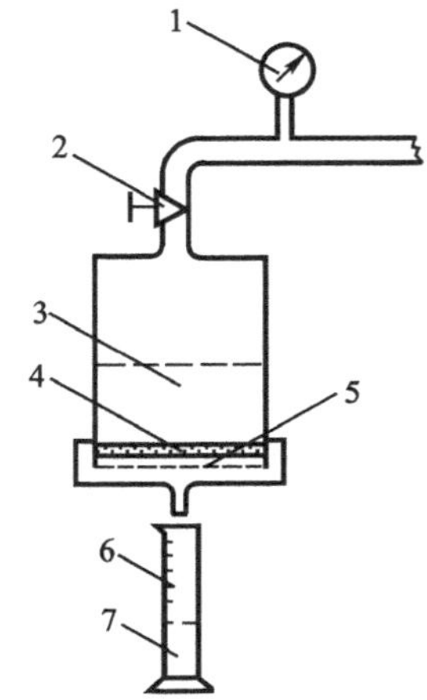

Fig. 2.3 Schematic diagram of conventional filter loss tester. 1—pressure gauge; 2—vent valve; 3—drilling fluid; 4—filter cake; 5—filter paper; 6—graduated cylinder; 7—drilling fluid filtrate

infiltration area of 45.8 cm^2, and the time period of 30 min) and high-temperature and high-pressure filtration loss (measured under the temperature of 150 ± 3 °C, the pressure difference of 3.45 MPa, the infiltration area of 45.8 cm^2, and the time period of 30 min). Figure 2.3 is the schematic diagram of drilling fluid conventional filter loss tester. The instrument can be used to measure the static filtration loss and the thickness of the filter cake under the conventional conditions.

The static filtration loss equation of drilling fluid can be deduced based on some boundary conditions before and after filtration and Darcy formula.

The volume of solid phase is constant before and after filtration, so the volume fraction of solid phase in the drilling fluid remains the same. The following equation can be obtained.

$$(q_{\mathrm{FL}} + AL)\varphi_{\mathrm{SM}} = AL\varphi_{\mathrm{SC}} \tag{2.1}$$

In Eq. 2.1, q_{FL} is drilling fluid filtration loss, m^3; A is infiltration area, m^2; L is thickness of filter cake; φ_{SM} is the volume fraction of the solid phase in the drilling fluid; φ_{SC} is the volume fraction of the solid phase in the filter cake.

Equation 2.1 can be transformed to the following form.

$$q_{\mathrm{FL}} = AL(\varphi_{\mathrm{SC}}/\varphi_{\mathrm{SM}} - 1) \tag{2.2}$$

The following formula is the Darcy formula.

$$\frac{\mathrm{d}q_{\mathrm{FL}}}{\mathrm{d}t} = \frac{kA\Delta p}{\mu L} \tag{2.3}$$

In the formula, $\frac{\mathrm{d}q_{\mathrm{FL}}}{\mathrm{d}t}$ is the infiltration rate of drilling fluid filtrate, m^3/s; k is filter cake permeability, m^2; Δp is the pressure difference on both sides of the filter cake, Pa; M is the viscosity of the filtrate, Pa · s.

By substituting Eq. 2.2 into Eq. 2.3 and integrating, you can obtain

$$q_{\mathrm{FL}} = A^{1/2}\left[\frac{2k\Delta p(\varphi_{\mathrm{SC}}/\varphi_{\mathrm{SM}} - 1)t}{\mu}\right]^{1/2} \tag{2.4}$$

Equation 2.4 is known as drilling fluid static filtration loss equation.

From the static filtration loss equation, it can be seen that the filtration loss of drilling fluid is proportional to the infiltration area. It is also proportional to the square root of infiltration time, filter cake permeability, solid content factor ($\varphi_{\mathrm{SC}}/\varphi_{\mathrm{SM}} - 1$), and the difference of pressures on both sides of filter cake (also referred as filter pressure difference). Drilling fluid's filtration loss is inversely proportional to the viscosity of the filtrate.

For drilling fluid with certain flow rate, after flowing for certain period of time, its scouring rate to the wall filter cake can be equal to the deposition rate of the solid particles on the filter cake, indicating that the thickness of the cake tends to be constant. So it can be assumed that k and L are both constants. Under these conditions, by integrating the Eq. 2.3, you can get dynamic filtration loss equation:

$$q_{\mathrm{FL}} = \frac{kA\Delta p}{\mu L} \tag{2.5}$$

It can be seen from Eqs. 2.4 and 2.5 that the factors affecting the static filtration loss also influence the dynamic filtration loss, however with different impact.

Although the temperature factor is not directly reflected in the static filtration loss equation and dynamic filtration loss equation, it functions by affecting the viscosity of the filtrate.

2.4.2 Control of the Filtration Property of Drilling Fluid

The filter loss property of the drilling fluid has an important effect on the protection of the reservoir, the stability of the wellbore and the formation of thick filter cake in the infiltration surface of high-permeability layers.

The filtration loss of drilling fluid can be controlled using filtrate reducer. Chemical agents that can reduce the filtration loss of drilling fluid are called drilling fluid filtrate reducer.

2.4.2.1 Classification of Drilling Fluid Filtrate Reducer

Drilling fluid filtrate reducers can be classified into the following categories:

Modified Lignite (also known as modified humic acid)
Lignite is a class of coal, which contains 20–80% of humic acid. Humic acid is not a single compound, but a mixture of compounds of different sizes and different compositions. These compounds have a skeleton containing an aromatic ring, and

the aromatic ring in the framework may be linked by an alkylene, carbonyl, ether, or amino group. Aromatic rings are surrounded by many carboxyl groups, hydroxyl groups, and sometimes methoxy groups. The following is a hypothetical illustration of the molecular structure of humic acid:

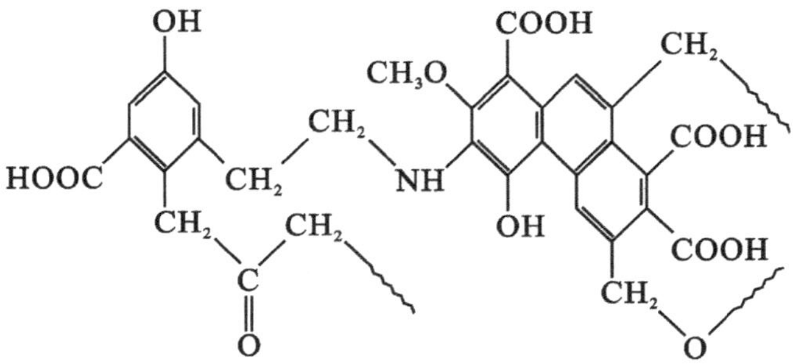

The relative molecular mass of humic acid lies in the range of 10^2–10^6. Humic acid is insoluble in water, yet it can react with alkali to produce water-soluble humate salts. Its water solubility can also be improved by means of nitrification or sulfomethylation. Therefore, the following modified lignite can be used as drilling fluid filtrate reducer.

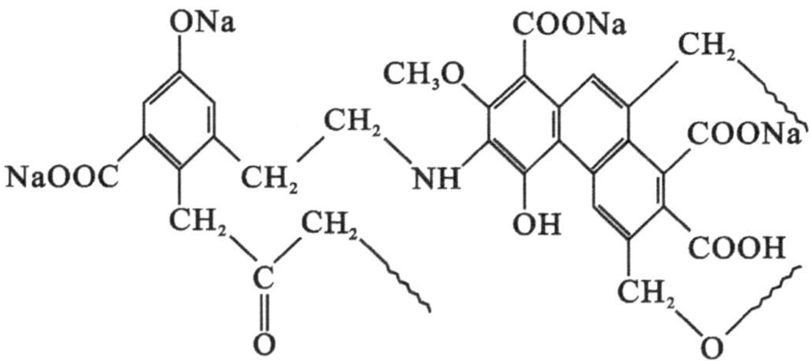

(Humic acid sodium, Coal alkali agen, Na-Hm)

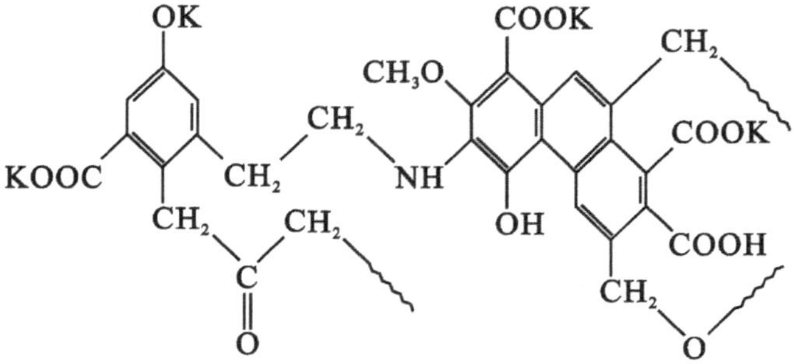

(Potassium humate, K-Hm)

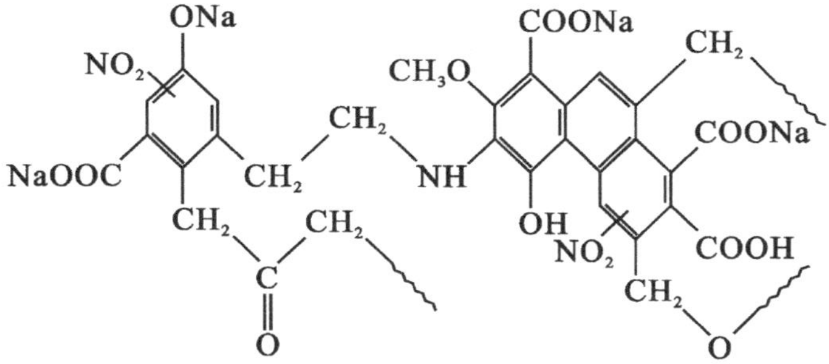

(Nitro-humic acid sodium, Na-NHm)

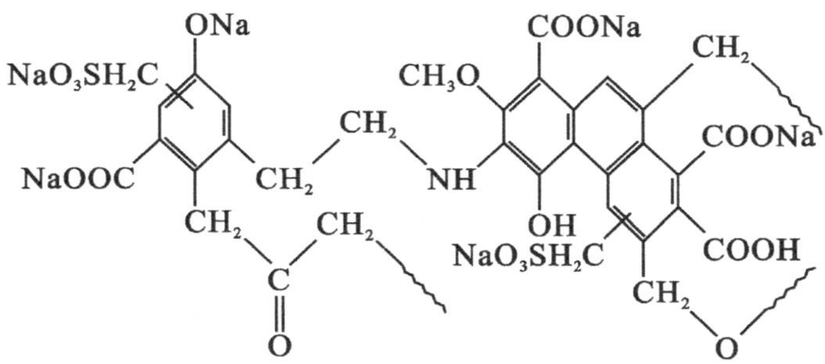

(Sodium sulfomethyl humate, Na-SMHm)

The above-mentioned modified lignite (modified humic acid) each has its own characteristic. The humic acid sodium has good thermal stability (up to 180 °C) while is sensitive to high salinity. The humic acid potassium can tolerate high temperature but not high salinity, and it bears simultaneously the functions of inhibiting shale expansion and dispersion. Nitro-humic acid and sodium sulfomethyl humate are both resistant to high temperature (up to 200 °C) and high salinity (up to 3×10^4 mg/L), and their tolerance to calcium and magnesium ions reaches up to 500 mg/L.

Modified Starch

Starch is a natural macromolecule composed of amylose and amylopectin. Amylopectin is a soluble starch and amylose is an insoluble starch (Chen 1980). In corn, potato, and other starch, the content of amylose is between 20 and 30%, while the content of amylopectin is between 70 and 80%. In order to use starch as a drilling fluid filtrate reducer, starch can be modified in various chemical ways such as alkalization, carboxymethylation, hydroxyethylation, and quaternization, to introduce hydrophilic groups into starch (Wang 1993).

The following modified starches can be used as drilling fluid filtrate reducer.

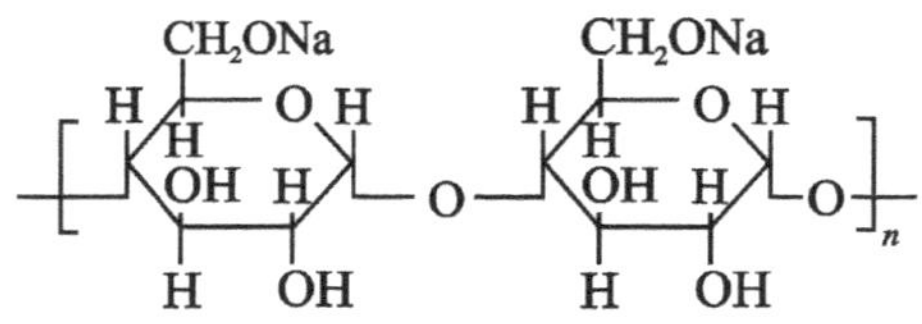

(Alkaline starch, Pregelatinized starch)

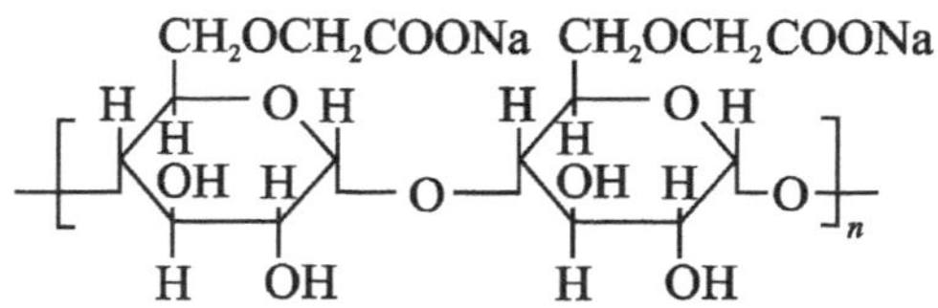

(Sodium carboxymethyl starch, Na-CMS)

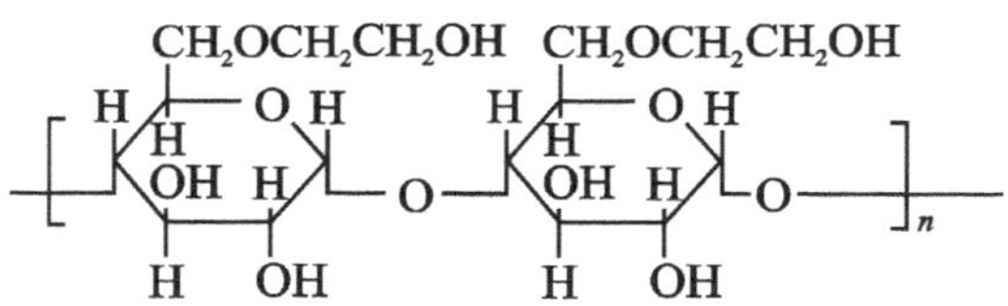

(Hydroxyethyl starch, HES)

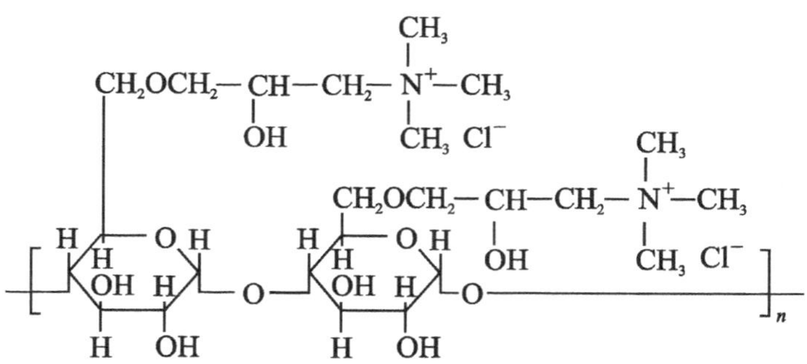

(The reaction product of epoxypropyltrimethyl ammonium chloride with starch)

(Wang Z and Wang A 1994; Jackson and Hartfiel 1978; Zhang 1998)

Modified starch can tolerate high temperature up to 120 °C. As the modified starch molecular chain is rigid, it has good salt resistance and can be used in saturated brine. The main drawback of modified starch is poor biostability.

Modified Cellulose (Wu and Yang 1992; Li et al. 1987)
Cellulose is also a natural polymer. Since the hydroxyethyl groups on the molecular chain can form intramolecular and intermolecular hydrogen bonds to produce crystals, it is insoluble in water. It is also possible to chemically modify the cellulose by carboxymethylation, hydroxyethylation and the like, to introduce hydrophilic groups into the cellulose so that it can be dissolved in water.

The following modified celluloses can be used as drilling fluid filtrate reducer.

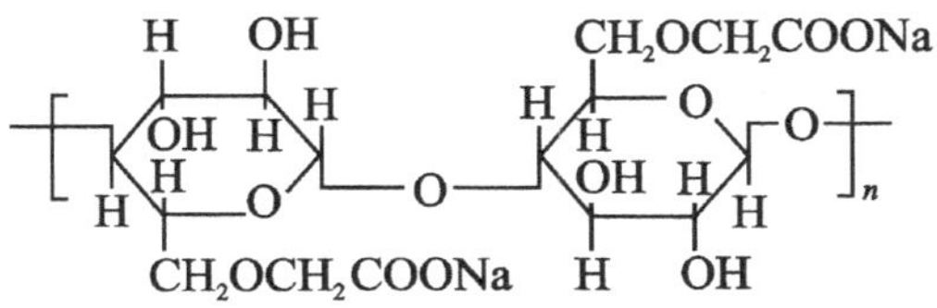

(Sodium carboxymethyl cellulose, Na-CMC)

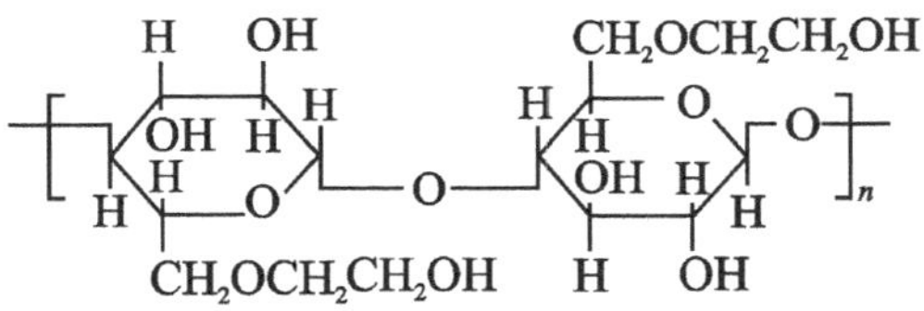

(Hydroxyethyl cellulose, HEC)

Modified cellulose can tolerate temperature as high as 130 °C. Meanwhile, it has good salt resistance, so it can be used in saturated brine. The biocompatibility of modified cellulose is better than that of modified starch.

Modified Resin
The resin here mainly refers to phenolic resin. The phenol used to synthesize these resins could be phenol, or phenols in phenol propane units of lignin (referring to Fig. 2.4) and phenol in the lignite (humic acid).

The following modified resins can be used as drilling fluid filtrate reducer.

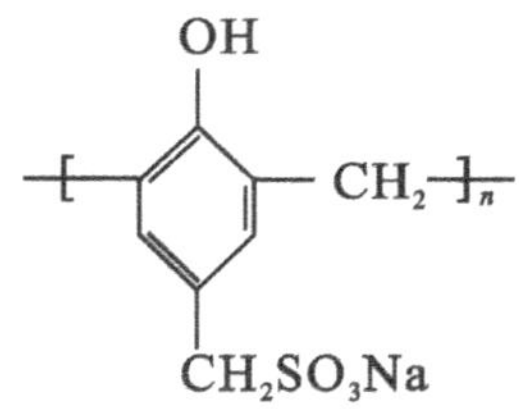

(Sulfomethyl phenolic resin, SMP)

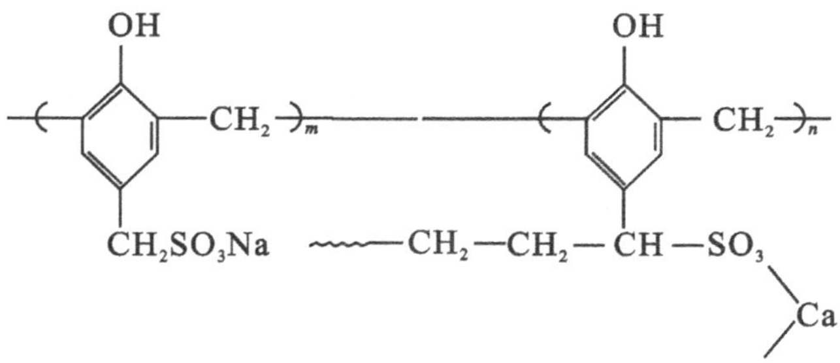

(Condensation compound of sulfomethyl phenolic resin and sulfonated lignin resin, SLSP)

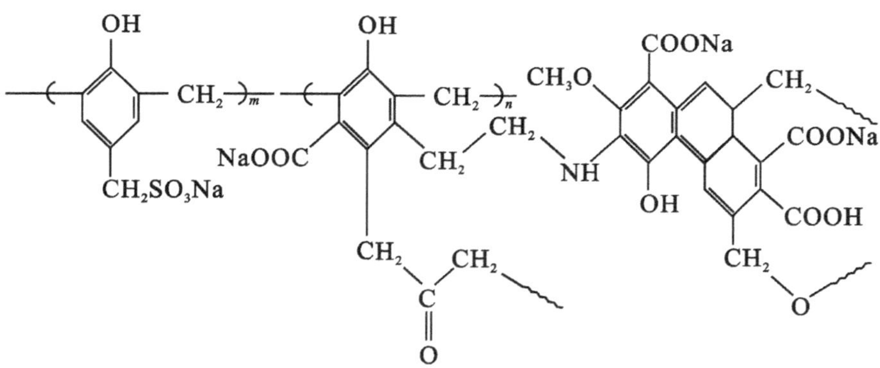

(Condensation compound of sulfomethyl phenolic resin and lignite resin)

For the above-mentioned modified resins, as their main chains contain aromatic rings as well as strong hydrophilic groups ($-SO_3Na$), they have good resistance to high temperature, high salinity, and high concentration of calcium and magnesium.

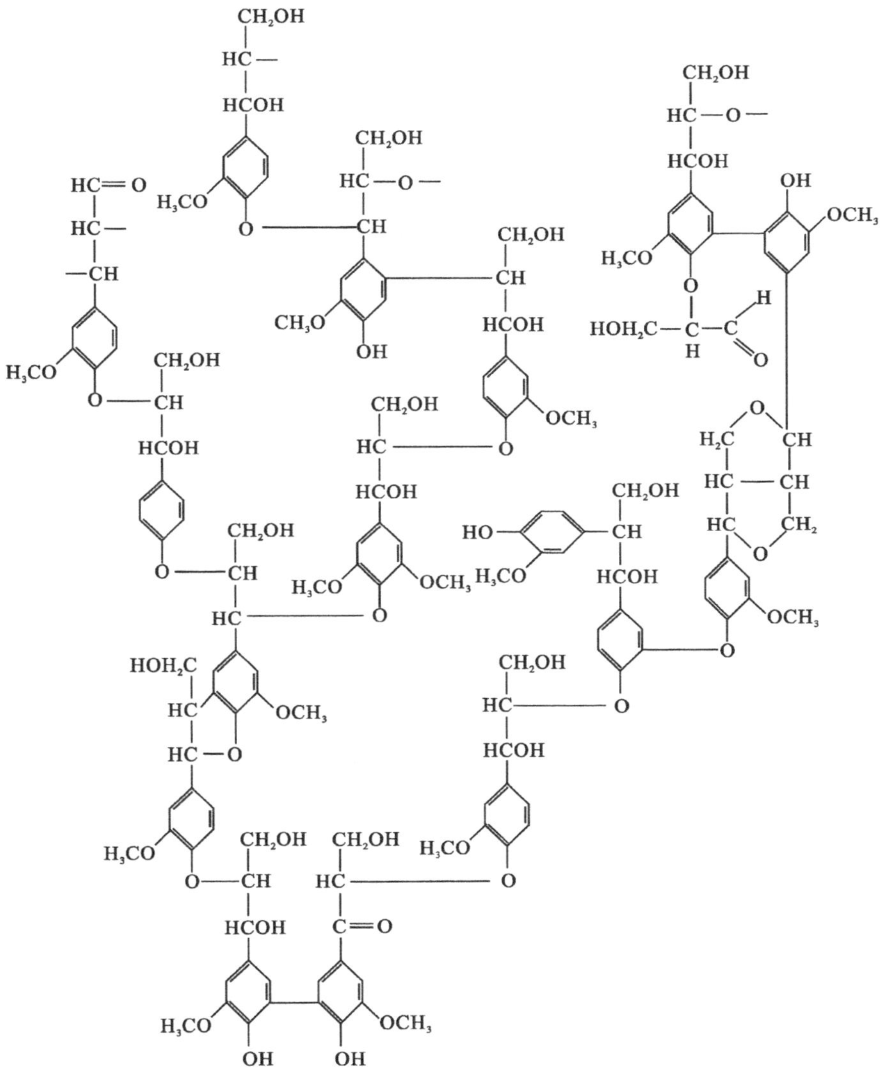

Fig. 2.4 A lignin structure from wood (Adler 1977)

For example, sulfomethyl phenolic resin can tolerate high temperature up to 200 °C, high concentration of calcium up to 2×10^3 mg/L, and it can also be used in saturated brine.

The modified resins have foaming problem, which can be addressed using defoamers such as pentanol, polydimethylsiloxane.

Olefin Monomer Polymer (Perricone 1986; Plank 1992; Audibert 1995)
The olefin monomers include acrylonitrile, acrylamide, acrylic acid, (2-acrylamido-2-methyl) propylsulfonate, and *N*-vinylpyrrolidone. They can form binary copolymers or terpolymers by copolymerization. These copolymers can be hydrolyzed and/or chemically modified to enhance their water solubility.

The following olefin monomer polymers can be used as drilling fluid filtrate reducer.

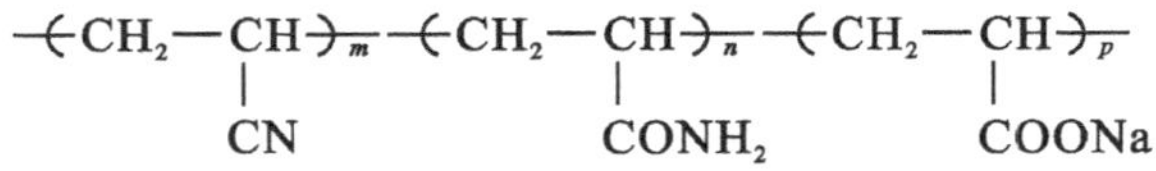

(Partially hydrolyzed polyacrylonitrile sodium salt, Na-HPAN)
(Fan et al. 2005)

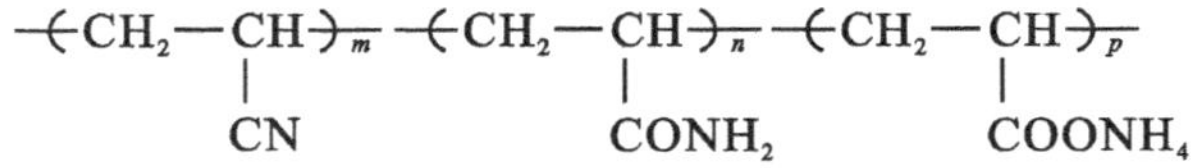

(Partially hydrolyzed polyacrylonitrile ammonium salts, NH_4-HPAN)

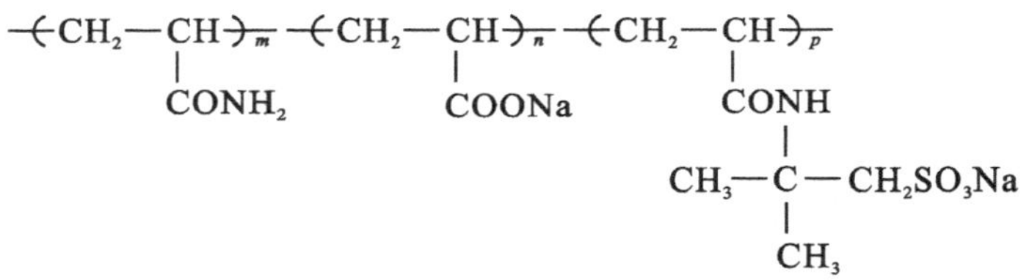

[Copolymer of acrylamide, sodium acrylate and (2-acrylamido-2-methyl) propyl sodium sulfonate, AM-AA-AMPS]

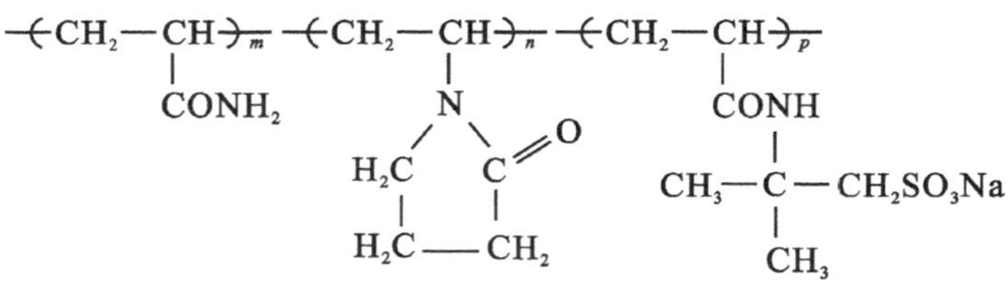

[Copolymer of acrylamide, N-vinylpyrrolidone and (2-acrylamido-2-methyl) propyl sodium sulfonate, AM-VP-AMPS]

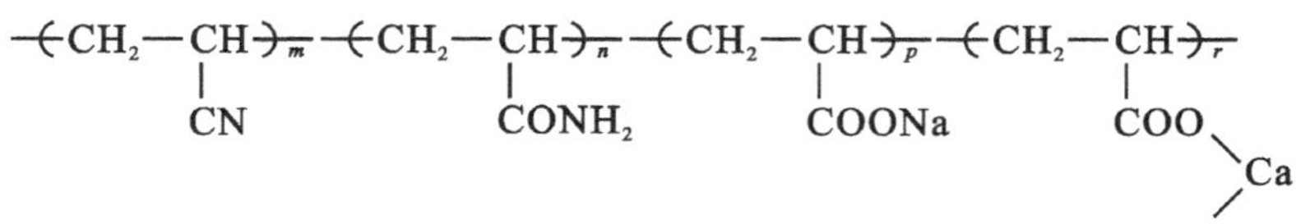

(Partially hydrolyzed polyacrylonitrile calcium salt, Ca-HPAN)

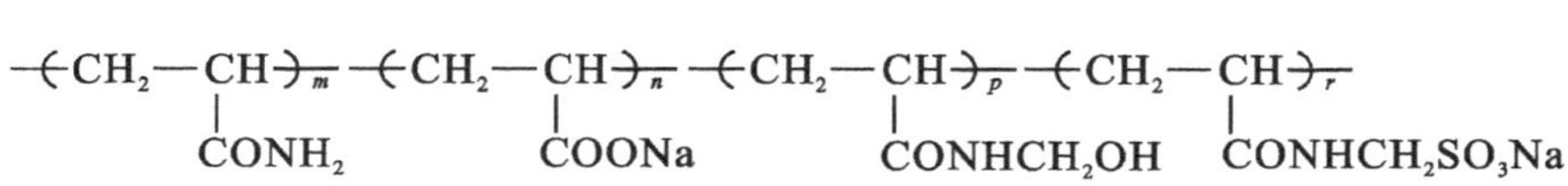

(Partially hydrolyzedsulfonated polyacrylamide, SHPAM)

In these olefinic monomer copolymers, the introduction of different monomers allows the copolymers to have different properties. The chains of acrylonitrile, (2-acrylamido-2-methyl) propyl sulfonate, and *N*-vinyl pyrrolidone endow the copolymer with high-temperature resistance and high-salinity resistance. The chains of acrylamide endow the copolymer with good adsorption performance. The chains of sodium acrylate endow the copolymer with good water solubility, however reduces the calcium resistance of the copolymer.

2.4.2.2 Function Mechanisms of the Drilling Fluid Filtrate Reducer

The above-mentioned five types of drilling fluid filtrate reducers function through the following mechanisms.

Thickening Mechanism
The above-mentioned five types of drilling fluid filtrate reducers are all water-soluble polymers which dissolve in drilling fluids to increase their viscosity. It can be observed from Eqs. 2.4 and 2.5 that the increase of the viscosity of the drilling fluid can alleviate the filtration loss.

Adsorption Mechanism
The above-mentioned five types of drilling fluid filtrate reducers can all adsorb on the surfaces of clay particles through hydrogen bonding, which will increase the electronegativity and thicken the hydration layer on the clay particle surface. Thusly, the agglomeration stability of clay particles is improved, with the clay particles maintaining relative small particle size and reasonable particle size distribution. In this way, high-quality filter cake which is thin and tough and has dense structure can be produced, reducing the permeability of the filter cake. It can be observed from Eqs. 2.4 and 2.5 that the decrease of the permeability would result in reduced filtration loss.

Capture Mechanism
Trapping refers to the phenomenon that the irregular polymer coils (or solid particles) remains in the pores through bridging effect. If we assume the diameter of the irregular polymer coil (or solid particles) is d_c and the diameter of the pore is d_p, trapping occurs when the following condition is satisfied.

$$d_c = \left(\frac{1}{3} \text{ to } 1\right) d_p$$

The above-mentioned five types of drilling fluid filtrate reducers are all polymers which consist of many substances with different molecular weight. These substances twist in water to form irregular coils with different sizes. The irregular coils would be retained in the pores of filter cake when their diameters meet the aforementioned trapping conditions, which will reduce the permeability of the filter cake and diminish the filtration loss of drilling fluid.

ocking Mechanism

· polymer coils with (or solid particles) with a d_c greater than d_p, they the pores of the filter cake. However, they can reduce the filtration loss of drilling fluid by blocking the inlet of the filter cake pores. This mechanism of reducing the filtration loss of drilling fluid is called the physical blocking mechanism, which is different from the trapping mechanism.

2.5 The Rheological Property of Drilling Fluid and Its Adjustment

The rheological property of drilling fluid refers to its flow and deformation property which is characterized by the relationship between shear stress and shear rate. The rheological property of drilling fluid is directly related to its washing capacity to the well bottom, its carrying and suspending capability of drilling cuttings, its power transfer capacity and the stability of the borehole wall.

2.5.1 Basic Concept

2.5.1.1 Flow State

Fluid flow state can be divided into two types, laminar flow and turbulent flow.

The laminar flow means fluid particles flow is laminar. The flow rates in each layer are different; however, they are all parallel to the direction of flow.

Turbulent flow means fluid particles flow in irregular form. The whole volume of fluid is filled with small eddies, and the macrospeeds of the particles are basically the same.

For drilling fluid, due to the existence of the dispersed phase structure, there is a plug flow state between the static and the laminar flow.

The plug flow means the fluid flows like a plug, and the velocity of each particle is the same.

In between different flow states lies a transitional state which is called transition flow.

Therefore, various flow states of drilling fluid are shown in Fig. 2.5.

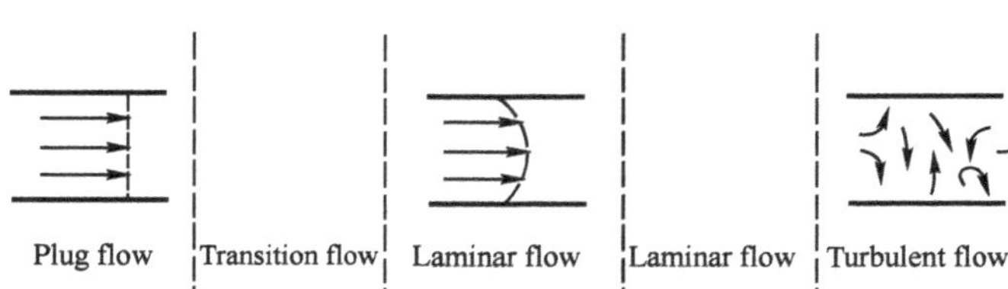

Fig. 2.5 Flow states of drilling fluid

2.5.1.2 Shearing Stress

When the flow state is in the laminar flow, the flow velocities in adjacent flow layers are different, therefore internal friction force or shear force exists. If the shear force is divided by the contact area of the adjacent flow layers, the shear stress is obtained which can be expressed in the following definition formula, where τ is the shearing stress, F is the shear force, and A is the contact area between adjacent flow layers.

$$\tau = \frac{F}{A} \quad (2.6)$$

2.5.1.3 Shear Rate

In a laminar flow, if the velocity difference between the adjacent flow layers is divided by their vertical distance, the shear rate is obtained which can be expressed in the following definition formula, where $\dot{\gamma}$ is the shear rate, dv is the velocity difference between the adjacent flow layers and dZ is the vertical distance between adjacent flow layers.

$$\dot{\gamma} = \frac{\mathrm{d}v}{\mathrm{d}Z} \quad (2.7)$$

The shear rate is different at different positions in the circulation of drilling fluid.

- Grit chamber, 10–20 s^{-1}
- Annular space, 15–150 s^{-1}
- Drill pipe, 10^2–10^3 s^{-1}
- Bit nozzle, 10^4–10^5 s^{-1}

2.5.1.4 Newton Viscosity and Apparent Viscosity

The law of Newton internal friction is

$$\tau = \mu\dot{\gamma} \quad (2.8)$$

or

$$\mu = \frac{\tau}{\dot{\gamma}} \quad (2.9)$$

In these equations, μ is the Newton viscosity.

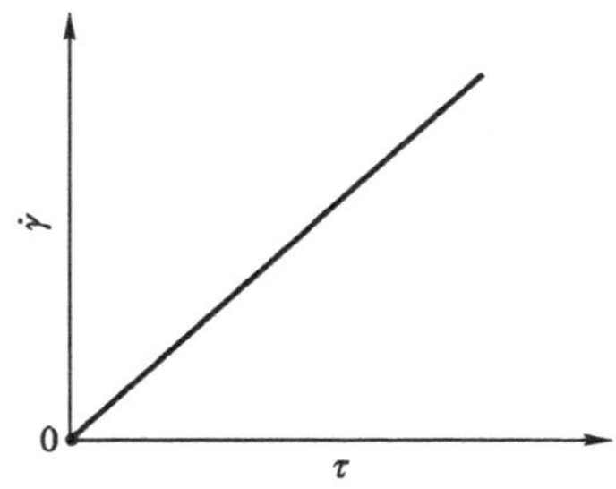

Fig. 2.6 Relationship between shear stress and shear rate

Newton fluid refers to the fluid that meets the law of Newton internal friction. Figure 2.6 shows the relationship between the shear stress and the shear rate of the Newton fluid. It can be seen from Fig. 2.6 that the relation line between shear stress and shear rate is a straight line passing through the origin which means the shear stress is proportional to the shear rate. Newton viscosity is constant at different shear rates.

Fluid which does not conform to the law of Newton internal friction is called non-Newtonian fluid. The viscosity of the non-Newtonian fluid varies with the shear rate. Drilling fluid is a non-Newtonian fluid. The viscosity of the non-Newtonian fluid defined by Eq. 2.9 is called the apparent viscosity. Therefore, for non-Newtonian fluid,

$$\mu_{AV} = \frac{\tau}{\dot{\gamma}} \tag{2.10}$$

In the equation, μ_{AV} is the apparent viscosity.

Since the apparent viscosity of drilling fluid varies with shear rate, when the performance of the drilling fluid is evaluated, the apparent viscosity is generally referred to as the apparent viscosity at the shear rate of 1,022 s^{-1}.

2.5.1.5 Thixotropy

Thixotropy refers to the property that non-Newtonian fluid gets thinning (or thickening) under mechanical actions, while gets thickening (or thinning) after elimination of the mechanical actions.

Drilling fluid possesses the property of thixotropy, which is determined by rotational viscometer. During the measurement, the drilling fluid is first placed in a rotation viscometer at 600 r/min (equivalent to a shear rate of 1,022 s^{-1}) for 10 s. Then, the shear stress of drilling fluid for being static for 10 s and 10 min, respectively, at 3 r/min (equivalent to a shear rate of 5.11 s^{-1}) is measured, whose difference is applied to determinate the thixotropy of drilling fluid.

2.5.2 *Rheological Models of Drilling Fluid*

The rheological property of drilling fluid is characterized by the relationship between shear stress and shear rate of drilling fluid. The curve of the relationship is called the rheological curve. The mathematical formula of the rheological curve is called the rheological model. The constant in the rheological model is called rheological parameter.

The following rheological models apply to drilling fluids.

2.5.2.1 Bingham Model

The rheological curve shown in Fig. 2.7 (straight line) can be expressed in the following Bingham model, where τ_d is the intercept on the τ axis and μ_{PV} is the reciprocal of the straight slope.

$$\tau = \tau_d + \mu_{PV}\dot{\gamma} \tag{2.11}$$

Fluid which accords with Bingham model is called Bingham fluid (also known as plastic fluid). The characteristic of Bingham fluid is that it starts flowing only when shear stress exceeds a certain value. The minimum shear stress which makes fluid flow (i.e., the intercept τ_d on the τ axis) is called the yield value. The reciprocal of the slope of Bingham fluid rheological curve (straight line) is called plastic viscosity, which reflects the magnitude of internal friction of a Bingham fluid in the flowing state.

By plugging Eq. 2.11 into Eq. 2.10, the following conclusion can be drawn:

$$\mu_{AV} = \mu_{PV} + \frac{\tau_d}{\dot{\gamma}} \tag{2.12}$$

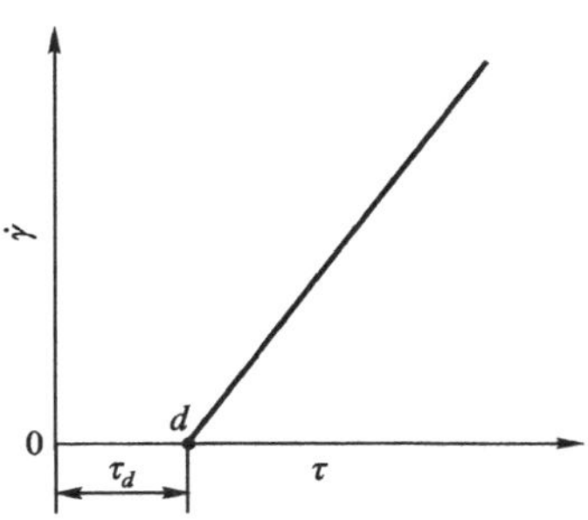

Fig. 2.7 Rheological curve of Bingham fluid

It can be seen from Eq. 2.12 that the apparent viscosity of Bingham fluid is composed of two parts, plastic viscosity and $\tau/\dot{\gamma}$. As $\tau/\dot{\gamma}$ depends on the structure strength in Bingham fluid, it is called structural viscosity.

The rheological properties of drilling fluid made up of bentonite generally conform to the Bingham model.

Figure 2.8 shows the rheological curve of this kind of drilling fluid.

It can be seen from rheological curve of the drilling fluid (Fig. 2.8) that the flow process can be divided into five stages ($0 \rightarrow 1 \rightarrow 2 \rightarrow 3 \rightarrow 4 \rightarrow 5$).

- The static stage ($0 \rightarrow 1$): when the shear stress is less than τ_1, the drilling fluid does not flow.
- The plug flow stage ($1 \rightarrow 2$): when the shear stress is larger than τ_1, only the structure of the drilling fluid close to the pipe wall will be destroyed. The drilling fluid is in plug flow.
- The plug–laminar transition flow stage ($2 \rightarrow 3$): when the shear stress continues to increase ($\tau_2 \rightarrow \tau_3$), the internal structure of the drilling fluid is gradually destroyed. The drilling fluid is in a transition state between plug flow and laminar flow.
- The laminar flow stage ($3 \rightarrow 4$): when the shear stress is greater than τ_3, the destruction and recovery of the internal structure of the drilling fluid is in a state of equilibrium. The drilling fluid is in laminar flow then.
- Turbulent stage ($4 \rightarrow 5$): when the shear stress is greater than τ_4, the drilling fluid gets into a turbulent state, from which Bingham model is no longer suitable for describing the rheological properties of drilling fluid.

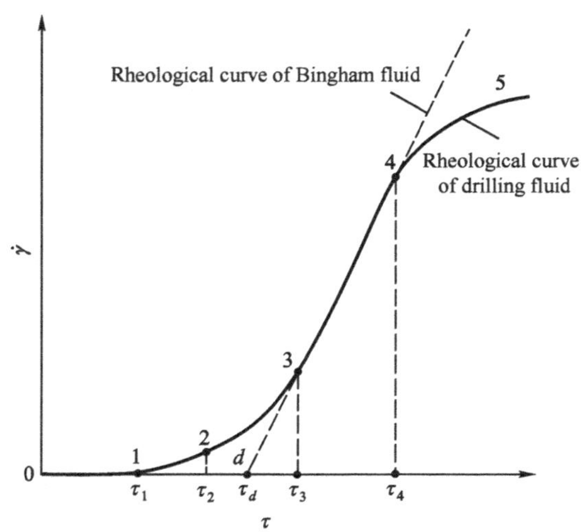

Fig. 2.8 Rheological curve of drilling fluid

Figure 2.8 shows that the rheological property of drilling fluid is consistent with the rheological model of Bingham fluid only in laminar flow.

The τ_1 in Fig. 2.8 is the minimum shear stress needed to move drilling fluid from static stage to flow state, known as static shear force (denoted by τ_g), reflecting the structural strength of drilling fluid in stationary state. Accordingly, the shear force indicated as point *d*, which is the intersection of the extension line of the rheological curve 3 → 4 and the τ axis in Fig. 2.8, is called dynamic shear force (denoted by τ_d), reflecting the structural strength of drilling fluid in flow state.

2.5.2.2 Power Low Model

The rheological curve, as shown in Fig. 2.9, can be expressed by a power-law model, where *K* is the consistency coefficient and *n* is the flow index.

$$\tau = K\dot{\gamma}^n \ (n < 1) \tag{2.13}$$

The fluid that conforms to the power-law model is called power-law fluid. If the fluid index of power-law fluid is less than 1, it is called pseudo-plastic fluid. The characteristic of a power-law fluid is that it starts flowing immediately when exposure to shear stress, thusly its rheological curve passes through the origin.

The flow index in the power-law model is concerned with the structure strength of the fluid. A stronger structure will lead to a smaller flow index. The viscosity coefficient is concerned with the internal friction in the fluid. The higher the internal friction between adjacent layers is, the larger the viscosity coefficient is.

The following equation can be obtained by taking logarithm on both sides of Eq. 2.13.

$$\lg \tau = \lg K + n \lg \dot{\gamma} \tag{2.14}$$

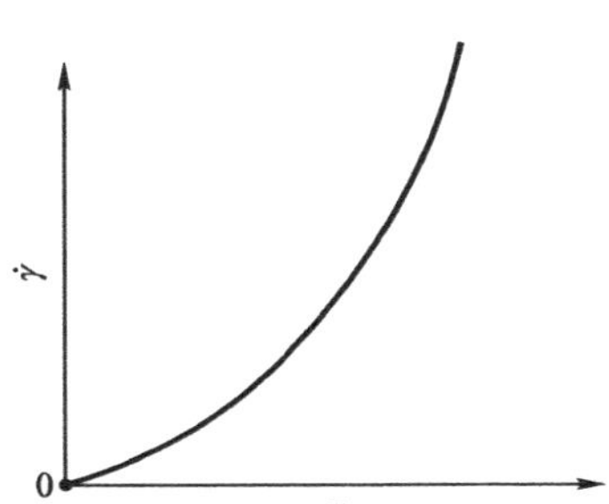

Fig. 2.9 Rheological curve of power-law fluid ($n < 1$)

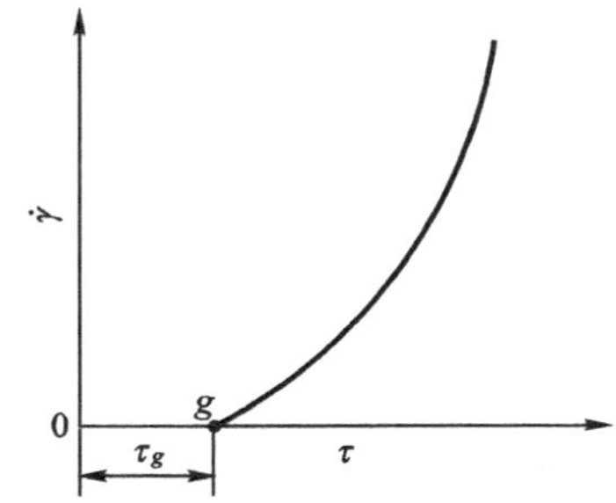

Fig. 2.10 Rheological curve of a power-law fluid with yield value

Eq. 2.14 shows that a linear relation exists between lg τ and lg γ. The slope of the line is the flow index, and the intercept is lg K.

Most of the drilling fluids prepared with polymers are in accord with the power-law model.

Figure 2.10 shows the rheological curve of a power-law fluid ($n < 1$) with yield value. This rheological curve can be expressed by the following modified power-law model, where τ_g is the yield value (also called static shear force).

$$\tau = \tau_g + K\dot{\gamma}^n \tag{2.15}$$

2.5.2.3 Casson Model

The rheological curve shown in Fig. 2.11 can be represented by the following Casson model.

$$\tau^{1/2} = \tau_c^{1/2} + \tau_\infty^{1/2}\dot{\gamma}^{1/2} \tag{2.16}$$

The following equation can be obtained if both sides of Eq. 2.16 are divided by $\dot{\gamma}^{1/2}$, where τ_c is the yield value and μ_∞ is the ultimate shear viscosity.

$$\mu_{\mathrm{AV}}^{1/2} = \mu_\infty^{1/2} = \tau_{\mathrm{c}}^{1/2}\dot{\gamma}^{-1/2} \tag{2.17}$$

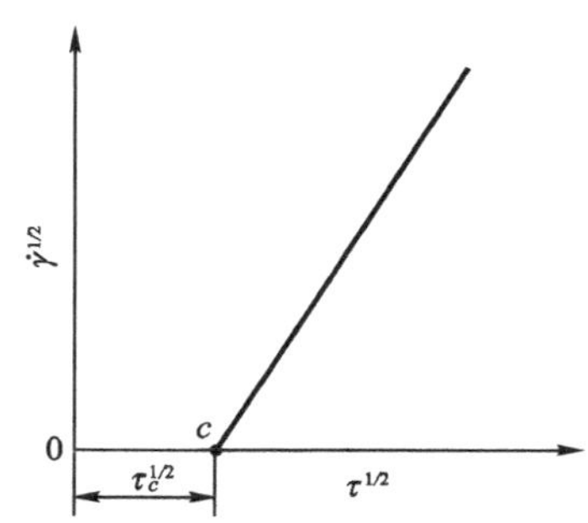

Fig. 2.11 Rheological curve of a Casson fluid

Fluid that is consistent with the Casson model is called Casson fluid. Casson yield value in Casson mode reflects the strength of fluid structure while the ultimate shear viscosity reflects the magnitude of the internal friction in high shear rate. Casson yield value and ultimate shear viscosity can be obtained by graphic method.

Compared with Binham mode and power-law model, Casson mode can be applied in a wider range of shear rate.

2.5.3 *Modification of the Rheological Properties of Drilling Fluid*

The modification of the rheological properties of drilling fluid is achieved mainly through the adjustment of drilling fluid viscosity (apparent viscosity) and shear force (static shear force and dynamic shear force).

In the drilling process, too high or too small drilling fluid viscosity and shear force will all result in adverse effects. Unreasonably, high drilling fluid viscosity and shear force will cause issues such as large flow resistance, high-energy consumption and generate serious impact on drilling rate. Furthermore, it will cause drill mud bag, sticking of tool, difficulty in drill cuttings removal and drilling fluid degassing. On the other hand, if the viscosity and shear force of the drilling fluid is too small, it will affect the cuttings carrying capability of drilling fluid and the wellbore stability.

The viscosity and shear force of drilling fluid can be modulated by adjusting the solid content in the drilling fluid. Although this is the first method in consideration, its application is limited because the changes of solid content will affect other properties of drilling fluid.

Based on the adjustment of solid content, rheology adjusting agents can also be used to adjust the drilling fluid viscosity and shearing force.

Rheology adjusting agents can be classified into the following two types.

2.5.3.1 Viscosity Depressant

Viscosity depressant refers to the rheology adjusting agents which can reduce the viscosity and shearing force of drilling fluid.

According to the chemical composition, viscosity depressant can be further classified as follows.

Modified Tannin

Gallnut tannin is the mostly used tannin in the modification, which is made from the esterification of five double gallic acids with glucose (reference figure is shown in Fig. 2.12.).

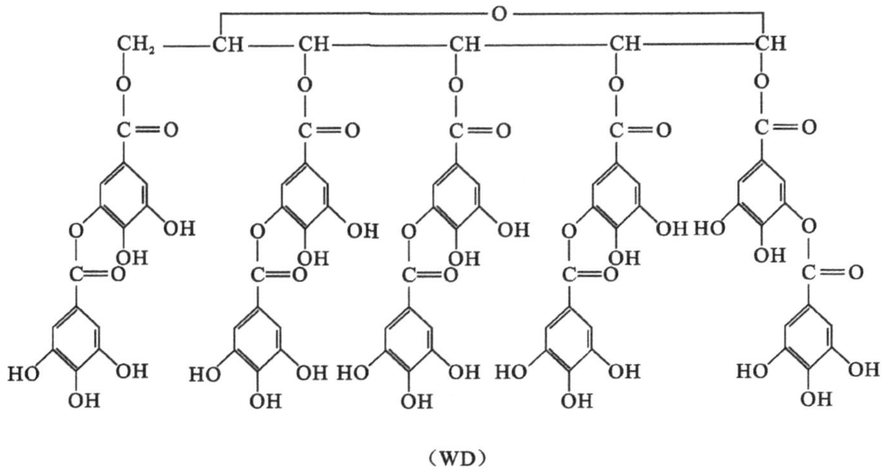

Fig. 2.12 Structure of Gallnut tannin

Gallnut tannin can generate double gallic acids and glucose by hydrolysis in water.

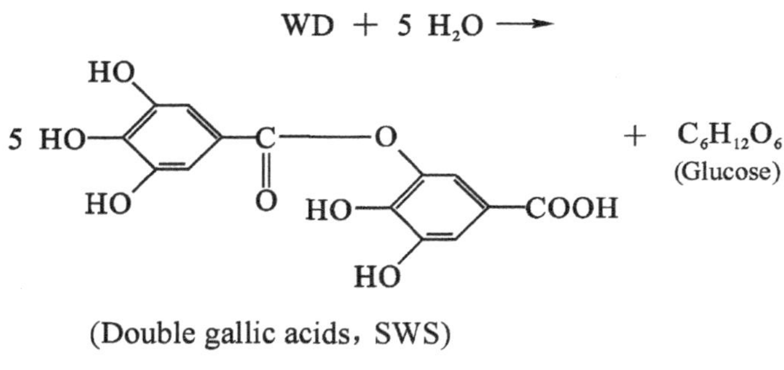

Double gallic acid can further hydrolyze into gallic acid.

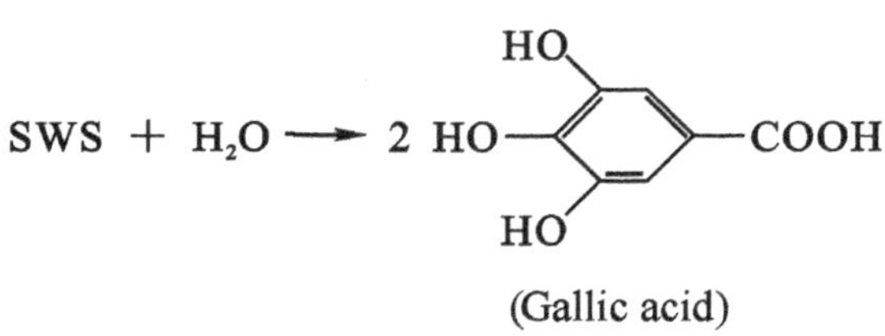

There are three kinds of modified tannin. One is tannin lye prepared by tannin and NaOH. Its main components are shown as follows.

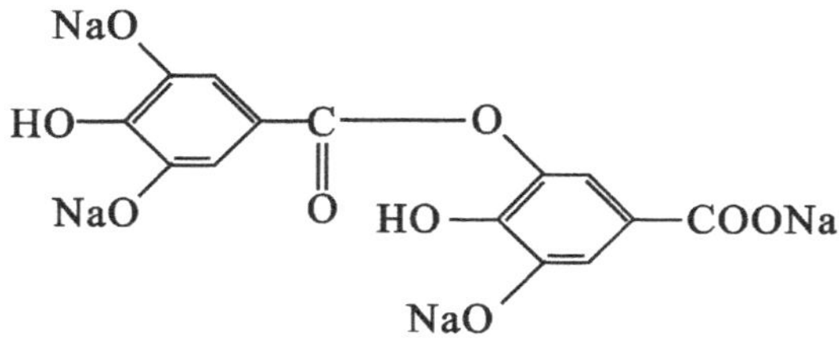

(Double sodium gallic acid)

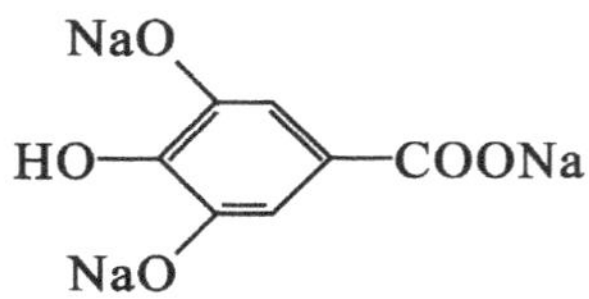

(Sodium gallic acid)

They can be collectively referred to as sodium tannate (NAI). Another kind is tanning extract lye. Though it comes from tanning extract (prepared from lixivium of red willow root or oak bark), its main component is still sodium tannate. The last kind is sulfomethylated tannin (SMT) which is prepared from the reaction of tannin with formaldehyde and sodium bisulfite under alkaline conditions. Its main components are shown as follows.

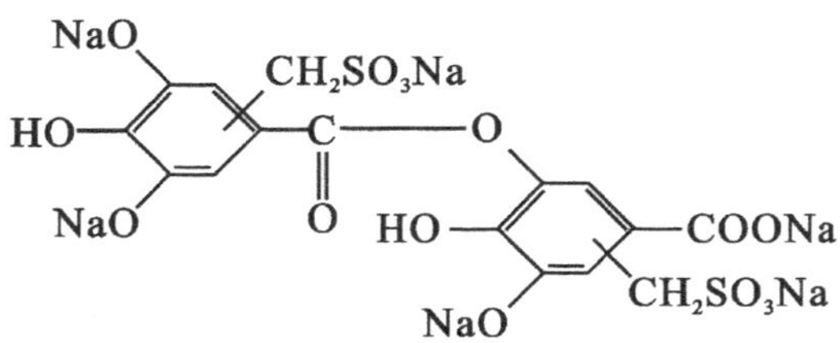

(Double sulfonated sodium gallic acid)

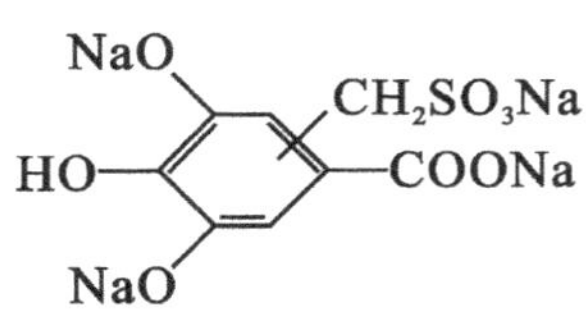

(Sulfonated sodium gallic acid)

The modified tannin can adsorb on the surface of clay particles through hydrogen bonds among the hydroxyl groups in the modified tannin and the hydroxyl groups on the clay surface. The polar groups such as –COONa, $–SO_3Na$, and –ONa in the modified tannin can dissociate in water to generate diffused double layer, which will improve the electronegativity of clay particle surface and increase the thickness of hydration layer. In this way, the structures of clay particles are broken up; therefore, the viscosity and shear force are reduced. Modified tannin can function as viscosity reducer for the drilling fluid with clay particles' structures.

Modified Lignin Sulfonate

Lignosulfonate is a kind of by-product generated in sulfite papermaking method, its basic structure unit can be shown by the following structure formula.

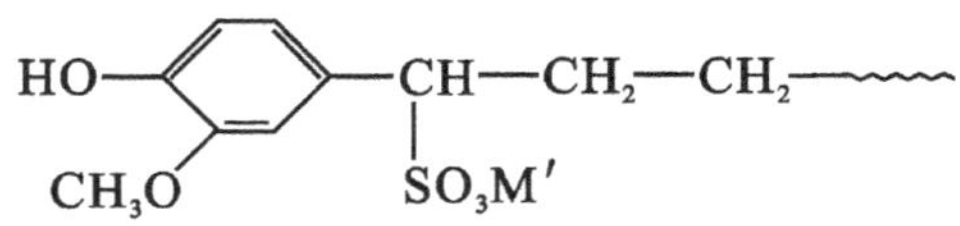

M′: Na or 1/2Ca

Trivalent metal ions M^{3+} such as Fe^{3+} and Cr^{3+} can form polynuclear olation complex ions by following reactions in water.

- Complexation

$$M^{3+} + 6H_2O \rightarrow [(H_2O)_6M]^{3+}$$

- Hydrolyzation

$$[(H_2O)_6M]^{3+} \rightarrow [(H_2O)_5M(OH)]^{2+} + H^+$$

- Olation

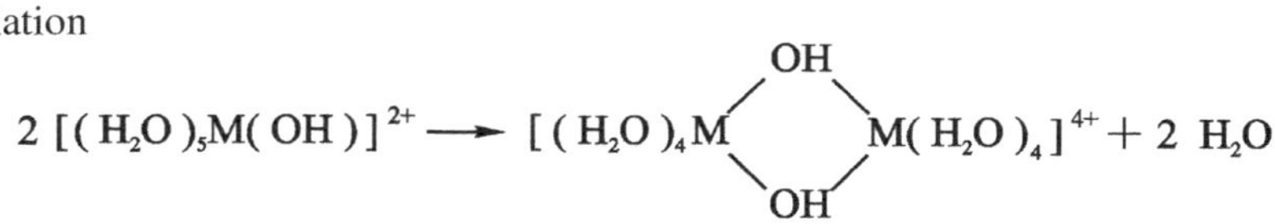

- Further hydrolyzation and olation

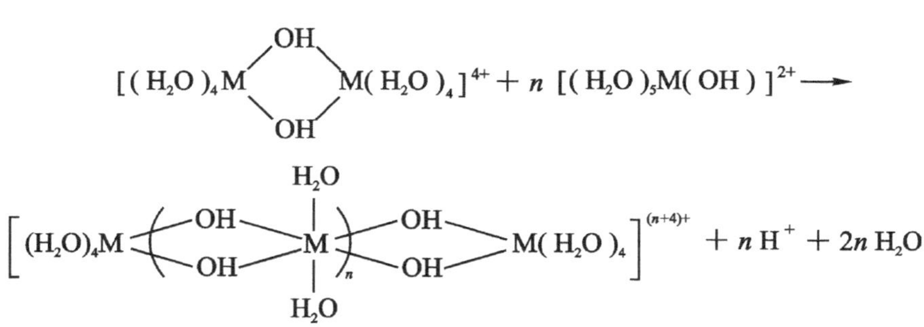

(Polynuclearolation complex ions generated by trivalent metal)

Lignin sulfonate can coordinate with polynuclear olation complex ions generated by trivalent metal. If the trivalent metal ions are Fe^{3+} or Cr^{3+}, their polynuclear olation complex ions can modify the lignin sulfonate to form the following structure.

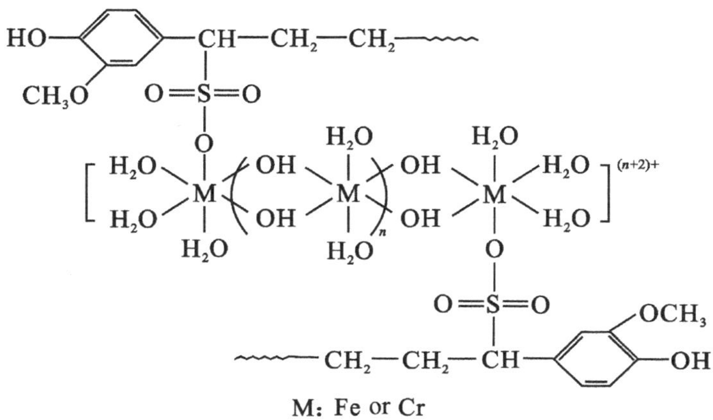

(Iron or chromium lignosulfonate, FCLS)

Iron or chromium lignosulfonate adsorb on clay particle surface through hydrogen bonds, which will improve the electronegativity of clay particle surface and increase the thickness of hydration layer. In this way, the structures of clay particles are broken up; therefore, the viscosity and shear force are reduced.

Compared with modified tannin, iron and chromium lignosulfonate possess more polar groups including salt and temperature-tolerant sulfonate groups. Therefore, they exhibit better performance in reducing viscosity and shear force and obtain good thermal resistance and salt tolerance properties.

Foaming problems still exist in iron and chromium lignosulfonate; therefore, defoamer is needed in practical application.

Considering the problem of chromium pollution, a series of chromium-free lignin sulfonate such as iron, zirconium, titanium lignin sulfonate (Park 1988; Miano and Lockhart 1994) have been developed.

Vinyl Monomer Oligomer

Oligomer refers to the polymer with relatively low molecular weight (1×10^2 to 6×10^2). The following vinyl monomer oligomers can be used as viscosity reducer.

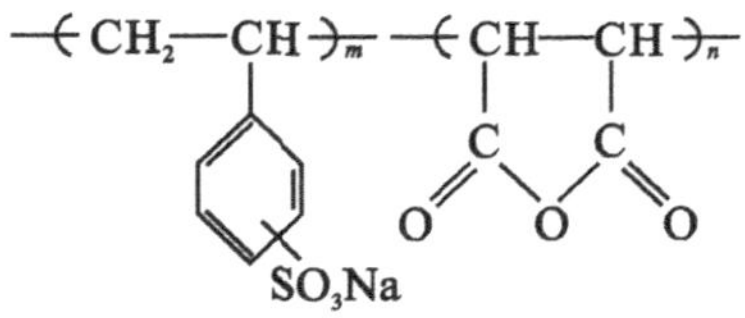

(Oligomer of sodium styrene sulfonate and maleic anhydride, SSMA)
(Hale 1998)

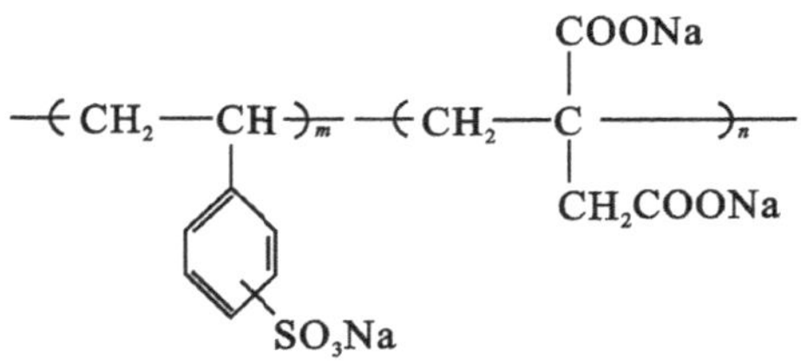

(Oligomer of sodium styrene sulfonate and itaconic acid)
(Zhao et al. 2009)

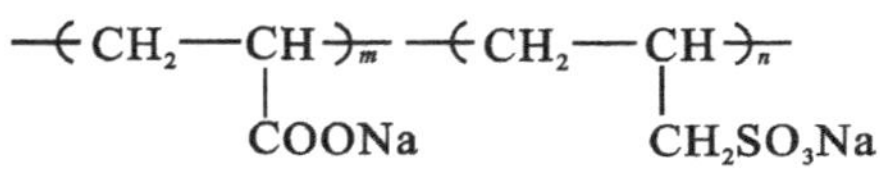

(Oligomer of sodium acrylate and sodium allylsulfonate)

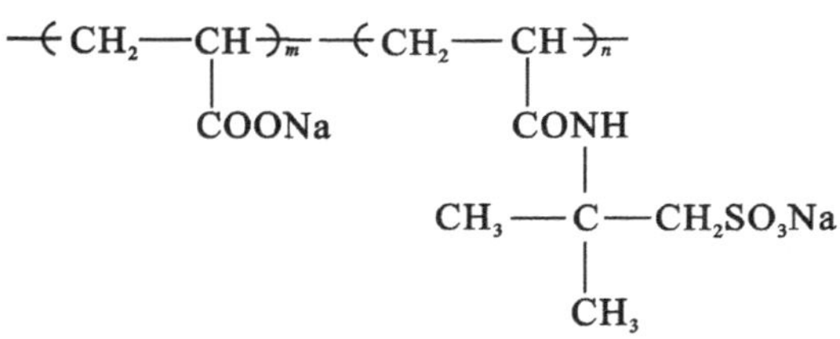

[Oligomer of sodium acrylate and (2-acrylamide-2- methyl)
sodium propyl sulfonate]

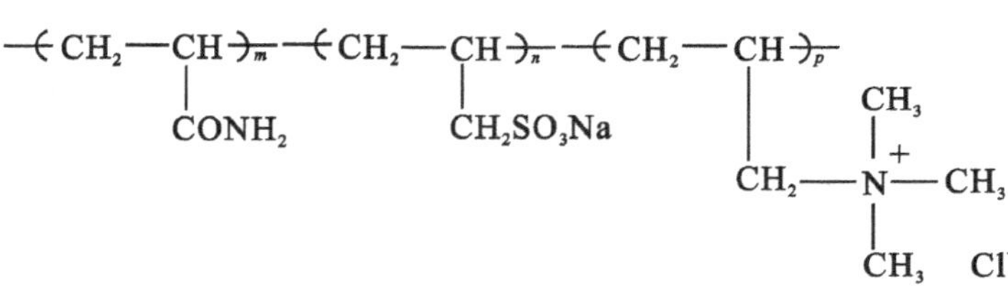

(Oligomer of acrylamide, sodium allylsulfonate and allyltrimethyl ammonium chloride)

Vinyl monomer oligomers can adsorb on the surface of clay particles by hydrogen bonding. If there are cationic chains in the oligomers, they may also absorb to the negatively charged surface of clay particles. Other polar groups which are not adsorbed on clay particles help with increasing the electronegativity on the surface and water layer thickness; in this way, the connection structure of clay particles are broken up, reducing the viscosity and shear force of the drilling fluid. In addition, for drilling fluid prepared with polymers, the addition of oligomers facilitates the desorption of polymers on the clay particles through competitive adsorption, destroying the structure constructed by polymers and clays and thusly reducing the viscosity and shear force (Zhang et al. 1988). Therefore, for drilling fluid prepared with polymers, vinyl monomer oligomer is particularly suitable as viscosity reducer to reduce the viscosity and shear force, which modified tannin and lignin fail to accomplish.

2.5.3.2 Tackifier

Tackifier refers to the rheological modifier which can improve the viscosity and shearing force of drilling fluid.

There are three kinds of important tackifier.

Modified Cellulose

The modified cellulose which can function as drilling fluid tackifier mainly refers to sodium carboxymethyl cellulose and hydroxyethyl cellulose. Modified cellulose with higher degree of polymerization and substitution can function better as tackifier.

Modified cellulose can improve the viscosity and shearing stress through the following mechanisms.

- Through the hydration of polar groups and the entanglement among molecules, the viscosity of water in the drilling fluid is enhanced.
- Through the adsorption on the surface of clay particles, the volume of the clay particles is increased to enhance the flow resistance.
- Through bridging adsorption, structure can be generated among clay particles, resulting in the corresponding structural viscosity.

Xanthan

Xanthan (also known as XC biopolymer) is produced from the fermentation of carbohydrates by xanthomonas bacteria (Wernau 1981). Its relative molecular weight is around 2×10^6 while in some cases as high as 13×10^6 to 15×10^6. The molecular formula of xanthan is shown below.

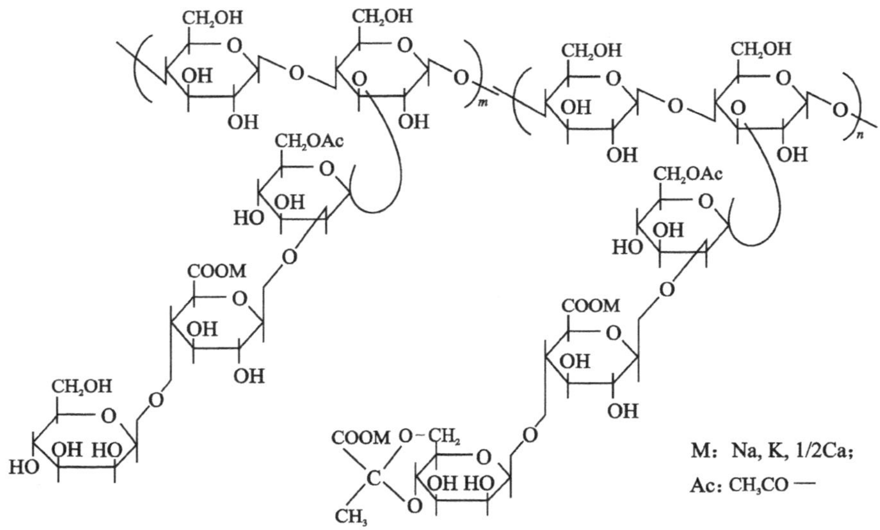

In improving the viscosity and shearing force, xanthan functions in similar mechanism with cellulose. As the long branches in xanthan molecule prevent it from adopting the curled conformation, xanthan exhibits better performance than modified cellulose does in improving the viscosity and shear force and shows enhanced salt resistance. Thusly, xanthan can be applied for preparing tackifier for saturated saltwater drilling fluid. Xanthan can tolerate high temperature up to 93 °C. Its main disadvantage is that it is subject to bacterial degradation, so fungicide (such as glutaraldehyde) should be added during application.

MMH (Burba et al. 1988; Burba et al. 1990; Polnaszek and Fraser 1991; Zhang et al. 1993)

MMH is a kind of tackifier prepared by mixing metal salt solutions and precipitating the metal ions using precipitant.

Available metal salts include divalent metal salts (such as magnesium chloride) and trivalent metal salts (such as aluminum chloride).

Available precipitant is generally ammonia water.

When using precipitants to precipitate metal ions, divalent metal salts and trivalent metal salts undergo different reactions.

For divalent metal salts, when the concentration of precipitating agent reaches the solubility product of the corresponding metal hydroxide, the divalent metal will precipitate. The precipitates generated follows the law of Fajans by preferentially adsorbing divalent metal ions and other metal ions with high valence states, and forming diffused double layer with the anions to generate positively charged precipitates.

For trivalent metal salts, their complex ions with water molecules can go through hydrolysis and olation by the addition of precipitating agent to form polynuclear olation complex ions.

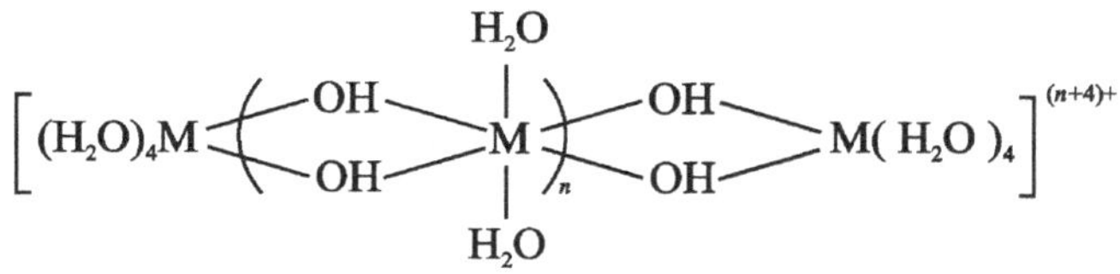

If more precipitating agents are added, the nuclear number and valence number in polynuclear olation complex ions are increased, until the solubility product of the trivalent metal hydroxide is reached and precipitation is generated. Following the Fanjans' law, the surface of precipitate can adsorb high-valence metal ions in the solution (especially polynuclear olation complex ions), rendering the surface of precipitate positively charged.

If the mixed metal salts are precipitated with precipitating agent, we can get positively charged mixed metal hydroxide (MMH).

Diffused double layer ions around MMH are hydrated. Water molecules in the hydration layer are directionally arranged according to their polarity, with the positive end outward. So the surface of MMH hydration layer is electropositive; it can bind with negatively charged clay particle surface through electrostatic interaction to form structure. The generated structural viscosity helps with the enhancement of the drilling fluid viscosity and shear force.

Characteristics of MMH in improving viscosity and shearing stress of drilling fluid are summarized as follows.

- MMH can only form structure with clay particles. No electrical neutralization exists therefore the coagulation of clay particles will not occur.
- Under the action of shear stress, the structure is fragile. Therefore, the shear thinning property of drilling fluid is more prominent. MMH is particularly suitable to function as tackifier of water-based drilling fluid.

2.6 The Solid Phase and Its Content Control in Drilling Fluid

2.6.1 Solid Phase in Drilling Fluid

The solid phase in drilling fluid can be classified according to different standards.

- If classified by source, the solid phase can be divided into slurry clay, cuttings, density adjustment materials, and the solid phase in the treatment agents, etc.
- If classified by density, the solid phase can be divided into high-density solid phase (≥ 2.7 g/cm^3) and low-density solid phase (<2.7 g/cm^3). For example, barite is high-density solid phase (density in the range of 4.2–4.6 g/cm^3), while

bentonite and cuttings belong to the low-density solid phase (density in the range of 2.4–2.7 g/cm^3).

- If classified by the chemical activity of the surfaces, the solid phase can be divided into surface active solid phase and surface inert solid surface. For surface active solid phase such as bentonite, its surface can react easily with water and some treating agents. While for the surface inert solid phase such as barite, its surface is not easy to react with water and treating agents.
- If classified by usefulness in the drilling fluid, the solid phase can be divided into useful solid phase and useless solid phase. The clay and the density-adjusting materials are all useful solid phase, while the cuttings are useless.
- If classified by particle diameter, the solid phase can be divided into colloidal particles (<2 μm), mud (2–74 μm), sand (>74 μm), etc.

2.6.2 Control Method for Solid Content in Drilling Fluid

The solid content in drilling fluid refers to the mass of solid matter in unit volume drilling fluid, with a unit of kg/m^3 or g/cm^3.

Solid content has important influence on the properties of drilling fluid. If the clay content is too high, the viscosity and shear force of the drilling fluid would be increased. If the debris content is too high, the permeability of filter cake would be increased to result in greater filter loss and thickened filter cake, which would cause sticking accidents of drilling tools. Therefore, the solid content of drilling fluid must be strictly controlled.

The following methods are used to control the solid content in the drilling fluid.

2.6.2.1 Sedimentation Method

When the drilling fluid is circulated to the ground, it will pass through a large pool in which larger solid particles can settle down. This method is called the sedimentation method, which is commonly applied to control the solid content when drilling occurs in the upper formation.

2.6.2.2 Dilution Method

Dilution method refers to the method of adding dispersion medium (such as water and oil) into the drilling fluid to reduce the solid content of the drilling fluid. As the addition of dispersion medium can also affect other properties of the drilling fluid, this method is seldom used.

2.6.2.3 Mechanical Method

Mechanical method refers to the method of separating large solid paticles using mechanical equipments (such as vibration sieve, desander, centrifuge, etc.).

2.6.2.4 Chemical Control Method

By adding flocculant, the solid particles in the drilling fluid can aggregate to form larger pieces, which would facilitate their removal through further sedimentation method or mechanical method. This method is called chemical control method. It can remove solid particles with a diameter below 5 μm, while sedimentation or mechanical method can only remove solid particles with a diameter above 5 μm.

2.6.3 *Drilling Fluid Flocculant*

The drilling fluid flocculant is a chemical agent that can make the solid particles aggregate to form larger pieces in the drilling fluid.

The drilling fluid flocculant is mainly water-soluble polymer. Some examples are given below.

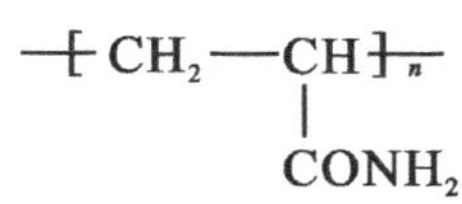

(Polyacrylamide, PAM)

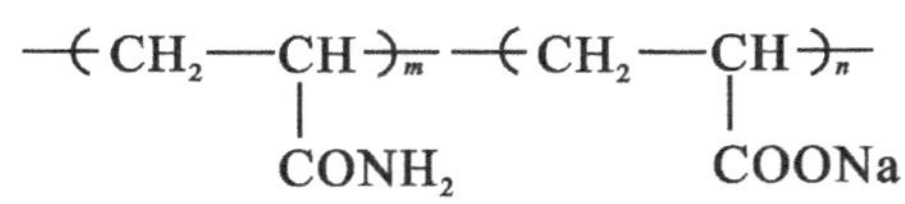

(Partially hydrolyzed polyacrylamide, HPAM)

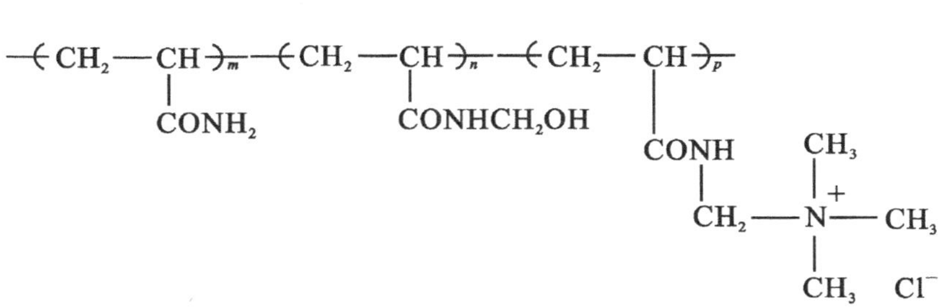

(Copolymer of lcrylamide, hydroxymethyl acrylamide and acrylamide based methylene trimethyl ammonium chloride, CPAM)

In these polymers, nonionic polymer (e.g., PAM) and anionic polymer (e.g., HPAM) can perform the function of flocculation by bridging–curling mechanism. Polymer molecules can simultaneously adsorb on two or more particle surfaces through bridging. Then, through the curling of the polymer chains, these particles can be flocculated. Beside this bridging–curling mechanism, cationic polymer (e.g., CPAM) can also function through electrical neutralization mechanism to achieve better flocculation effect.

Although the above-mentioned polymers all have flocculation effect, they have their own characteristics. PAM is a non-selective flocculant. It can flocculate both inferior soil (e.g., the debris whose surface bears fewer negative charges in water) and also high-quality soil (e.g., bentonite whose surface bears more negative charges in water), thusly, it belongs to total flocculant. HPAM is a selective flocculant because it contains negatively charged ($-COO^-$) chains. It can only adsorb on inferior soil (with fewer negative charges) through hydrogen bonding and perform the flocculation, while has no effect on the flocculation of high-quality soil. For cationic and nonionic chains of CPAM, the amide and hydroxyl methyl groups can adsorb on the hydroxyl groups on the clay surfaces via hydrogen bonding, and the cationic groups can adsorb on the negatively charged clay surfaces via electrostatic interaction. Therefore, compared with PAM and HPAM, CPAM exhibits more efficient and faster flocculation effects.

HPAM is the most commonly used drilling fluid flocculant. The main factors that affect the flocculation of HPAM are given below.

2.6.3.1 Relative Molecular Mass

The higher the relative molecular mass, the better the flocculation effect, because polymer with longer molecular chains can flocculate solid particles at a larger distance. HPAM will not have flocculation effect until its relative molecular mass surpasses 1×10^6. When used as drilling fluid flocculant, HPAM is required to have a relative molecular mass of more than 3×10^6.

2.6.3.2 Degree of Hydrolysis

There is an optimal degree of hydrolysis (30%) for HPAM to achieve the most desirable flocculation effect. If the degree of hydrolysis is too low, it will affect the stretching of HPAM molecular chains, reducing the flocculation effect. While if the degree of hydrolysis is too high, it will affect the adsorption of HPAM to the negatively charged clay surfaces, also resulting in weakened flocculation effect.

2.6.3.3 Concentration

There is also an optimal concentration for HPAM to achieve the most desirable flocculation effect. If the concentration is too low, flocculation will not be conducted completely. While if the concentration is too high, HPAM can form network structures with clay particles, which is not conducive to flocculation.

2.6.3.4 PH Value

The pH value can influence the flocculation effect by affecting the existential states of HPAM and clay particles.

If the pH value is too low, a greater number of $-COO^-$ in HPAM would combine with H^+ to form –COOH. The molecular chains would become more curled due to the reduced electrostatic repulsion among molecular chains, thusly reducing the flocculation effect. While if the pH value is too high, the clay particles tend to be dispersed, also not conducive to the flocculation.

For HPAM, the most desirable pH value of flocculation lies in the range of 7–8.

2.7 The Lubricity of Drilling Fluid and Its Improvement

2.7.1 Drilling Fluid Lubricity

In the drilling process, the existence of drilling fluid turns the dry friction between the drill string and the shaft wall to the wet friction between the drill string and the borehole wall (covered by filter cake). This friction-reducing property of the drilling fluid is called the lubricity of drilling fluid.

The lubricity of drilling fluid can be measured using the friction coefficient. Friction coefficient refers to the ratio of the friction generated between objects (solid) in relative motion to the force perpendicular to the friction surface under certain conditions. The above-mentioned conditions mainly refer to the movement speed of the objects in relative motion and whether there is liquid between them. The friction coefficient of the drilling fluid is obtained in mimicking the wet friction between drill string and the borehole wall (using the filter cake obtained in measuring the filtration loss). The smaller the frictional coefficient, the better the lubricating property of drilling fluid.

The lubricating property of drilling fluid has important influence on reducing the wear of drill string and improving the drilling speed.

2.7.2 Improvement of Drilling Fluid Lubrication

Lubricant can be added to the drilling fluid to improve the lubrication of drilling fluid. The material that can improve the lubricating property of drilling fluid is called drilling fluid lubricant.

There are two types of drilling fluid lubricant.

2.7.2.1 Liquid Lubricants

Liquid lubricants mainly refer to oil including vegetable oil (such as soybean oil, cottonseed oil, and nettle oil), animal oil (such as lard), and mineral oil (such as kerosene, diesel oil, and mechanical lubricating oil).

Because the viscosity of oil is higher than that of water, it is not easy to be squeezed out from the friction surface during the friction between drill string and borehole wall, so the lubricating property of drilling fluid can be improved.

To facilitate the formation of a uniform oil film on the friction surface, surface active agent can be added into the drilling fluid. The surface active agent can form an adsorption layer on the friction surface. As the surfaces of the drill string (due to the existence of oxide film) and the borehole wall are both hydrophilic, the surface active agents adsorbed based on the rule of close polarity can turn these hydrophilic surfaces to lipophilic surfaces. Therefore, oil can form uniform films on the surface of drill string and borehole wall, strengthening its lubricating effect.

The available surfactants are mainly water-soluble surfactants. Some examples are given below.

$$C_{12}H_{25}-SO_3Na$$

(Dodecyl sulfonic acid sodium)

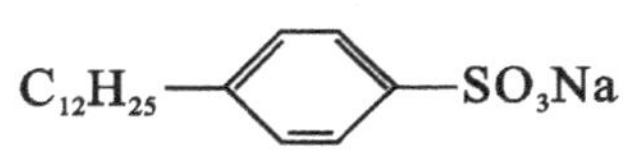

(Dodecyl benzene sulfonic acid sodium)

$$CH_3-(CH_2)_7-CH=CH-(CH_2)_7-COONa$$

(Sodium oleic acid)

$$CH_3-(CH_2)_5-\underset{\displaystyle OH}{\underset{|}{CH}}-CH_2-CH=CH-(CH_2)_7-COONa$$

(Sodium ricinate)

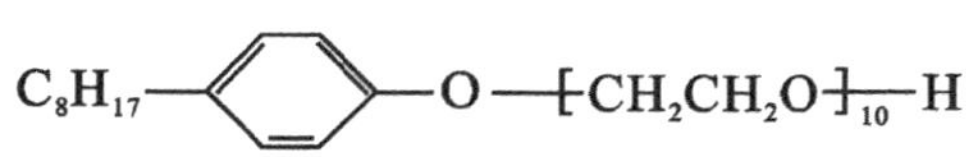

(Octyl phenol polyoxyethylene ether-10,OP-10)

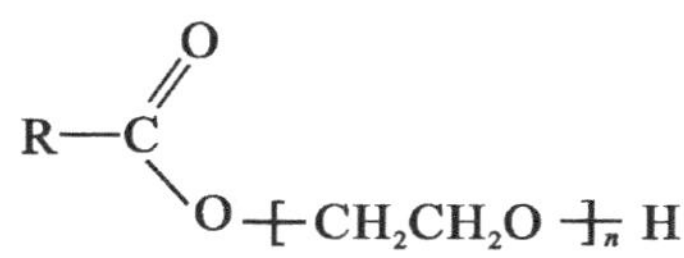

R: C_{21}–C_{25}; *n*: 20

(Polyoxyethylene high carbon carboxylic ester)
(Sun et al. 2002)

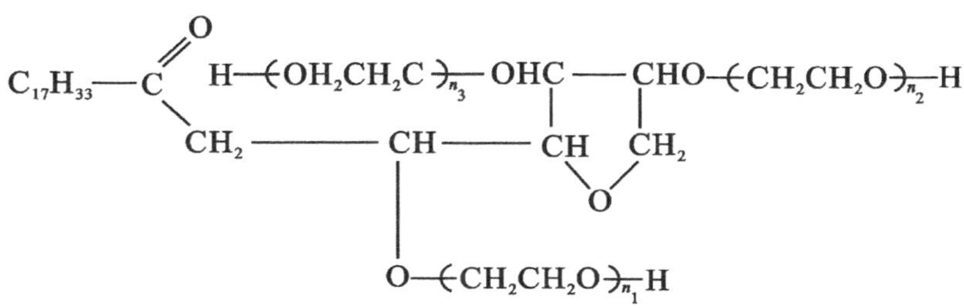

$n_1+n_2+n_3$: 21–26

(Sorbitan monooleate polyoxyethylene ether, Tween 80)

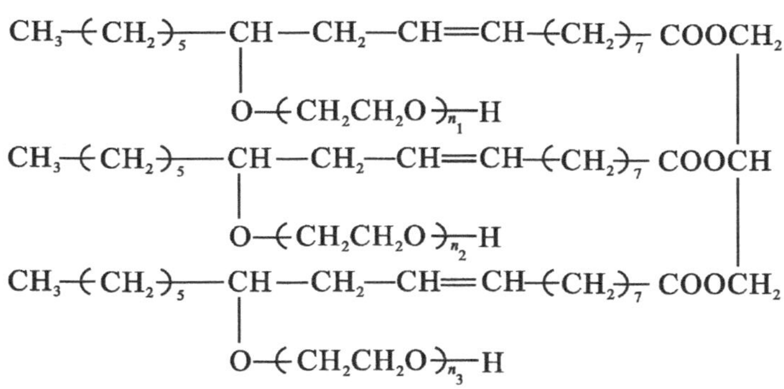

$n_1+n_2+n_3$: 40

(Polyoxyethylene castor oil, EL-40)

Water-soluble surface active agents can be used as emulsifier in preparing oil-in-water emulsion. Therefore, the oil acting as the drilling fluid lubricant, the surface active agent used for strengthening the oil lubrication and water together can be prepared into oil-in-water emulsion first before they are added in drilling fluid.

Although adding surface-active agent alone into the drilling fluid can also improve the lubricity of drilling fluid, its effect is far inferior to that of using oil and surface active agents simultaneously.

2.7.2.2 Solid Lubricant

There are mainly two types of solid lubricants.

Solid Pellet
The commonly used solid pellets refer to the plastic pellets and the glass pellets.

The plastic pellets include polyamide (nylon) pellets and polystyrene and divinylbenzene copolymer pellets. These pellets have various advantages such as high-temperature resistance, compression resistance, and chemical inertness and can be applied as lubricants for various types of drilling fluids.

Glass pellets can be made of glasses with different components (such as sodium glass and calcium glass) and have advantages of high-temperature resistance and chemical inertness. For glass pellets, their cost is lower than that of their plastic counterparts; however, their compression resistance is inferior and is more easily to sink.

The solid lubricant can reduce the friction between the drill string and the borehole wall by turning the sliding friction to rolling friction.

Graphite
Graphite is a flake crystalline form of carbon with high melting point, low hardness, and chemical inertness. Dispersed in the drilling fluid, when it forms filter cake on the borehole wall through filtration, it can turn the friction between string and borehole wall into friction from the relative motion between low-hardness graphite crystal sheets or between the drill columns and low-hardness graphite crystal sheets, thus reducing the friction.

Graphite can be used as lubricant for various types of drilling fluids.

2.8 The Borehole Wall Stability and Its Control

2.8.1 Borehole Wall Stability

Borehole wall stability refers to its capability to maintain its original state. If the borehole wall can keep its original state, it is called stable borehole. While if the wall cannot maintain its original state, it is called unstable borehole wall. In order to keep the stability of the borehole wall, we must understand the factors that affect the stability of the borehole wall (Bol et al. 1992; Wong and Kenter 1993; Chee et al. 1996; Santos 1998).

2.8.1.1 Geological Factors

The instability of borehole wall can be caused by many geological factors such as the stress release of high-pressure formation, the stress release of high tectonic stress formation, the collapse of loose strata, and the plastic deformation of salt rock strata.

2.8.1.2 Engineering Factors

Many engineering factors can also result in instability of borehole wall such as the collision of drill to the borehole wall during the trips, the excessive scouring of the borehole wall due to excess flow of annulus drilling fluid, and the pressure fluctuation due to over quick drilling processes.

2.8.1.3 Physical and Chemical Factors

The contact of shale and water can cause instability of borehole wall. Shale beds are high in clay. If shale beds consist mainly of expansive clays, the contact with water can lead to the swelling and dispersion of shale. While if the shale beds contain mainly non-swelling clays such as illite and kaolinite, the contact with water can cause clay exfoliation, resulting in strata unstability.

2.8.2 Control Method for Borehole Wall Stability

Because the factors that cause the instability of borehole wall are different, the control methods are thusly different.

If the instability of borehole wall is caused by geological factors, it can be addressed by suitably increasing the density of drilling fluid or consolidating the wall using chemical method (water glass can react with the exchangeable calcium and magnesium ions in minerals of the borehole wall to generate calcium silicate and magnesium silicate which will contribute to the well cementation).

If the borehole wall instability is caused by engineering factors, it can be addressed by improving the drilling technology.

If the borehole wall instability is caused by physical and chemical factors, it can be addressed by improving the properties of drilling fluid, such as adjusting the density of drilling fluid and adding shale inhibitors. The method for adjusting the density of the drilling fluid has been introduced previously, while the method of adding shale inhibitors will be introduced as follows.

2.8.3 *Shale Inhibitor*

Chemical agents that inhibit the swelling and/or dispersion (including spalling) of shale are called shale inhibitors (Federici et al. 2008).

Shale inhibitors are classified as follows.

2.8.3.1 Salt

The salts here mainly refer to the inorganic salts (such as sodium chloride, ammonium chloride, potassium chloride, calcium chloride) and organic salts (such as sodium formate, formic acid potassium, sodium acetate, potassium acetate). The above-mentioned are all water-soluble salts. Beyond certain concentrations, all water-soluble salts can function to stabilize the shale. The salts function by compressing the thickness of the diffused double layer on the shale surface and reducing the ζ potential (Martin and Vincent 1998).

Although all water-soluble salts can help with the stabilization of the shale, they have different effects.

Among water-soluble salts, potassium (Sun et al. 2005) and ammonium salts work best for the stabilization of the shale. This is because the diameters of their cations (see Table 2.3) are close to those of the hexagonal oxygen ring (0.288 nm) formed by oxygen atoms on the bottom surface of the clay silicon oxygen tetrahedron, so that they can enter the clay crystal layer while are not easy to release. Therefore, the clay tablets whose surface negative charges are neutralized can link together to effectively inhibit the shale swelling.

2.8.3.2 Cationic Surfactants

The following cationic surfactants (Beihoffer 1990) can stabilize the shale.

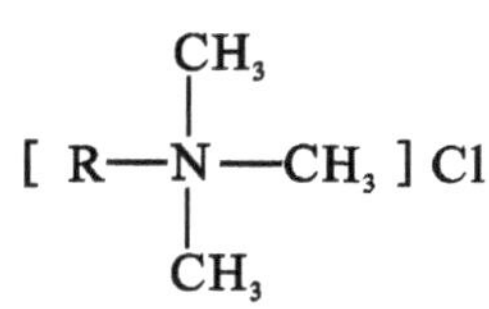

R: C_{12}–C_{18}

(Alkyl trimethyl ammonium chloride)

Table 2.3 Ion diameter of some cations

Ion	Ion diameter (nm)	Ion	Ion diameter (nm)
Li^+	0.120	Cs^+	0.340
Na^+	0.196	NH_4^+	0.286
K^+	0.266	Ca^{2+}	0.212
Rb^+	0.300	Mg^{2+}	0.156

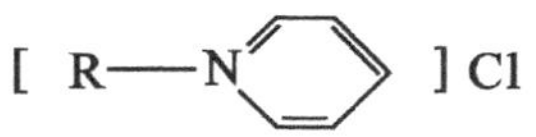

R: C_{12}–C_{18}

(Alkyl chloride pyridine)

The stabilization of shale by cationic surfactants is achieved mainly through the adsorption of the active cations on the surface of shale, which will neutralize the electronegativity on the shale surface and convert it to lipophilic surface.

Some organic cationic compounds with low molecular weights (some examples are shown below) can also stabilize shale by neutralizing the electronegativity on the shale surface.

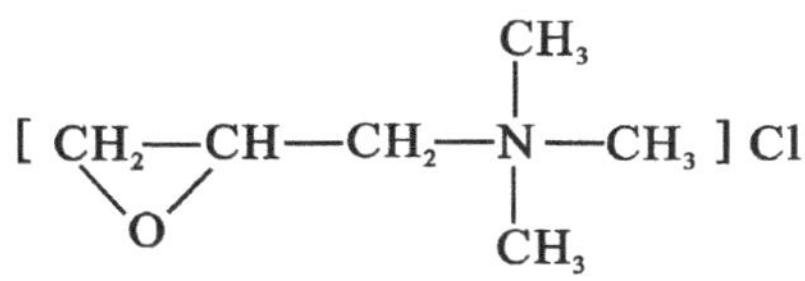

(Glycidyltrimethyl ammonium chloride)

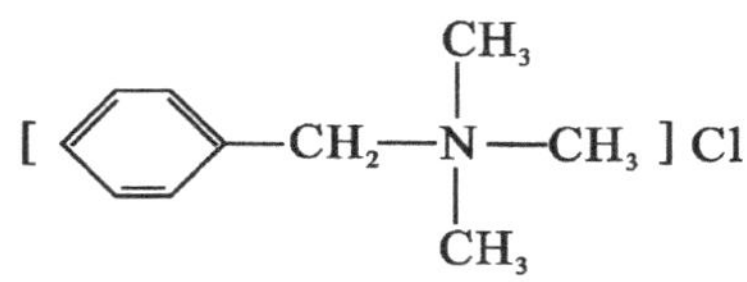

(Benzyl trimethyl ammonium chloride)

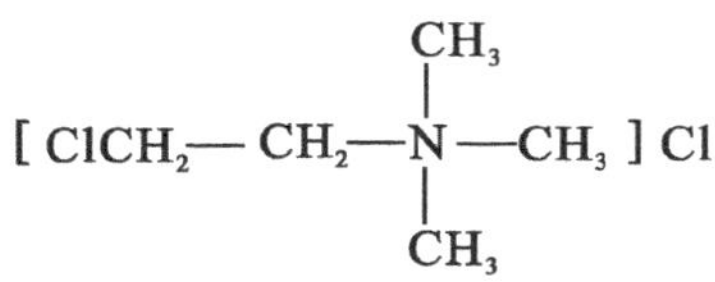

(2-chlorocholine chloride)
(Zhu et al. 2003)

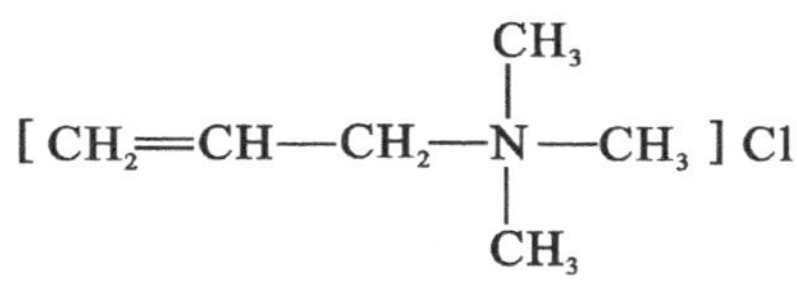

(Allyltrimethyl ammonium chloride)
(Yu et al. 2008)

2.8.3.3 Cationic Polymers

The following cationic polymers (Homer and Jimmie 1982; Jimmie 1984; Retz et al. 1991; Ma 1995; Li et al. 2009; Ma and Zhao 1995) have the function of stabilizing the shale.

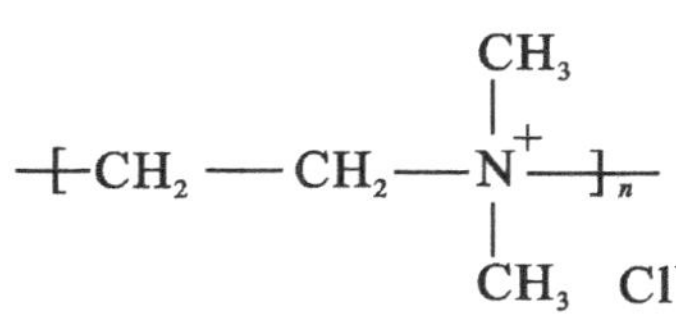

[Poly(1,2- ethylidene dimethyl ammonium chloride)]

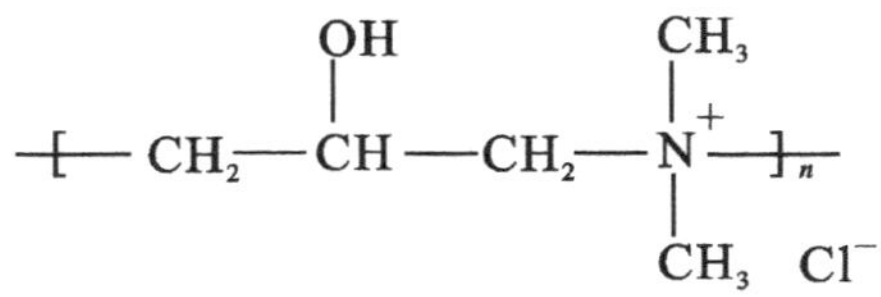

[Poly(2-hydroxyl-1,3- propylidene dimethyl ammonium chloride)]

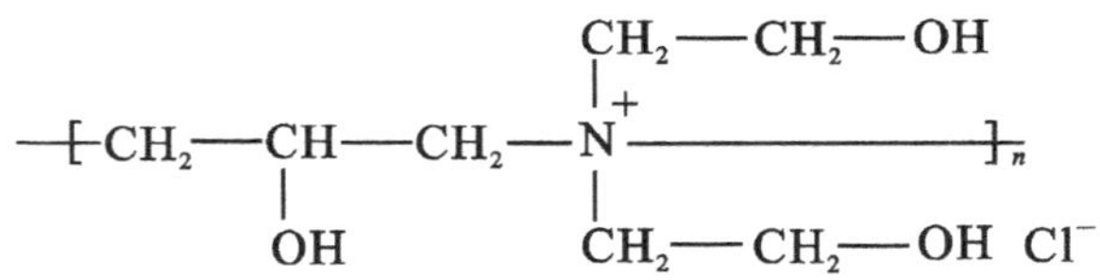

[Poly(2- hydroxyl-1,3- propylidene dihydroxyethyl ammonium chloride)]

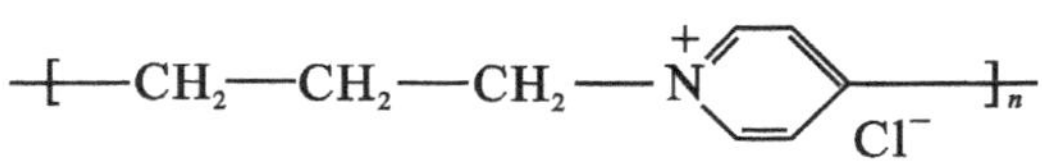

[Poly(1,3-propylidene chloride pyridine)]

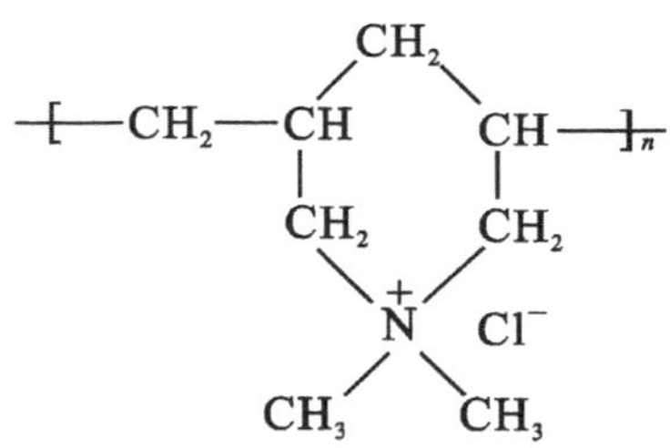

[Poly(diallyldimethyl ammonium chloride)]

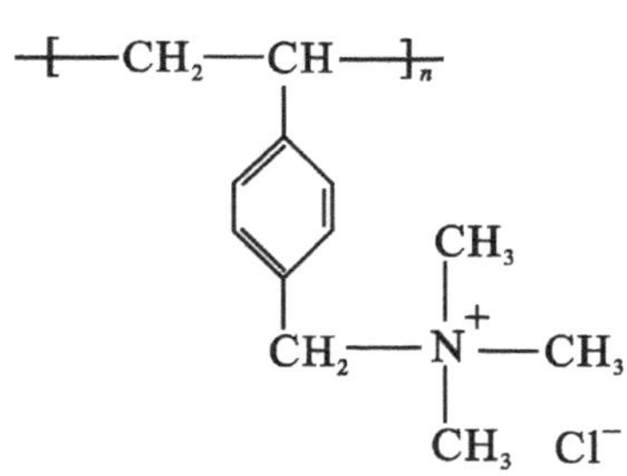

[Poly(benzylvinyl trimethyl ammonium chloride)]

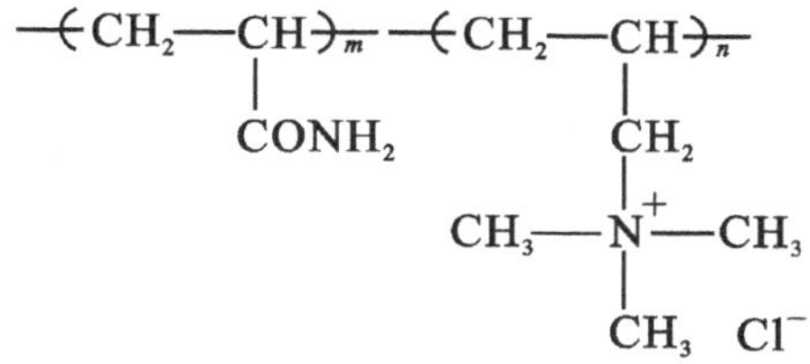

(Copolymer of acrylamide and allyltrimethyl ammonium chloride)

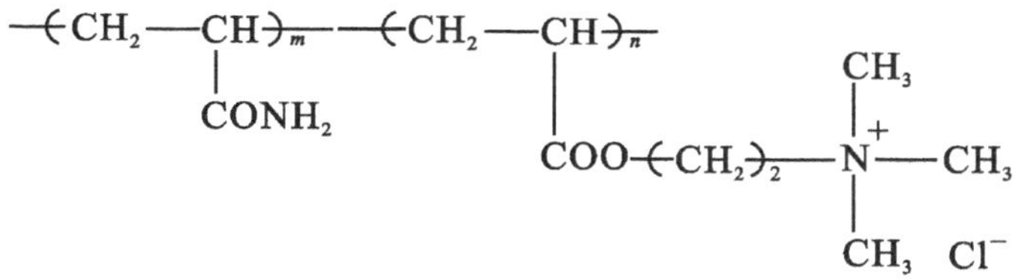

(Copolymer of acrylamide and acrylic acid-1,2- carbethoxy trimethyl ammonium chloride)

Cationic polymers stabilize the shale mainly through neutralizing the electronegativity on the shale surface and adsorptive bridging among clay sheets.

2.8.3.4 Nonionic Polymers

Ether-based polymers (Chenecert 1989; Perricone 1989) (some examples are shown below) are the main nonionic polymers used as shale inhibitors.

(Polyethylene glycol)

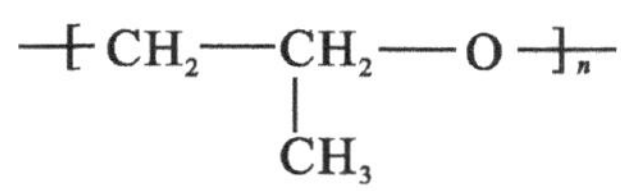

(Polypropylene glycol)

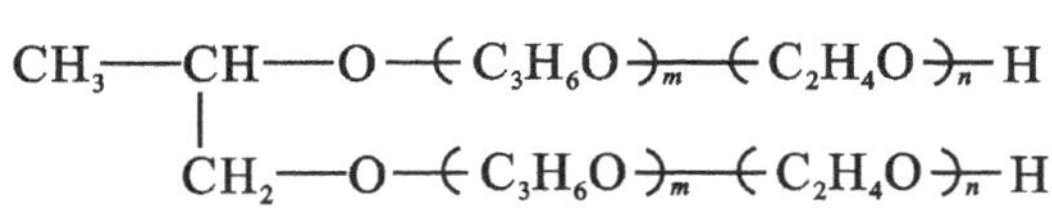

(Polyoxyethylene polyoxypropylene propylene glycol ether)

These ether polymers are dissolved in water by forming hydrogen bonds (expressed by –) with water molecules.

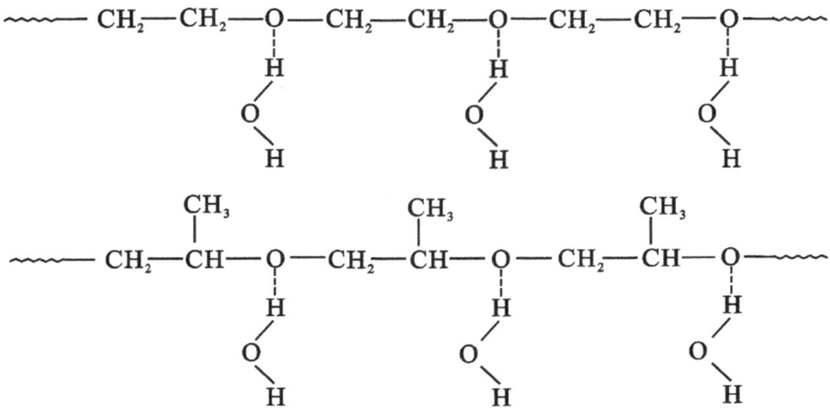

Ether polymers with certain concentrations are water-soluble under the ground temperature. However, when the temperature rises to certain value, which means polymers circulate to stratum with certain depth, the ether polymers will experience a saturated precipitation due to the weakening hydrogen bonds. These precipitated ether polymers can adhere to the shale surface, blocking the pores of the shale and reducing the contact between the shale and water to stabilize the shale (Downs et al. 1993; Reid and Dolan 1995; Bland and Smith 1995).

2.8.3.5 Modified Asphalt

Asphalt is composed of a small amount of hydrocarbon compounds (containing carbon and hydrogen elements only) and a large number of nonhydrocarbon compounds (containing elements including oxygen, sulfur, and nitrogen besides carbon and hydrogen). Asphalt can be divided into three types, which are natural asphalt, petroleum asphalt, and tar asphalt. Natural asphalt is formed by petroleum under natural conditions. Petroleum asphalt is generated by the refining and processing of petroleum. Tar asphalt is obtained by the destructive distillation of coal, timber, etc.

There are two important modified products of asphalt (Yao 1989).

One is oxidized asphalt, which is produced by the air oxidation of residual oil from atmospheric distillation, vacuum distillation, or cracking process under high temperature (200–220 °C). Oxidized asphalt is mainly used in oil-base drilling fluid.

The other is sulfonated asphalt, which is produced by the sulfonation of asphalt using sulfonating agents (such as concentrated sulfuric acid, fuming sulfuric acid, and sulfur trioxide). The major component of sulfonated asphalt is asphalt sulfonate and it can be used in water-base drilling fluid.

Modified asphalt can help stabilize the shale through its adhesion on the shale surface, which will block the pores of the shale, form hydrophobic oil film and reduce the contact of shale with water.

2.9 The Sticking and Stuck-Freeing of Drilling Tool

The sticking of drilling tool refers to the phenomenon that the drilling tool gets stuck in the borehole and cannot function normally, which is a complicated situation encountered in the drilling process. The pipe-sticking related to the properties of drilling fluid is mainly adhesive pipe-sticking (also called differential pressure sticking).

Differential pressure sticking is caused by the differential pressure between drilling fluid and formation, after the adhesion of filter cake on the drill string (Fig. 2.13).

The release of the sticking problem is called the stuck-freeing of drilling tool. The lift force required to release the adhesive pipe-sticking shown in Fig. 2.13 can be estimated using the following equation.

$$\text{Lift force} = \text{Adhesion area} \times \text{Adhesion pressure difference} \times \text{Frictional coefficient}$$

$$F = r\theta L(\rho_{\mathrm{m}}gh - p)f \tag{2.18}$$

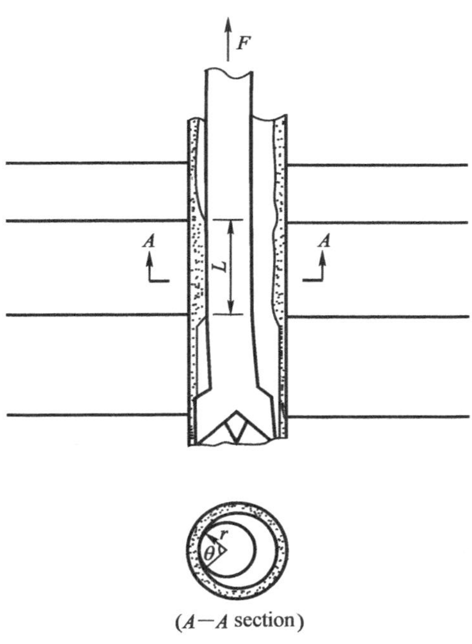

Fig. 2.13 Sticking of drilling tool

In this equation, F is the lift force, r is the radius of drill string where sticking occurs, θ is the wrap angle of filter cake adhering to the drill pipe, L is the adhesion length, ρ_m is the drilling fluid density, g is the acceleration of gravity, h is the well depth where sticking occurs, P is the formation pressure, and f is the friction coefficient between filter cake and drill pipe.

It can be seen from Eq. 2.18 that the lifting force required in the releasing of adhesive pipe-sticking can only be reduced by two methods, which are decreasing the drilling fluid density and decreasing friction coefficient between the filter cake and drill pipe. The latter method is the mainly used one which requires the usage of pipe-freeing agent. Pipe-freeing agent is a chemical agent that can reduce the friction coefficient between the filter cake and the drill pipe so as to achieve the purpose of stuck-freeing. The aforementioned drilling fluid lubricant can be used as pipe-freeing agent. Diesel thickened by asphalt is often used as drilling fluid lubricant. In order to dissolve the asphalt-thickened diesel in water, compound emulsifiers including water-soluble surface active agent (e.g., polyoxyethylene alkyl ether) and oil-soluble surface active agent (e.g., calcium oleate) are used.

Because the fluorescent material contained in diesel and asphalt will affect the interpretation of logging data, surface active agent itself can also be used as pipe-freeing agent. For example, alkyl sodium sulfonate and alkyl benzene sodium sulfonate whose hydrophile–lipophile balance properties are adjusted by alcohol with low molecular weight (such as ethanol, propanol) and salt (sodium chloride) can also be used as pipe-freeing agent; however, its effect is inferior than that of the agent with thickened diesel.

2.10 The Leakage of Drilling Fluid and the Plugging of Adsorbent Formation

2.10.1 The Leakage of Drilling Fluid

In the drilling process, the phenomenon that large amounts of drilling fluid in the borehole flow into the formation is defined as the leakage of drilling fluid.

According to the features of leakage formations, the leakage of drilling fluid can be divided into three categories.

2.10.1.1 Permeable Leakage

The leakage of drilling fluid caused by the high permeable sandstone formation or conglomerate formation (as shown in Fig. 2.14a) is defined as permeable leakage. The feature of the permeable leakage is that the leakage rate is not high (0.5–10 m^3/h), manifested by the slow declining of the level of the drilling fluid pond.

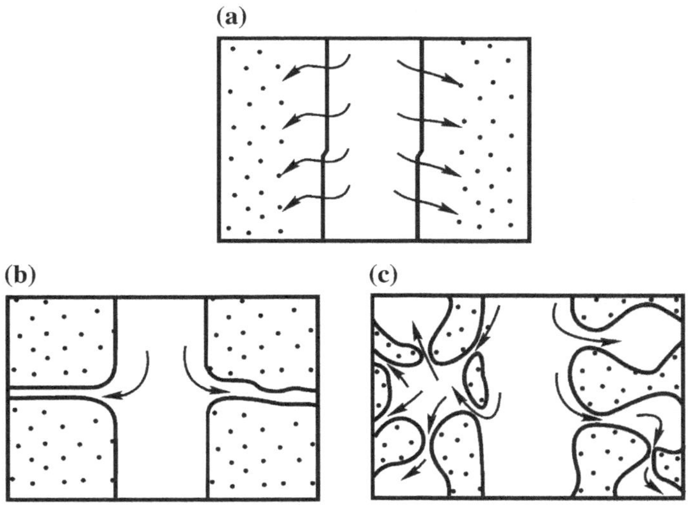

Fig. 2.14 Different types of drilling fluid leakage. **a** Permeable leakage, **b** fractured leakage, **c** karst cave leakage

2.10.1.2 Fractured Leakage

The leakage of drilling fluid caused by the fractured formation is called the fractured leakage (as shown in Fig. 2.14b). The fractures leading to the drilling fluid loss include the natural fracture in the limestone and sandstone formations and the artificial fracture formed by the high pressure of drilling fluid to the limestone and sandstone formations. The feature of this fractured leakage is that the leakage rate is high (10–100 m^3/h), manifested by the rapid drop of the level of the drilling fluid pond.

2.10.1.3 Karst Cave Leakage

The leakage of drilling fluid caused by the karst cave formation is called the karst cave leakage, which usually occurs in the limestone formation. The feature of this karst cave leakage is that the leakage rate is very high (>100 m^3/h), and the drilling fluid will be leaked completely into the formation.

2.10.2 The Plugging of Leakage Formation

The materials used for plugging of the leakage formation is called plugging materials. Different plugging materials should be used in different leakage formations (Verret 2005; Xiang 2005; Hoskins 2007).

2.10.2.1 The Plugging of Permeable Formation

The materials used for plugging of permeable formation are shown as follows.

Silicic Acid Gel
Firstly, the sodium silicate is added into the hydrochloric acid to form the silicic acid sol. Then, the silicic acid sol is injected into the leakage formation. After a certain time, the silicic acid gel is formed to plug the leakage formation.

Chromium Gel
This plugging material is prepared by mixing HPAM, sodium dichromate, and sodium sulfite in the aqueous solutions. After the mixture is injected into the leakage formation, the Cr^{6+} in the sodium dichromate can be reduced to Cr^{3+} by sodium sulfite, and then form polynuclear olation complex ions which can crosslink the HPAM into chromium gels for plugging the leakage formation.

Phenolic Resin
This material is prepared by the pre-polycondensation of phenol and formaldehyde. If excess amount of formaldehyde is used in the pre-polycondensation, this kind of material can undergo further polycondensation to form insoluble and infusible phenolic resin after injected into the formation, to plug the leakage formation.

Urea resin can also be used to plug permeable leakage formation.

The plugging materials used for the fractured or karst cave formations could also be applied for the extra-high-permeability formations.

2.10.2.2 The Plugging of the Fractured or Karst Cave Formations

Fiber Material
The plant fiber (e.g., linter) or mineral fiber (e.g., asbestos fiber) can be applied for plugging the fractured or karst cave formation. The fiber materials can be suspended in the carrying medium (such as water, gelled water, and suspension of sodium bentonite) before injected into the leakage formation. They can plug the formation by forming filter cake in the narrow region of crevice or inlet of the karst cave.

Particle Material
The plant material (such as walnut shell, peanut shell, and corncob) and mineral material (such as clay, diatomite, perlite, and limestone) ground to certain particle size can be used as plugging materials. When these plugging materials are injected into the formation, if the particle diameter is larger than 1/3 of the width of the crevice narrow region or 1/3 of the inlet diameter of the karst cave, the particle will be detained in the pore throats by bridging, forming filter cake to plug the leakage formation.

Cement is a special mineral material in the particle material. After forming the filter cake in the leakage formation, the cement can solidify by a series of hydration reactions to produce high-intensity plugging for the leakage formation.

Cement can be mixed with other mineral plugging materials for application. Acting as inorganic cementing agent, cement can improve the plugging strength of other mineral plugging materials.

2.11 Drilling Fluid System

The drilling fluid system refers to all kinds of drilling fluids used in the general formation and special formation (such as rock salt stratum, gypsum stratum, shale stratum, and high-temperature stratum).

Drilling fluid system can be divided into three categories based on the dispersion medium, water-base drilling fluid, oil-base drilling fluid, and air drilling fluid.

These three types of drilling fluids can be further divided according to other standards. For example, the water-base drilling fluid can be divided into inhibiting drilling fluid and noninhibiting drilling fluid based on its inhibition to shale. Based on the treating agents used, the inhibiting drilling fluid can be divided into calcium-treated drilling fluid, potassium salt drilling fluid, saltwater drilling fluid, silicate drilling fluid, polymer-base drilling fluid, and cationic sol drilling fluid (Gulf Publishing Company 1998).

The classification and reclassification of drilling fluid system based on the above-mentioned standards are shown in Fig. 2.15.

The various drilling fluids listed in Fig. 2.15 are introduced in the following part.

2.11.1 Water-Base Drilling Fluid

Using water as the dispersion medium, this kind of drilling fluid is prepared by water, bentonite clay, and treatment agents. Water-base drilling fluid can be classified further into inhibiting drilling fluid and noninhibiting drilling fluid, or oil-in-water drilling fluid, and (water-base) foam drilling fluid.

2.11.1.1 Noninhibiting Drilling Fluid

This is a water-base drilling fluid using the viscosity reducer as the main treatment agent. Because the viscosity reducer decreases the viscosity and shear stress by breaking up the structure among clay particles, the viscosity reducer is also called dispersant. The water-base drilling fluid using the viscosity reducer as the main treatment agent is also called dispersible drilling fluid.

This kind of drilling fluid bears high density (over 2 g/cm^3), dense and tough filter cake, low filtration loss and high-temperature resistance.

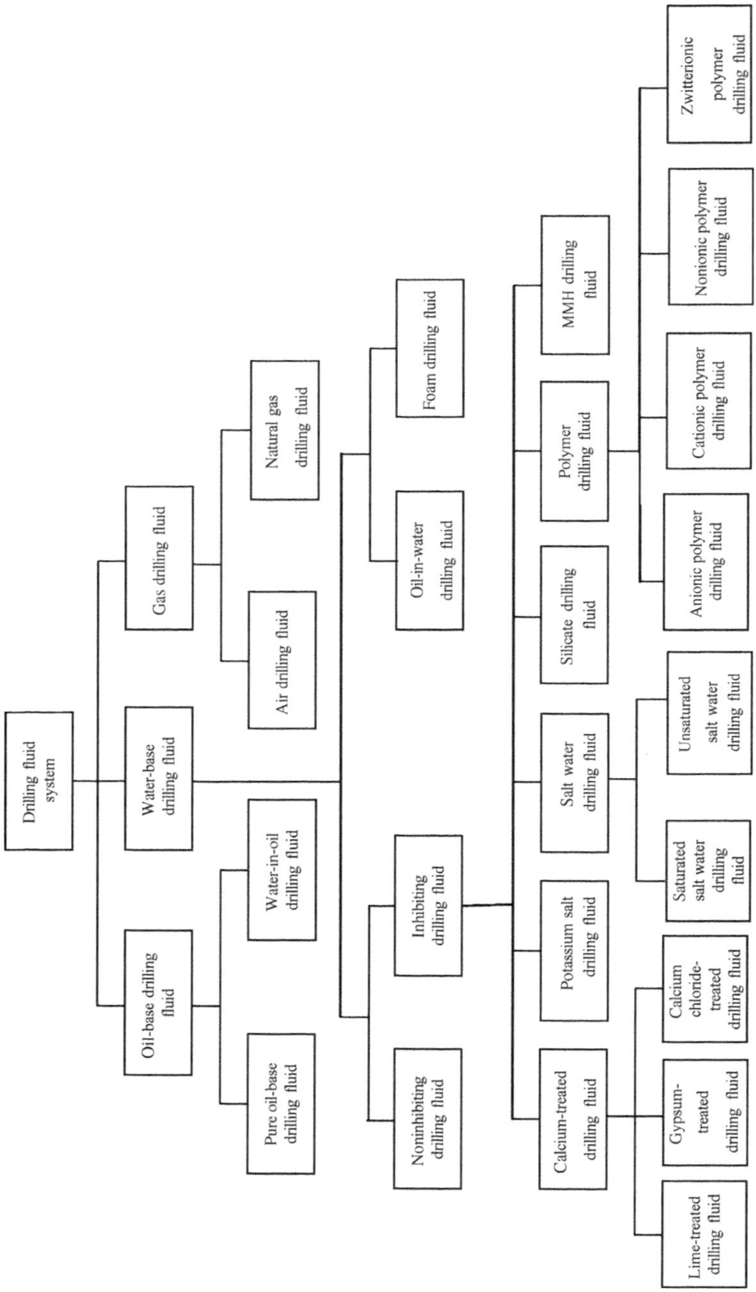

Fig. 2.15 Drilling fluid system

This drilling fluid contains high content (over 70% of solid content) of clay submicron particles (diameter < 1 μm), which are not good for increasing drilling velocity.

To guarantee the satisfactory performance of this drilling fluid, the content of bentonite in the drilling fluid is controlled within 10% and decreases accordingly with the increase of density and temperature. At the same time, the salt content in the drilling fluid is required to be less than 1% and pH value to be above 10, so that the viscosity reducer can properly function.

This kind of noninhibiting drilling fluid is applicable to drilling deep well (>4500 m) and high-temperature well (>200 °C) in the general formation, but is not applicable to rock salt stratum, gypsum stratum, and shale stratum.

2.11.1.2 Inhibiting Drilling Fluid

Inhibiting drilling fluid is a kind of water-base drilling fluid using shale inhibitor as the main treatment agent. Because the shale inhibitor can keep the clay particles in coarse-grained state, inhibiting drilling fluid is also called coarse disperse drilling fluid, which is developed for the overcoming the deficiencies of non-inhibiting drilling fluid (high content of submicron particles and poor salt resistance).

This kind of drilling fluid can be classified according to different shale inhibitors.

Calcium-treated Drilling Fluid

This is a kind of water-base drilling fluid using calcium as main treating agent.

The calcium treating agents include lime, gypsum, and calcium chloride. The drilling fluids using these treating agents are called lime-treated drilling fluid (lime drilling fluid), gypsum-treated drilling fluid (gypsum drilling fluid), and calcium chloride-treated drilling fluid (calcium chloride drilling fluid), respectively.

Calcium-treated drilling fluid can resist the calcium contamination, stabilize shale, and control the clay dispersion in the drilling system.

To guarantee good performance of this kind of drilling fluid, the pH of lime-treated drilling fluid is required to be over 11.5 and the concentration of Ca^{2+} ranges between 0.12 and 0.20 g/L. The mass concentration of excess amounts of lime (compensating for the Ca^{2+} consumed by the ion exchange with drilling cuttings) should be in the range of 3–6 g/L. The pH value of gypsum-treated drilling fluid is required to range between 9.5 and 10.5, the concentration of Ca^{2+} between 0.6 and 1.5 g/L, and the mass concentration of excess amounts of gypsum (compensating for the Ca^{2+} consumed by the ion exchange of drilling cuttings) varies from 6 to 12 g/L. The pH value of calcium chloride-treated drilling fluid is required to be in the range of 10–11, and the concentration of Ca^{2+} ranges in the range of 3–4 g/L.

A small amount of viscosity reducers need to be added into the calcium-treated drilling fluid to keep the drilling fluid particles in the disperse state, thereby maintaining stable usage performance of the drilling fluid.

This kind of drilling fluid is especially applicable for the drilling in gypsum stratum.

Potassium Salt Drilling Fluid

Potassium salt drilling fluid is a kind of water-base drilling fluid using polymer potassium salt or polymer–ammonium salt and potassium chloride as the main treatment agents.

Potassium-treated drilling fluid can inhibit the swelling and dispersing of shale, control the formation mud making, prevent the formation collapse and reduce the content of submicron particles in the drilling fluid.

To guarantee good performance of this drilling fluid, the pH value of lime-treated drilling fluid is required to be in the range of 8–9 and the mass concentration of K^+ is over 18 g/L.

Potassium-treated drilling fluid is mainly applied in the drilling in shale stratum.

Saltwater Drilling Fluid

Saltwater drilling fluid is a kind of water-base drilling fluid using sodium chloride as the main treating agent. The mass concentration of salt ranges from 10 g/L (the mass concentration of Cl^{-1} is around 6 g/L) to saturation (the mass concentration of Cl^{-1} is around 180 g/L). In saltwater drilling fluid, the drilling fluid in which the salt content reaches saturation is called saturated saltwater drilling fluid.

Saltwater drilling fluid has characteristics of salt resistance, Ca^{2+} and Mg^{2+} tolerance, strong shale stability and low damage to the formation.

To guarantee good performance of this drilling fluid, salt-resistant clay (e.g, sepiolite) and salt-resistant and Ca^{2+} and Mg^{2+} resistant treating agents are preferably used in preparing drilling fluid. In order to prevent the corrosion of drilling tool, the corrosion inhibitor is required. For the saturated saltwater drilling fluid, salt crystallization inhibitors are also required to avoid the salt crystallization.

Saltwater drilling fluid is applicable to the drilling in the offshore and other regions which lack freshwater. Saturated saltwater drilling fluid is mainly applied in the rock salt stratum, shale stratum, and rock salt–gypsum mixed stratum.

Silicate Drilling Fluid (Guo and Yan 2003; Lan and Qiu 2007; Liang and Pu 2005)

Silicate drilling fluid is a kind of water-base drilling fluid prepared using silicate as the main treating agent.

Because the silicate ions in silicate can react with Ca^{2+} and Mg^{2+} in the borehole wall surface and in the formation water to form precipitates of calcium silicate and magnesium silicate. These precipitates can deposit on the surface of the borehole wall to form protective layer. Therefore, the silicate drilling fluid can resist calcium contamination and control the swelling and dispersion of shale. The pH value of drilling fluid is required to be in the range of 11–12 during the practical application. When the pH value is lower than 11, the silicate can transform to silicic acid to deactivate the treating agents.

Silicate drilling fluid is especially applicable to the drilling in gypsum stratum and mixed stratum of shale and gypsum.

Polymer Drilling Fluid (Strickland 1994)
Polymer drilling fluid is a kind of water-base drilling fluid using polymer as the main treating agent.

Clay particles in the drilling fluid maintain in the coarse state due to the bridging of polymers. Meanwhile, the protective layers generated by the adsorption of polymers on the surfaces of drillings protect them from being dispersed into finer particles. Therefore, higher drilling speed can be obtained using polymer drilling fluid.

Polymer drilling fluid is also called non-dispersed drilling fluid.

This kind of drilling fluid can be further classified into four types according to the polymer used.

- Anionic polymer drilling fluid

This is a kind of water-base drilling fluid using anionic polymer as the main treating agent. This kind of drilling fluid has strong cuttings carrying capability, low content of submicron clay particles, and low nozzle viscosity (the viscosity of fluid at bit port under high shear rate). Besides, it can stabilize borehole wall and protect the hydrocarbon reservoir. To guarantee good performance of this drilling fluid, the solid content in the drilling fluids is controlled less than 10% (preferably below 4%). The mass ratio of the cuttings to bentonite in the solid phase is controlled in the range of 2:1–3:1. The ratio between the dynamic shear force and plastic viscosity of drilling fluid is controlled at about 0.48 Pa /(mPa · s). This drilling fluid is applicable to the reservoir with well depth lower than 3,500 m and well temperature lower than 150 °C.

- Cationic polymer drilling fluid

This is a kind of water-base drilling fluid using cationic polymer as the main treating agent. Because cationic polymers have bridging effect and can neutralize the electronegativity on the shale surface, they have strong ability of stabilizing shale. In addition, cationic surfactant can be added to disperse into the clay crystal layer which polymer cannot enter. Cationic polymer drilling fluid is applicable to the drilling in shale stratum.

- Zwitterionic polymer drilling fluid (Research Group of Amphoteric Polymer Mud, China National Petroleum Corporation 1994)

This is a kind of water-base drilling fluid using zwitterionic polymer as the main treating agent. The cationic groups in the zwitterionic polymer can stabilize shale like cationic polymer, and the anionic groups in the zwitterionic polymer can enhance the stability of drilling fluid by the hydration effect. What's more, zwitterionic polymer has good compatibility with other treating agents. Therefore, zwitterionic polymer drilling fluids bear excellent properties.

- Nonionic polymer drilling fluid (Dorothy and Martin 1992; Halliday 1993)

This is a kind of water-base drilling fluid using ether type polymer as the main treating agent. This kind of drilling fluid has characteristics of non-toxicity, non-pollution, good lubricating performance, and strong capacity of preventing bit balling and sticking. It can also stabilize the borehole wall and protect the oil reservoirs. Therefore, nonionic polymer drilling fluid is especially applicable to the drilling in the offshore and shale stratum.

- MMH drilling fluid

This is a kind of water-base drilling fluid using MMH as the main treating agent. This drilling fluid system has characteristics of good cuttings transportation, wellbore stability, and oil layer protection. It is especially applicable to drilling horizontal wells and opened hydrocarbon reservoir.

2.11.1.3 Oil-in-Water Drilling Fluid

Oil-in-water drilling fluid is prepared by adding oil and oil-in-water type emulsifier into drilling fluid (Ma et al. 2006).

The oil used in preparing the oil-in-water drilling fluid can be mineral oil or synthetic oil. The former are mainly diesel or machine oil in which the fluorescent substance (aromatic hydrocarbon) affecting the logging can be removed by the sulfuric acid refining. The latter are mainly organic compounds free of fluorescent substance (Growcock et al. 1994; Friedheim and Comn 1996; Zhang 1998); some examples are given below.

$$CH_3 - (CH_2)_n - CH_3$$

(Linear paraffin)

$$CH_3 - (CH_2)_{n_1} - CH{=}CH - (CH_2)_{n_2} - CH_3$$

(Linear olefine)

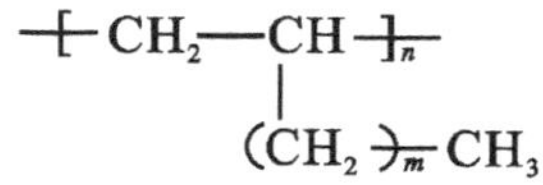

(Poly(α-olefine))

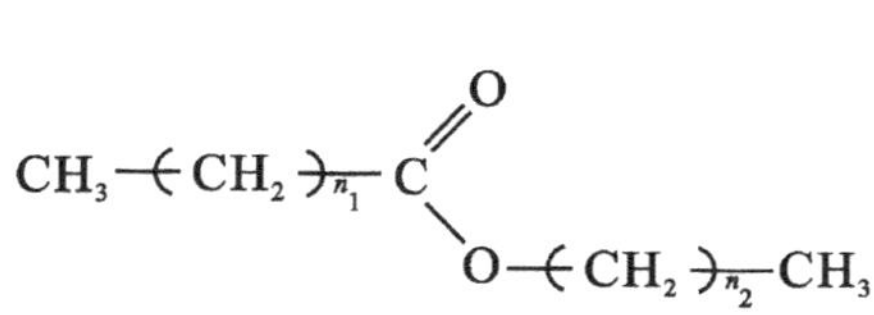

(Reaction product of fatty acid and alcohol)

The property of synthetic oil is required to be similar to that of mineral oil. At 25 °C, the density ranges from 0.76 to 0.86 g/cm^3 and viscosity ranges from 2 to 6 mPa · s.

The emulsifier used in preparing the oil-in-water drilling fluids is water-soluble surfactant. Some examples are given below.

$$R{-}SO_3Na$$

(Alkyl sodium sulfonate)

$$R{-}OSO_3Na$$

(Alkyl alcohol sulphate sodium salt)

$$R{-}O{+}CH_2CH_2O{+}_nH$$

(Polyoxyethylene alkyl alcohol ether)

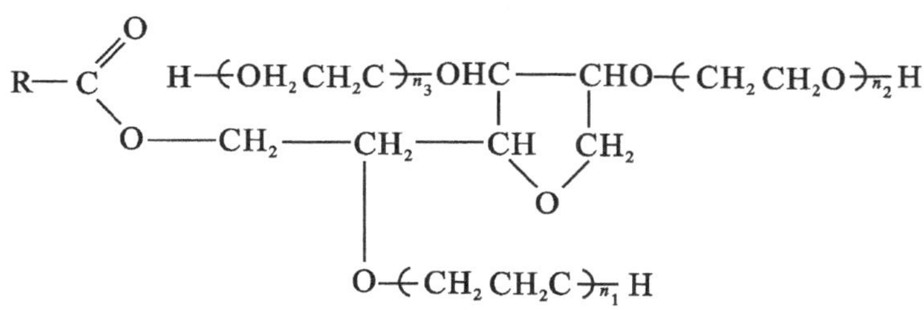

(Sorbide carboxylic ester polyvinylether)

The oil-in-water drilling fluid has characteristics of good lubrication, low filtration loss, and hydrocarbon reservoir protection effect.

The oil-in-water drilling fluid is applicable to the drilling in formations which are easy to generate bit balling and sticking.

2.11.1.4 Foam Drilling Fluid

The foam drilling fluid is prepared by injecting the foaming agents and gas into the water-base drilling fluid. Because water is used as the dispersion medium, foam drilling fluid belongs to the water-base drilling fluid.

The gas used for preparing foam drilling fluid can be nitrogen and carbon dioxide.

The foaming agents used for preparing the foam drilling fluid are water-soluble surfactants, such as alkyl sodium sulfonate, sodium alkyl benzene sulfonate, alkyl sulfate sodium salt, polyoxyethylene alkyl alcohol ether, polyoxyethylene alkyl alcohol ether sulfate sodium salt (Kuhlman et al. 1995; Xu and Yang 1995; Argillier et al. 1998).

The bentonite content in the drilling fluid is determined by the well depth and formation pressure.

The system has good prop-carrying capacity, low frictional resistance, and protective effect for the oil reservoirs.

To guarantee good performance of this drilling fluid, the up-hole velocity of drilling fluid in the annulus is required to be over 0.5 m/s.

Foam drilling fluid is mainly used in low-pressure leaking formations.

2.11.2 *Oil-Base Drilling Fluid*

This drilling fluid is composed of oil, organic soil, and treatment agents, using oil as dispersion medium (Kirsner et al. 2005). As this drilling fluid contains water, it can be classified into two categories according to the water content.

2.11.2.1 Pure Oil-Base Drilling Fluid

The oil-base drilling fluid with the water content lower than 10% is called pure oil-base drilling fluid.

The oil used for preparing the pure oil-base drilling fluid can be mineral oil (such as diesel oil and machine oil) and synthetic oil (such as linear paraffin, linear olefin, and poly-α-olefin).

Organic soil used for preparing the oil-base drilling fluid is prepared by treating the bentonite using the following quaternary ammonium salt surfactants.

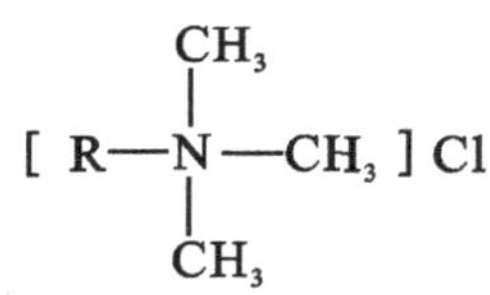

R: C_{12}–C_{18}

(Alkyl trimethyl ammonium chloride)

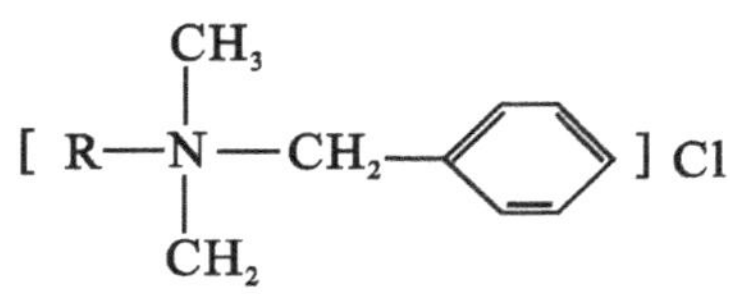

R: C_{12}–C_{18}

(Alkyl benzyl dimethyl ammonium chloride)

Quaternary ammonium salt surfactants can absorb on the surface of bentonite particles and transform the surface into oil-wet, making the bentonite particles more easily dispersed in the oil.

The treating agents used for preparing the oil-base drilling fluid are mainly filtrate reducer (e.g., oxidized asphalt) and emulsifier (e.g., calcium stearate).

The advantages of pure oil-base drilling fluid are temperature tolerance, anti-collapse performance, anti-sticking performance, good lubrication performance, protective effect for the oil reservoirs, and so on, while the deficiencies are high cost, environment pollution, and poor safety.

Pure oil-base drilling fluid can be applied in the shale stratum, rock salt stratum, and gypsum stratum and is especially applicable to the high-temperature stratum and opening oil–gas reservoirs.

2.11.2.2 Water-in-Oil Drilling Fluid

The water-in-oil drilling fluid is prepared by adding water (content higher than 10%) and water-in-oil emulsifier to the pure oil-base drilling fluid.

The surfactants used for preparing drilling fluid are oil-soluble surfactants. Some examples are given below.

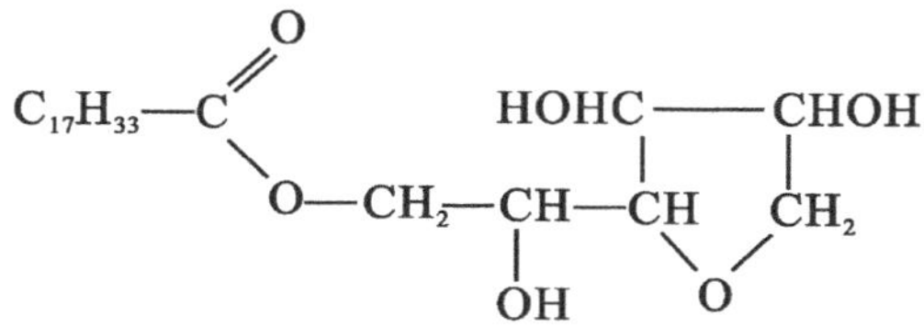

(Sorbide monooleate, Span 80)

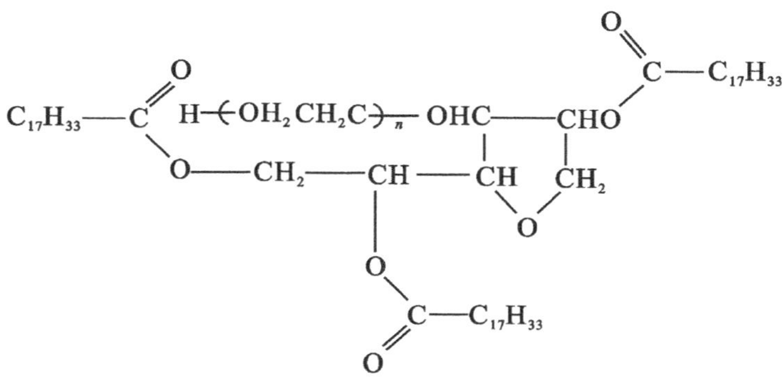

(Sorbide trioleate polyoxyethylene ether, Span 85)

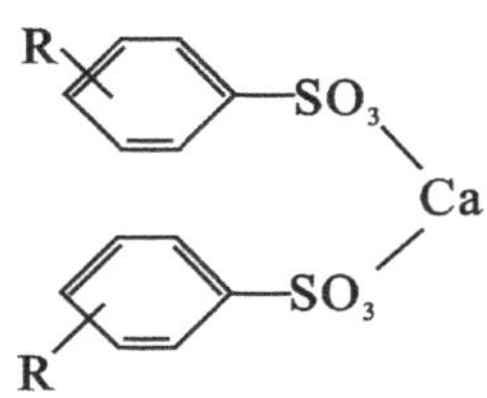

(Calcium alkylbenzene sulfonate)

Oil-in-water drilling fluid has the same characteristics as the pure oil drilling fluid; however, its cost is lower than pure oil drilling fluid. Oil-in-water drilling fluid has the same range of application as the pure oil drilling fluid.

2.11.3 Gas Drilling Fluid

Gas (such as air and natural gas) used as drilling fluid is called gas drilling fluid.

To ensure the carrying of the cuttings, the up-hole velocity of gas drilling fluid in the annulus should be over 15 m/s.

Gas drilling fluid can enhance the drilling rate and protect the reservoirs, but it has the drawbacks such as dry fiction, easily catching fire or exploding. Pipe-sticking and borehole collapse can also occur easily after the water production in the formation.

Gas drilling fluid can be applied in the oil–gas reservoirs with low pressure, severe leakage, and water shortage.

References

Adler E (1977) Lignin chemistry—past, present and future. Wood Sci Technol 11(3):169–218

Argillier JF, Saintpere S, Herzhaft B (1998) Stability and flowing properties of aqueous foams for underbalanced drilling. SPE 48982

Audibert A (1995) Thermal stability of sulfonated polymers. SPE 28953

Beihoffer TW, Dorrough DS, Schmidt DD (1990) The development of an inhibitive cationic drilling fluid for slim-hole coring applications. SPE 19953

Bland RG, Smith GL (1995) Low salinity polyglycol water-based drilling fluids as alternations to oil-based muds. SPE 19378

Bleakley WB (1980) Clear completion and workover fluids—what they are, what they'll do. Pet Eng 52(10):19–22

Bol GM, Davidson CJ, Woodland DC et al (1992) Borehole stability in shales. SPE 24975

Burba JL, Holman WE, Crabb CR (1988) Laboratory and field evaluations of novel inorganic drilling fluid additive. SPE 17198

Burba JL, Tehan WF, Hamilton FD et al (1990) Field evaluations confirm superior benefits of MMLHC fluid system on hole cleaning, borehole stability, and rate of penetration. SPE 19956

Chen X (1980) Which starch fraction is water-soluble. Chemistry 11:173–174

Chenecert ME (1989) Glycerol mud additive provides shale stability. Oil Gas J 87(29):60–64

Chee P, Grian G, Rahman SS et al (1996) Managing physico-chemical wellbore instability in shales with the chemical potential mechanism. SPE 36971

Chilingarian GV (1981) Drilling and drilling fluids. Elsevier Scientific Publishing Company, New York, p 18

Dorothy P, Martin F (1992) An environmentally safe water-based alternative to oil muds. SPE Drill Eng 7(1):15–19

Downs JD, Van Oort E, Redman DI et al (1993) TAME: a new concept in water-based drilling fluids for shales. SPE 26699

Fan Z, Wang L, Sun M (2005) Preparation of heat resistant, salts tolerant and inhibitory filtrate loss reducer KWY for aqueous drilling fluids from acrylic fibrous waste. Oilfield Chem 22(1):10–12

Federici F, Merli L, Vigano L et al (2008) Swelling inhibitors for clay and shales. WO Patent 08/031806A1, 20 Mar 2008

Friedheim JE, Comn HL (1996) Second generation synthetic fluids in the North Sea: are they better? SPE 35061

Gulf Publishing Company (1998) Classifications of fluids systems. World Oil 219(6):106

Gulf Publishing Company (1998) Classifications of fluid systems. World Oil 219(6):103–106

Growcock FB, Andrews SL, Frederick TP (1994) Physicochemical properties of synthetic drilling fluids. SPE 27450

Guo J, Yan J (2003) Research advances in silicate-based drilling fluids and their applications. Oil Drill Prod Technol 25(5):20–24

Hale AH (1998) Well drilling fluids and process for drilling wells. US Patent 4,728,445, 1 Mar 1988

Halliday WS (1993) Polyglycol and aluminum chemistry drilling fluid helps operator reach project goals. SPE 25702

Homer C, Jimmie D (1982) Oil well treating method and composition. US Patent 4,366,071, 28 Dec 1982
Hoskins T (2007) Drilling fluid and method for reducing lost circulation. WO Patent 107015A1, 27 Sept 2007
Jackson JM, Hartfiel AH (1978) Water loss reduction agents. US Patent 4,090,968, 23 May 1978
Jimmie D (1984) Methods and hydrocarbon base treating fluids for stabilizing water sensitive clay containing formations. US Patent 4,460,483, 17 July 1984
Kirsner J, Miller J, Bracken J (2005) Additive for oil-based drilling fluids. EP Patent 1496096A3, 12 Jan 2005
Kuhlman ML, Lau HC, Falls AH (1995) Carbon dioxide foam with surfactants used below their critical micelle concentrations. SPE 28952
Lan Q, Qiu Z (2007) Laboratory studies on silicate drilling fluids. Drill Fluid Complet Fluid 24(1):20–22
Li Z, Zhang W, Lu P (1987) Carboxymethyl celluloses I. Synthesis. Oilfield Chem 4(4):318–326
Li B, Liu Z, Zhang L et al (2009) Synthesis and anti-swelling property of poly2-hydroxylproylidene dihydroxyethyl ammonium chloride. J Daqing Pet Inst 33(5):90–92, 102
Li K, Xie J, Li P et al (1990) Drilling manual (the first party). Petroleum Industry Press, Beijing, pp 602–603
Liang D, Pu X (2005) Inhibition and borehole stability mechanism of silicate. Drill Prod Technol 28(6):105–107
Ma X (1995) Synthesis and application of polyacrylamide cationic polymer. Chem World 36(11):588–591
Ma X, Zhao J (1995) Acrylamide/trimethylallylammonium chloride co-polymer as heat and salts tolerant clay stabilizer: synthesis and properties. Oilfield Chem 12(3):197–200
Ma Y, Cui M, Sun S (2006) Research and application of oil-in-water drilling fluids system. Fault-Block Oil Gas Field 13(1):4–6
Martin EC, Vincent P (1998) Control of shale swelling pressures using inhibitive water-base muds. SPE 49263
Miano F, Lockhart TP (1994) Chrome-free lignosulfonate muds: Zirconium additives provide extended high-temperature performance. SPE 28305
Park LS (1988) A new chrome-free lignosulfonate thinner performance without environmental concerns. SPE Drill Eng 3(3):311–314
Perricone AC, Enright DP, Lucas JM (1986) Vinyl sulfonate copolymers for high-temperature filtration control of water-based muds. SPE Drill Eng 1(5):358–364
Perricone AC, Clapper DK, Enight DP (1989) Modified non-polluting liquid phase shale swelling inhibition drilling fluid and method of using same. US Patent 4,830,765, 16 May 1989
Plank JP (1992) Water-based muds using synthetic polymers developed for high temperature drilling. Oil Gas J 90(9):40–45
Polnaszek SC, Fraser LJ (1991) Drilling fluid formulation for shallow offshore horizontal well applications. SPE 22577
Reid PI, Dolan B (1995) Mechanism of shale inhibition by polyols in water based drilling fluids. SPE 28960
Retz RH, Norge AS, Friednein J et al (1991) An environmentally acceptable and field-practical, cationic polymer and system. SPE 23064
Research Group of Amphoteric Polymer Mud, China National Petroleum Corporation (1994) Review of amphoteric polymer mud: Part I. Drill Fluid Complet Fluid 11(4):13–22, 27
Santos H (1998) Wellbore stability: a new conceptual approach based on energy. SPE 49264
Sawdon CA, Bradbury AJ (2006) An additive for increasing the density of a fluid and fluid comprising such additive. EP Patent 1626077A2, 15 Feb 2006
Strickland SD (1994) Polymer drilling fluids in the 1990s: will they replace oil-based muds? JPT 56(8):691–692
Sun J, Pan X, Liu J (2002) Development of a novel drilling fluid lubricant GXRH. Drill Fluid Complet Fluid 19(6):13–14

Sun M, Hou W, Sun D et al (2005) Mechanism study on the potassium ion stabilizing wellbore. Drill Fluid Complet Fluid 22(5):7–9
Verret RJ (2005) Method and composition for inhibiting lost circulation during well operation. US Patent 170973A1, 4 Aug 2005
Wang Z (1993) Modified starch additives for drilling fluid. Chem Eng Oil Gas 22(2):108–110
Wang Z, Wang A (1994) Synthesis and application of cationic starches. Chem World 35(10):524–528
Wernau WC (1981) Fermentation process for production of xanthan. US Patent 4,282,321, 4 Aug 1981
Wong SW, Kenter CJ (1993) Optimizing shale drilling in the Northern North Sea: Borehole stability considerations. SPE 26736
Wu L, Yang F (1992) Colloid chemistry principle of drilling fluids treating additive. Cheng Du Technology University Press, Cheng Du, pp 70–86
Xiang T (2005) Drilling fluid systems for reducing circulation losses. US Patent 221994A1, 6 Oct 2005
Xu R, Yang L (1995) A new binary surfactant mixture improved foam performance in gas well production. SPE 29004
Yao F (1989) Asphaltic additives in drilling fluids. Oilfield Chem 6(2):176–184
Yu J, Zhang Y, Hu S (2008) Synthesis of mini-cationic clay stabilizer and its performance evaluation. J Yangtze Univ (Natural Science Edition) 5(2):26–28
Zhang Y (1998) Review on the development of synthetic-based drilling fluids. Drill Fluid Complet Fluid 15(3):28–32
Zhang Y (1998) Preparation of cationic starch by one-step semi-dry process. Chem World 39(11):591–593
Zhang Z, Yan Z, Fan S (2005) Low density drilling fluid technology. China University of Petroleum Press, Dong Ying, pp 222–223
Zhang C, Hou W, Sun D (1993) MMH drilling fluid system & cationic treating additive. Drill Fluid Complet Fluid 10(3):1–19
Zhang C, Hou W, Wang G (1988) Sepiolite mud II. The thinning effect of SSMA on sepiolite/HPAM suspensions. Oilfield Chem 5(1):1–5
Zhao X, Hu Z, Zhang J et al (2009) Preparation of SSHIA as ultra-temperature viscosity reducer for drilling fluid. J Oil Gas Technol 31(5):105–108
Zhu W, Zhang R, Liu G et al (2003) Property of a clay stabilizer-CETA. Chin J Appl Chem 20(10):955–958

Chapter 3
Cement Slurry Chemistry

Cement slurry is the working fluid employed in well cementation. The well cementation mentioned here is a kind of operation during which the cement slurry is injected into the annular space between borehole wall and casing from the casing pipe and moved upwards to certain height before solidified to cement stone to consolidate the borehole wall and casing.

In this chapter, the function, composition, properties, and chemical treatment of cement slurry are introduced. The cement slurry system used in general and special circumstances are also introduced, corresponding to the drilling fluid system.

3.1 Function and Composition of Cement Slurry

3.1.1 Function of Cement Slurry

The function of cement slurry is well cementing, which can achieve the following purposes (Sabins and Autton 1982).

3.1.1.1 Fixing and Protecting Casing Pipes

All the casing pipes used in the process of drilling must be fixed through the well cementing operation. In addition, the cement stone outside the casing can reduce the extrusion of formation to casing, which plays the role of protecting casing.

C. Dai and F. Zhao, *Oilfield Chemistry*,
https://doi.org/10.1007/978-981-13-2950-0_3

3.1.1.2 Protecting High-Pressure Oil and Gas Layers

When drilling in high-pressure oil and gas layers, well blowout accidents are easily to take place. The density of drilling fluid needs to be increased to balance the formation pressure. Therefore after drilling, casing and cementing operations must be done to protect the high-pressure oil and gas layers.

3.1.1.3 Sealing Serious Leakage Layers and Other Complex Layers

When drilling in serious leakage layers, the methods of reducing the density of drilling fluid and/or adding plugging material can be adopted. After drilling in serious leakage layers, casing and cementing operations must also be done to seal the leakage layers, so that the subsequent well drilling will not be affected.

When drilling in other complex layers such as the sloughing formation, casing, and cementing can also be conducted after drilling.

3.1.2 Composition of Cement Slurry

The cement slurry is composed of water, cement, additive, and admixture.

3.1.2.1 Water

The water used for preparing cement slurry can be freshwater or brine (including seawater).

3.1.2.2 Cement

The cement used for preparing cement slurry is made of limestone and clay through calcination in 1,450–1,650 °C, cooling and levigation. It mainly contains the following silicate and aluminate.

Tricalcium Silicate ($3CaO{\cdot}SiO_2$)
Tricalcium silicate has the highest content in cement. Its hydration rate, strength increase rate, and final strength are all high.

Dicalcium Silicate ($2CaO{\cdot}SiO_2$)
In cement, the content of dicalcium silicate is lower than that of tricalcium silicate. The hydration rate and strength increase rate of dicalcium silicate are low, but the final strength is high.

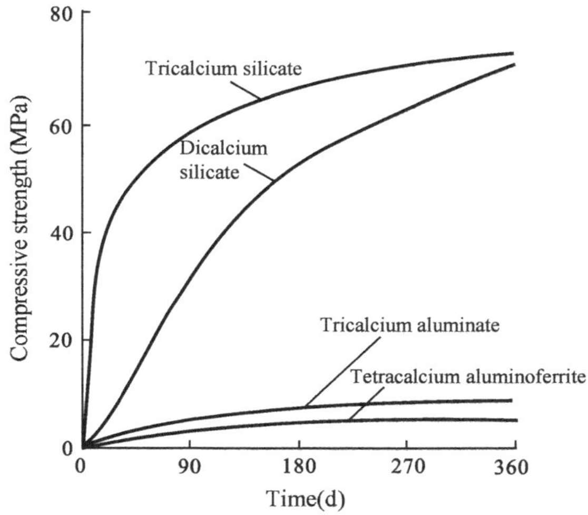

Fig. 3.1 Changes of compressive strength with time after hydration of each component in the cement (tricalcium silicate, dicalcium silicate, tricalcium aluminate, tetracalcium aluminoferrite)

Tricalcium Aluminate ($3CaO{\cdot}Al_2O_3$)
The content of tricalcium aluminates in cement is low. Although the hydration rate of tricalcium aluminate is high, the strength increase rate is low. Thus, the final strength is low.

Tetracalcium Aluminoferrite ($4CaO{\cdot}Al_2O_3{\cdot}Fe_2O_3$)
In cement, the content of tetracalcium aluminoferrite is slightly higher than that of tricalcium aluminate. However, the hydration rate, strength increase rate, and final strength of tetracalcium aluminoferrite are similar to those of tricalcium aluminate.

The intensity changes after hydration of the four kinds of compositions above are shown in Fig. 3.1 (Xu 1988). The compressive strength is determined by placing certain size of cement sample under vertical force in automatic loading strength-testing machine until failure.

As is shown in Fig. 3.1, the early strength after hydration of cement is mainly determined by tricalcium silicate, the late strength is mainly determined by tricalcium silicate and dicalcium silicate, while tricalcium aluminate and tetracalcium aluminoferrite have minor effect on the early strength and late strength.

Table 3.1 Mineral composition of API oil well cement

Mineral	w(mineral) (%)							
	A	B	C	D	E	F	G	H
Tricalcium silicate	53	47	58	26	26	50	50	50
Dicalcium silicate	24	32	16	54	54	30	30	30
Tricalcium aluminate	8	5	8	2	2	5	5	5
Tetracalcium aluminoferrite	8	12	8	12	12	12	12	12

In addition, there are gypsum, alkali metal sulfate, magnesium oxide, and calcium chloride in cement, which have certain effect on the hydration rate of cement and the properties after cement solidification.

The cement used in well cementing is called oil well cement (Brothers et al. 2003).

According to American Petroleum Institute (API) standard, oil well cement can be classified into nine grades, A, B, C, D, E, F, G, H, and J. The mineral compositions of some of these grades are shown in Table 3.1 (Liu 1988).

The applicable well depth of oil well cement is shown in Table 3.2.

According to Chinese petroleum industry (SY) standards, oil well cement can be divided into four grades, 45, 75, 95, and 120 °C, based on the applicable formation temperature. The mineral compositions of some grades of oil well cements are shown in Table 3.3. For comparison, the mineral composition of ordinary silicate cement is also listed in Table 3.3.

The applicable well depth of SY oil well cement is shown in Table 3.4.

Table 3.2 Applicable well depth of oil well cement

Grade	Applicable well depth (m)	Grade	Applicable well depth (m)
A	0–1,830	F	3,050–4,880
B	0–1,830	G	0–2,440
C	1,830–3,050	H	0–2,440
D	3,050–4,270	J	3,600–4,880
E	3,050–4,270		

Table 3.3 Mineral composition of SY oil well cement

Mineral	*w*(mineral) (%)			
	45 °C oil well cement	75 °C oil well cement	95 °C oil well cement	Ordinary silicate cement
Tricalcium silicate	49–57	49–57	21–41	38–60
Dicalcium silicate	18–23	18–23	35–58	15–38
Tricalcium aluminate	0.7–9	0.7–9	0–1.1	7–15
Tetracalcium aluminoferrite	1.8–16	1.8–16	1.5–17	10–18

Table 3.4 Applicable well depth of SY oil well cement

Grade	Applicable well depth (m)	Grade	Applicable well depth (m)
45 °C oil well cement	0–1,500	95 °C oil well cement	2,500–3,500
75 °C oil well cement	1,500–2,500	Ordinary Portland cement	3,500–5,000

3.1.2.3 Additives and Admixtures of Cement Slurry

In order to adjust the performance of cement slurry, some special materials need to be added. The material whose addition quantity no more than 5% of the cement mass is called additive, while the material whose addition quantity larger than 5% of the cement mass is called admixture (Gulf Publishing Company 1996, Allouche and Maroy 2003).

If classified by application, the additives and admixtures of cement slurry together can be divided into seven categories, which are coagulant, retarder, friction reducer, expansive agent, filtrate reducer, density adjustment admixture, and leak-proof admixture.

3.2 The Density of Cement Slurry and Its Adjustment

In well cementing, to guarantee that the drilling fluid between the well wall and the casing pipe is replaced completely by the cement slurry, the density of cement slurry is required to be greater than that of drilling fluid. However, the density of cement slurry should not be too high so that no leakage of formations would occur.

When preparing cement slurry, the ratio between water and cement is called water–cement ratio. The common water–cement ratio of cement slurry is in the range of 0.3–0.5, while the density of prepared cement slurry is in the range of 1.8–1.9 g/cm^3. According to the circumstances of different formations, the density of cement slurry should be adjusted to different ranges. If the density of cement slurry is not in the required range, it can be adjusted by density adjustment admixture. The density adjustment admixture for cement slurry can be divided into density-reducing admixture and density-increasing admixture.

3.2.1 Density-Reducing Admixture for Cement Slurry

The material that can reduce the density of cement slurry is called density-reducing admixture for cement slurry. In the cementing of low-pressure oil–gas reservoir or leaking formation, density-reducing admixture should be added into cement slurry.

There are several kinds of density-reducing admixtures for cement slurry.

3.2.1.1 Clay

The solid density of clay (2.4–2.7 g/cm^3) is lower than that of cement (3.1–3.2 g/cm^3). If part of cement is replaced by clay when preparing the cement slurry, the density of cement slurry can be reduced. This is the partial substitution mechanism of reducing the density of cement slurry by using solid with lower density.

In addition, compared with other low-density solids, clay has a unique density-reducing mechanism, thickening mechanism. Clay (especially sodium bentonite) has an excellent thickening effect on water which can greatly increase the water content in the cement slurry, therefore effectively reducing the density of the cement slurry.

Generally, the addition amount of clay in cement slurry is around 5–32% of cement mass, and it can be used to prepare the cement slurry with density in the range of 1.3–1.8 g/cm^3.

If the water used for preparing cement slurry is freshwater or low concentration brine, bentonite can be used as density-reducing admixture; if the water used for preparing cement slurry is high concentration brine, attapulgite or sepiolite should be used as density-reducing admixture.

3.2.1.2 Fly Ash

Fly ash is the hollow particles produced by the combustion of pulverized coal, consisting mainly of silica (as shown in Table 3.5). The solid phase density of fly ash (around 2.1 g/cm^3) is lower than that of cement.

If the fly ash is used as partial substitute of cement for preparing cement slurry, the density of cement slurry can be reduced. The cement slurry with density in the range of 1.6–1.8 g/cm^3 can be prepared with fly ash.

3.2.1.3 Expanded Perlite

Expanded perlite is the porous solid produced by perlite through high-pressure melting, rapid decompression and cooling. The pores in solids are generated by the vaporization of crystal water in perlite at high temperature and decompression. The solid phase density of expanded perlite (around 2.4 g/cm^3) is lower than that of cement.

If expanded perlite is used as partial substitute of cement for preparing cement slurry, the density of cement slurry can be reduced. The cement slurry with density in the range of 1.1–1.2 g/cm^3 can be prepared with expanded perlite.

Table 3.5 Analysis results of the composition of one kind of fly ash

Composition	w(composition) (%)	Composition	w(composition) (%)
SiO_2	55–65	CaO	1–3
Al_2O_3	25–35	MgO	<4
Fe_2O_3	3–5	Loss of ignition	<2

3.2.1.4 Hollow Glass Bead

Hollow glass beads are produced by spraying molten glass through a special nozzle. The diameter of hollow glass beads is in the range of 20–200 μm, the wall thickness is in the range of 0.2–0.4 μm, and the apparent density is in the range of 0.4–0.6 g/cm^3.

If the hollow glass bead is used as partial substitute of cement for preparing cement slurry, the density of cement slurry can be reduced. The cement slurry with density in the range of 1.0–1.2 g/cm^3 can be prepared with hollow glass bead.

In addition, hollow ceramic beads (with apparent density of around 0.7 g/cm^3) and hollow urea formaldehyde resin beads (with apparent density of around 0.5 g/cm^3) can be used to prepare the cement slurry with low density.

3.2.2 Density-Increasing Admixture for Cement Slurry

The material that can increase the density of cement slurry is called density-increasing admixture for cement slurry. In the well cementing of high-pressure oil–gas reservoir, density-increasing admixture should be added into cement slurry.

There are two types of density-increasing admixture for cement slurry.

3.2.2.1 High-Density Solid Powder

High-density solid includes barite, siderite, ilmenite, magnetite, and pyrite. If these high-density solids are ground to powder with certain particle size and are added into cement slurry, the density of cement slurry can be improved. The cement slurry with density in the range of 2.1–2.4 g/cm^3 can be prepared with high-density solid powder.

3.2.2.2 Water-Soluble Salt

Water-soluble salt can improve the density of cement slurry by increasing the density of water. Water-soluble salt mainly refers to sodium chloride. The density of cement slurry can be increased to 2.1 g/cm^3 by using sodium chloride.

In addition, under the premise of ensuring the rheological property of cement slurry, adding friction reducer (to be mentioned later) can greatly reduce the water–cement ratio of cement slurry, therefore increasing the density of cement slurry.

3.3 The Thickening of Cement Slurry and the Adjustment of Thickening Time

3.3.1 The Thickening of Cement Slurry

3.3.1.1 Behavior After the Mixture of Water and Cement

Behavior after the mixture of water and cement is mainly manifested by the gradual thickening of the cement slurry. The gradual thickening phenomenon of the cement slurry is called cement thickening. The degree of cement thickening is expressed by consistency. The consistency of cement slurry, with the unit of Bc, is obtained by measuring the resistance of a blade with certain rotational speed in the cement slurry through a thickening apparatus. The rate of cement thickening is represented by thickening time. Thickening time refers to the time required for the consistency of the mixture of water and cement reaching 100 Bc. In order to inject the cement slurry into the annular space between the wall and the casing smoothly, the thickening time is required to be equal to the construction time of injecting the cement slurry (i.e., the time from preparing the cement slurry to the cement slurry returning to a predetermined height) plus 1 hour. Cement slurry thickening time is determined by the curves of cement slurry consistency with time. A typical curve of cement slurry consistency with time is shown in Fig. 3.2.

3.3.1.2 Hydration Reactions of Different Components in Cement

Cement thickening is caused by the hydration of cement. In water, different components in cement can undergo the following hydration reactions.

$$3CaO \cdot SiO_2 + 2H_2O \rightarrow 2CaO \cdot SiO_2 \cdot H_2O + Ca(OH)_2$$

(Tricalcium silicate)

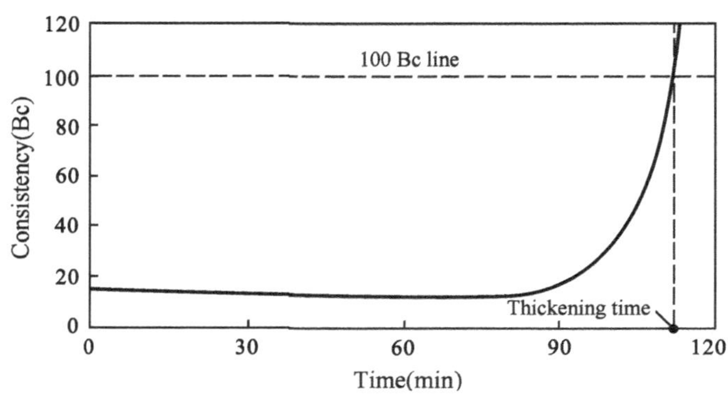

Fig. 3.2 Curve of the consistency of cement slurry with time

$$2CaO \cdot SiO_2 + H_2O \rightarrow 2CaO \cdot SiO_2 \cdot H_2O$$
(Dicalcium silicate)

$$3CaO \cdot Al_2O_3 + 6H_2O \rightarrow 3CaO \cdot Al_2O_3 \cdot 6H_2O$$
(Tricalcium aluminate)

$$4CaO \cdot Al_2O_3 \cdot Fe_2O_3 + 7H_2O \rightarrow 3CaO \cdot Al_2O_3 \cdot 6H_2O + CaO \cdot Fe_2O_3 \cdot H_2O$$
(Tetracalcium aluminoferrite)

$Ca(OH)_2$ produced by hydration can further react with $3CaO{\cdot}Al_2O_3$ and $4CaO{\cdot}Al_2O_3{\cdot}Fe_2O_3$, respectively.

$$3CaO \cdot Al_2O_3 + Ca(OH)_2 + (n-1)H_2O \rightarrow 4CaO \cdot Al_2O_3 \cdot nH_2O$$

$$4CaO \cdot Al_2O_3 \cdot Fe_2O_3 + 4Ca(OH)_2 + 2(n-2)H_2O \rightarrow 8CaO \cdot Al_2O_3 \cdot Fe_2O_3n \cdot 2nH_2O$$

3.3.1.3 The Stages of Hydration Process of Cement

Hydration process of cement can be measured by thermal method. The change of heat release rate with time during the hydration of cement is shown in Fig. 3.3 (Nelson 1990).

As is shown in Fig. 3.3, the hydration process of cement can be divided into five stages.

Pre-induction Stage

This stage refers to the several minutes after the mixture of water and cement. In this stage, lots of heat (including moist heat and reaction heat) is released as a result of the wetting of cement dry powder and hydration reaction. The hydrate generated

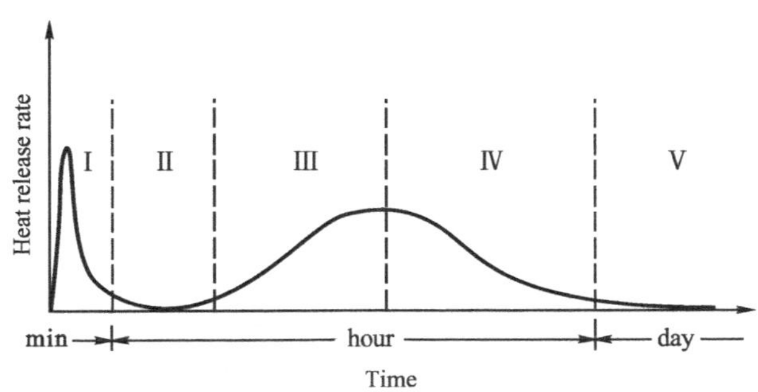

Fig. 3.3 Diagrammatic sketch of the change of heat release rate with time during the hydration of cement. I—Pre-induction stage; II—induction stage; III—curing stage; IV—hardening stage; V—terminating stage

from hydration reaction forms a saturated solution near the surface of the cement particles and precipitates on the surface, preventing the further hydration of cement particles. Therefore, the hydration rate of cement drops rapidly, and the hydration process enters the induction stage.

Induction Stage
In this stage, the hydration rate of cement is considerably low. However, due to the gradual dissolving of the hydrate precipitated on the cement surface (because it does not get saturated to aqueous phase of the cement slurry), the hydration rate increases in the late stage.

Curing Stage
In this stage, the rate of hydration increases. A large number of hydrates are produced from the hydration of cement, which are first dissolved in water followed by saturated precipitation. Network structure is formed among cement particles to cure the cement slurry.

Hardening Stage
In this stage, the network structure among cement particles becomes denser and the strength of cement stone gets higher. Therefore, the permeability is getting lower, affecting the contact of unhydrated cement particles with water, resulting in further reduced hydration rate.

Terminating Stage
In this stage, the amount of water permeating into cement stone is getting smaller and smaller until it cannot permeate anymore; thus, the hydration of cement comes to the end. The whole process of cement hydration is completed.

3.3.2 Adjustment of Cement Thickening Time

In order to meet the construction requirements, cement thickening time needs to be adjusted. The material which can adjust the cement thickening time is called setting conditioner for cement slurry. The setting conditioner can be divided into accelerator and retarder.

3.3.2.1 Accelerator for Cement Slurry

The agent which can shorten the thickening time of cement slurry is called accelerator for cement slurry (Lebo and Resch 2002).

Table 3.6 Influence of the addition of calcium chloride on the thickening time of cement slurry

$w(CaCl_2)$ (%)	Thickening time (min)		
	32 °C	40 °C	45 °C
0	240	180	152
2	77	71	61
4	75	62	59

Table 3.7 Influence of the addition of calcium chloride on the early compressive strength of cement stone

$w(CaCl_2)$ (%)	Compressive strength (MPa)		
	6 h	12 h	24 h
0	240	180	152
2	77	71	61
4	75	62	59

Calcium chloride is a typical kind of accelerator for cement slurry. The influence of its addition on the thickening time of cement slurry is shown in Table 3.6.

As is shown in Table 3.6, the addition of calcium chloride can significantly shorten the thickening time of cement slurry.

By compressing the diffused double layer of precipitated hydrate surface and promoting the formation of network structure with high permeability in cement particles, calcium chloride facilitates the infiltration of water and the progressing of hydration reaction. As a result, calcium chloride has good accelerating effects on cement slurry.

Table 3.7 suggests that apart from the accelerating effect, calcium chloride can also improve the early compressive strength of cement stone.

Other water-soluble salts (such as chloride, carbonate, phosphate, sulfate, aluminate, low molecular organic acid salt) all have accelerating effects similar to that of calcium chloride.

3.3.2.2 Retarder for Cement Slurry

The agent which can extend the thickening time of cement slurry is called retarder for cement slurry (Lebo and Resch 2002).

Classification of Retarder for Cement Slurry

- Boric acid and its salts

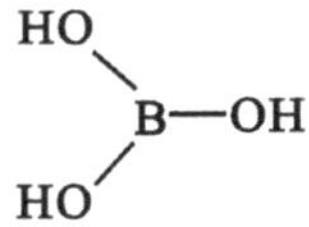

(Boric acid)

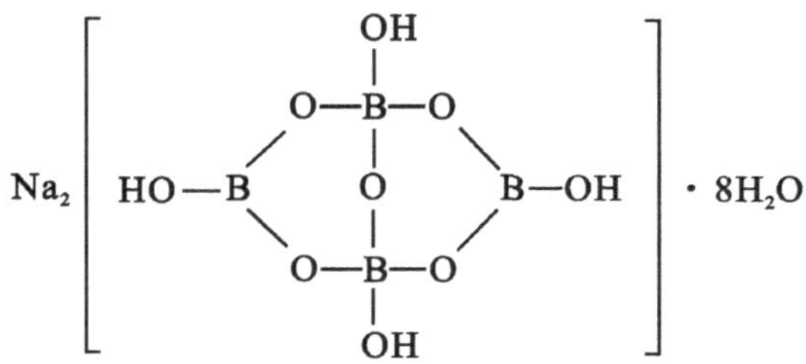

(Sodium tetraborate)

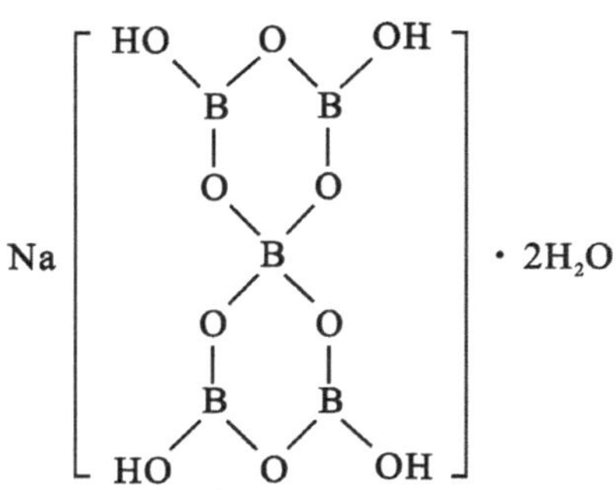

(Sodium pentaborate)

- Phosphoric acid and its salts

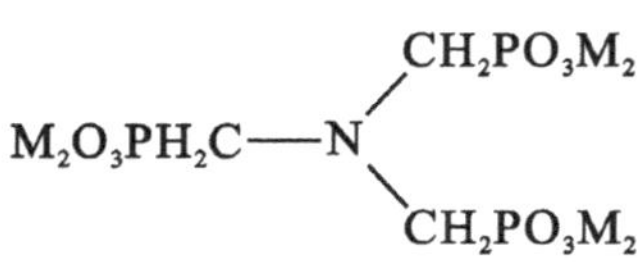

(Nitrilo-trimethylene phosphonic acid and its salts, ATMP)

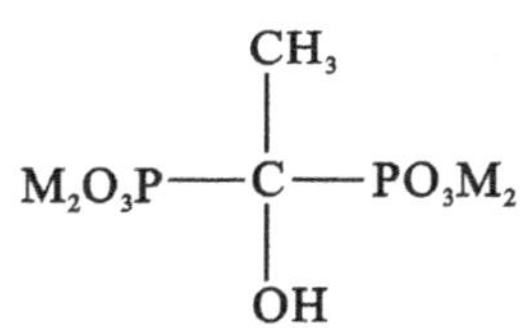

(Ethylidyne hydroxy diphosphate and its salts, HEDP)

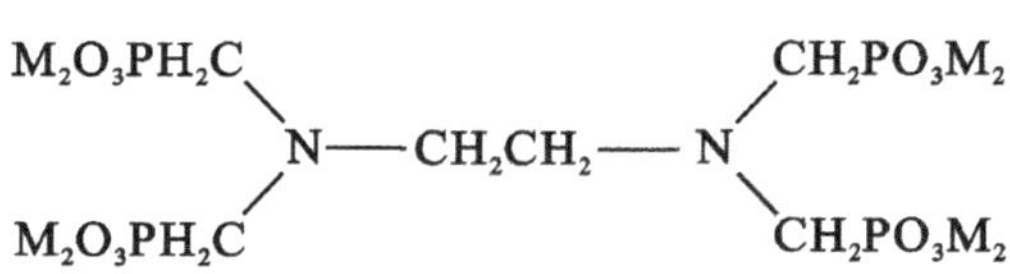

(Ethylenediamine tetramethylene phosphonic acid and its salts, EDTMP)

The M in the formula is H or Na, K, NH_4, etc.

- Hydroxy carboxylic acid and its salts

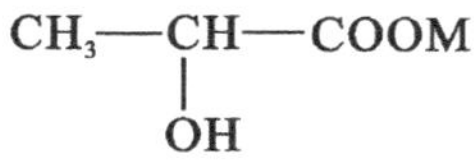

(Lactic acid and its salts)

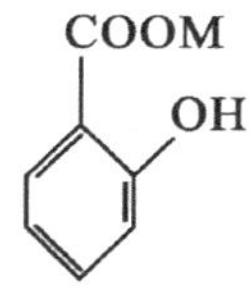

(Salicylic acid and its salts)

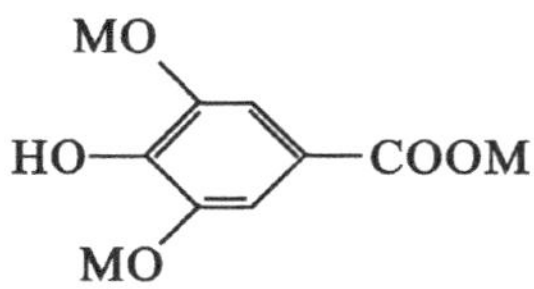

(Gallic acid and its salts)

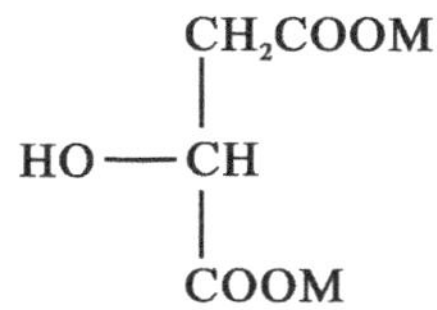

(Malic acid and its salts)

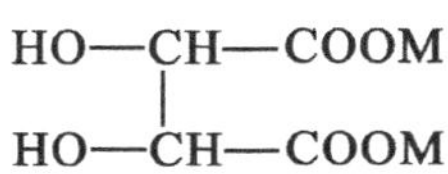

(Tartaric acid and its salts)

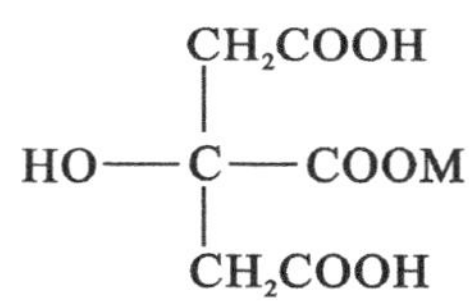

(Citric acid and its salts)

The M in the formula is H or Na, K, NH_4, etc.

- Lignin sulfonate and its modified products

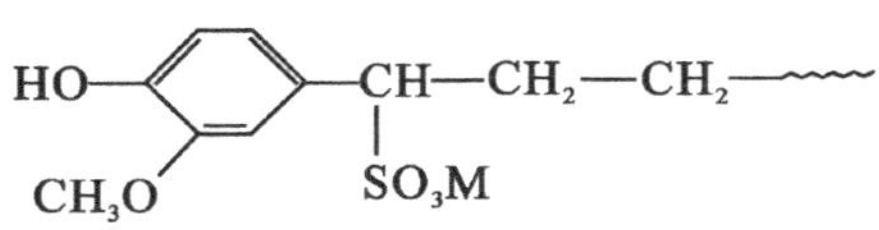

(M: Na or 1/2Ca)

(Sodium lignin sulfonate or calcium lignin sulfonate)

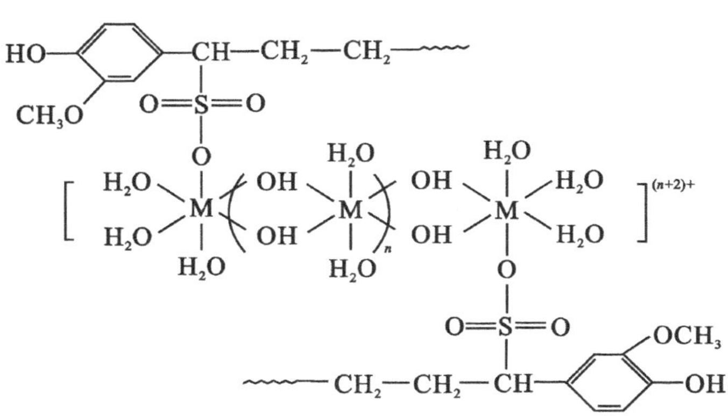

(M: Fe or Cr)

(Ferric chromium lignin sulfonate)

- Water-soluble polymer

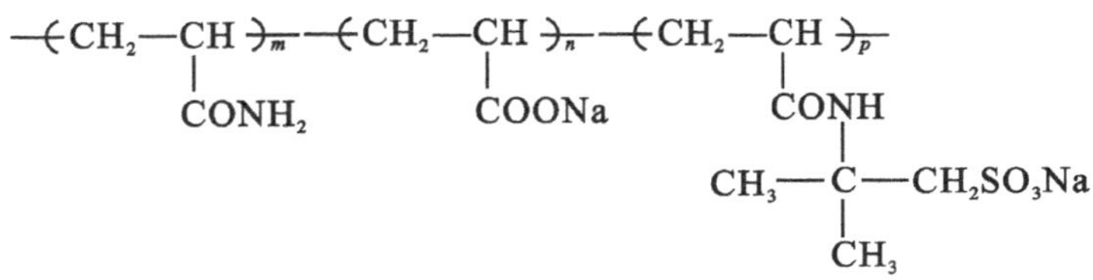

[Copolymerization of acrylamide, sodium acrylate and (2-acrylamide-2-methyl) sodium allylsulfonate] (Brothers and Rao 1985; Eoff and Buster 1995)

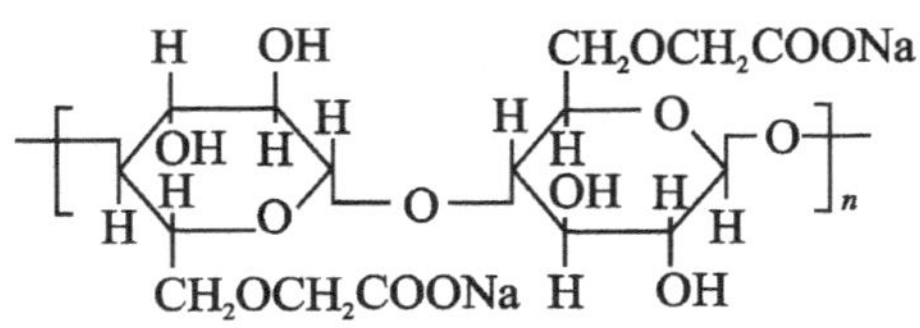

(Sodium carboxymethyl cellulose)

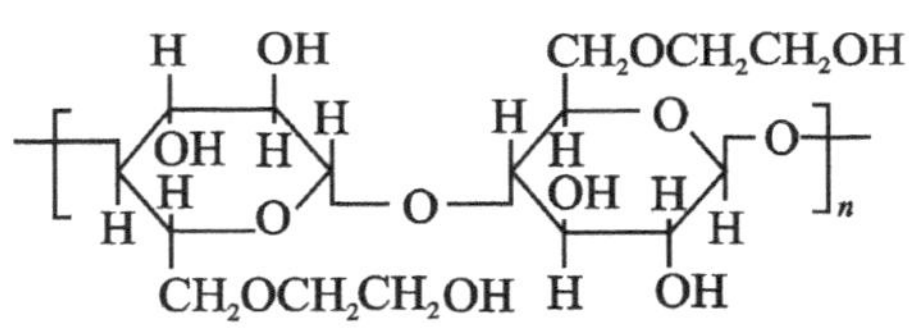

(Hydroxyethyl cellulose)

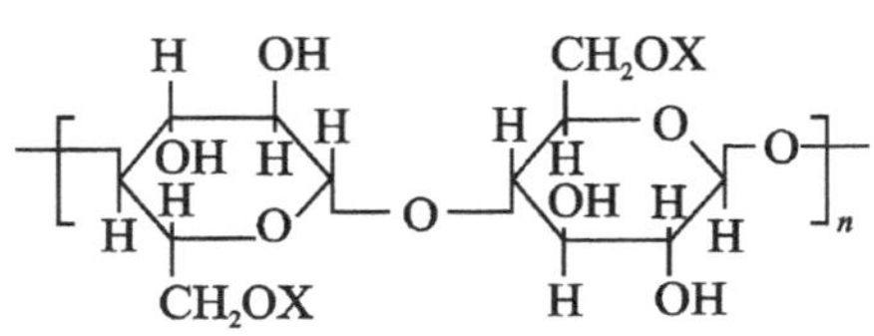

X: CH_2COONa or CH_2CH_2OH

(Sodium carboxymethyl hydroxyethyl cellulose)

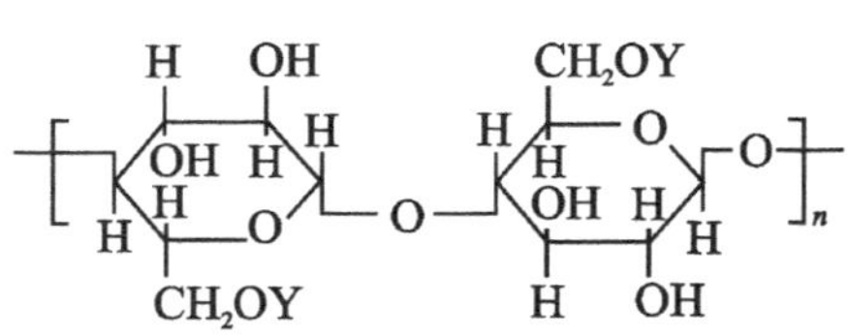

Y: CH_2COONa or CH_3CHCH_2OH

(Sodium carboxymethyl hydroxypropyl cellulose)

The Reaction Mechanism of Retarder for Cement Slurry

The retarder for cement slurry mentioned above can play the role of retarding mainly through two mechanisms.

- Adsorption mechanism

The retarder for cement slurry can adsorb on the surface of cement particle to hinder its contact with water. It can also adsorb on the surface of cement hydrates which are precipitated from the saturation, affecting the rate of forming network structure in curing stage and hardening stage. In these ways, the retarder can play the role of retarding.

Lignin sulfonate and its modified products and water-soluble polymer take effect mainly through this mechanism.

- Chelation mechanism

The retarder for cement slurry can form the following stable five or six-membered ring structure with Ca^{2+} through chelation, which can affect the rate of saturated precipitation of cement hydrate, playing the role of retarding.

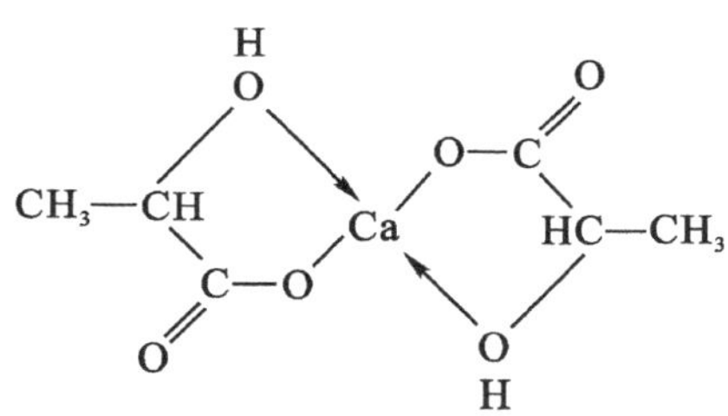

(Structure formed by lactic acid and Ca^{2+})

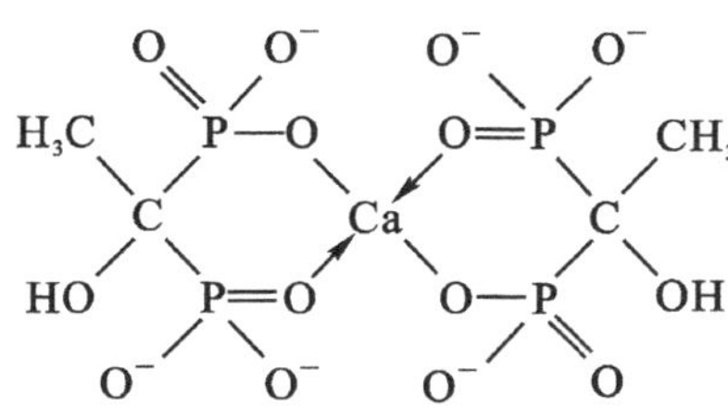

(Structure formed by HEDP and Ca^{2+})

Boric acid and its salts, hydroxy carboxylic acid and its salts, and phosphoric acid and its salts can play the role of retarding mainly through this mechanism.

3.4 The Rheological Property of Cement Slurry and Its Adjustment

3.4.1 The Rheological Property of Cement Slurry

The rheological property of cement slurry is related with its flow resistance at injection, its replacing efficiency to drilling fluid, and also the quality of cementing.

The rheological property of cement slurry is similar to that of drilling fluid, because the cement and clay share similar electrical property and cohesion in water.

The reason why the cement slurry and clay have similar properties in water can be attributed to two aspects. One is that the major component of cement slurry is clay, and the other is that the cement slurry particles can also generate hydroxyl surface in water.

Therefore, the rheological models of drilling fluid can also be used for describing the rheological properties of cement slurry.

3.4.2 Adjustment of the Rheological Property of Cement Slurry

Cement slurry has high flow resistance due to the high content of solid phase. Therefore, the adjustment of the rheological property of cement slurry is mainly to reduce the flow resistance of cement slurry. Drag reducer can be used to reduce the flow resistance of cement slurry.

The working mechanisms for cement slurry drag reducer and drilling fluid viscosity reducer are the same. Both of them can improve the electronegativity of the cement slurry particle surfaces by adsorption and increase the hydration layer thickness, so that the structure formed by cement slurry particles can be broken up to reduce the flow resistance.

Available cement slurry drag reducers can be classified into the following types.

3.4.2.1 Hydroxy Carboxylic Acid and Its Salts

Examples are lactic acid, gallic acid, citric acid, salicylic acid, malic acid, tartaric acid, and their salts. These drag reducers have high thermostability, strong salt-resistance, and good retarding effect.

3.4.2.2 Lignosulfonate and Its Modified Products

Examples are sodium lignosulfonate, calcium lignosulfonate, and Fe–Cr lignosulfonate. These drag reducers have similar characteristics as hydroxy carboxylic acid and their salts; however, defoamer is needed when in use.

3.4.2.3 Olefin Monomer Oligomers

Examples are polyethylene sulfonate sodium, polystyrene sulfonate sodium, copolymer of vinyl sulfonate sodium and acrylamide copolymer, copolymer of styrene sulfonate sodium and maleic acid sodium (Li and Wang 1998).

The relative molecular mass of oligomer mentioned above ranges from 2×10^3 to 6×10^3.

Olefin monomer oligomers have characteristics of high thermal stability, non-foaming, non-retarding, and good drag reduction effect.

3.4.2.4 Sulfonated Resin Condensation Polymer

An important sulfonated resin (Li 1987) is as below.

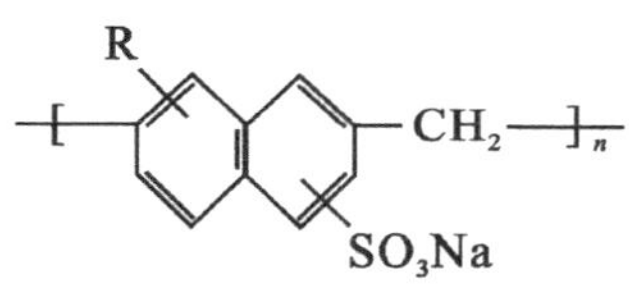

(Sulphonated alkylnaphthalene formaldehyde resin)

The relative molecular mass of this kind of sulfonated resin ranges from 2×10^3 to 4×10^3. It has similar characteristics with olefin monomer oligomer, but with certain retarding effect.

3.5 The Filtration Property of Cement Slurry and Its Control

3.5.1 The Filtration Property of Cement Slurry

Similar with drilling fluid, cement slurry also has filtration phenomenon.

The filtration rate of cement slurry is much higher than that of drilling fluid. The common filtrate volume of cement slurry without chemical treatment is more than 1,500 mL.

The filtration of cement slurry will lead to its poor mobility, while the filtrate flowing into the formation will cause damage to the formation.

For well cementation with different purposes, they have different requirements for the filtrate volume of cement slurry. For general well cementation, the filtrate volume is required to be smaller than 250 mL; for deep well cementation, the filtrate volume is required to be smaller than 50 mL; for oil–gas reservoir well cementation, the filtrate volume is required to be smaller than 20 mL.

Furthermore, formations with different permeabilities have different requirements for the filtration properties of cement slurries. The higher the permeability of the formation, the lower the filtrate volume of cement slurry is required.

The filtration theory and influence parameter of cement slurry are the same as those of drilling fluid.

3.5.2 The Control of the Filtrate Volume of Cement Slurry

To control the filtrate volume of cement slurry at a given range, filtrate reducer can be added into cement slurry (Cook and Commingham 1977).

Filtrate reducer for cement slurry can be divided into three categories.

3.5.2.1 Solid Particle

Bentonite, limestone, bitumen, and thermoplastic resin can be ground into particles with different grain sizes to function as filtrate reducer for cement slurry.

Trapping mechanisms and physical plugging mechanism are employed by solid particles to reduce the filtrate loss.

3.5.2.2 Latex

The latex is a kind of dispersion system generated by emulsion polymerization.

Emulsion polymerization is a kind of polymer preparation method, in which emulsifier is used to emulsify the water-insoluble monomers (or oligomers) in water by stirring.

Various homopolymers (such as polyvinyl chloride, polybutadiene, and polyvinyl acetate) and copolymers (such as copolymer of styrene and butadiene and copolymer of styrene and methyl methacrylate) can be prepared through emulsion polymerization.

The diameter of liquid bead in latex ranges from 0.05 to 0.5 μm.

The stable latex can reduce filtrate loss through superimposed Jamin effect generated by viscous liquid beads in the pore structure of the formation. While the liquid beads of unstable latex can form film on the formation pore surface, reducing the permeability of the formation and therefore the filtrate loss.

3.5.2.3 Water-Soluble Polymer

The water-soluble polymer listed below can be used as filtrate reducer for cement slurry (Mckenzie 1986; Rao and Burkhalter 1985).

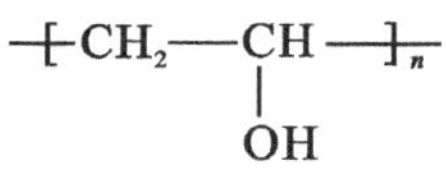

(Polyvinyl alcohol)

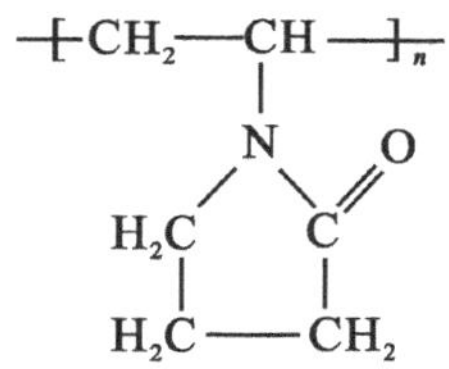

(Poly *N*-vinyl pyrrolidone)

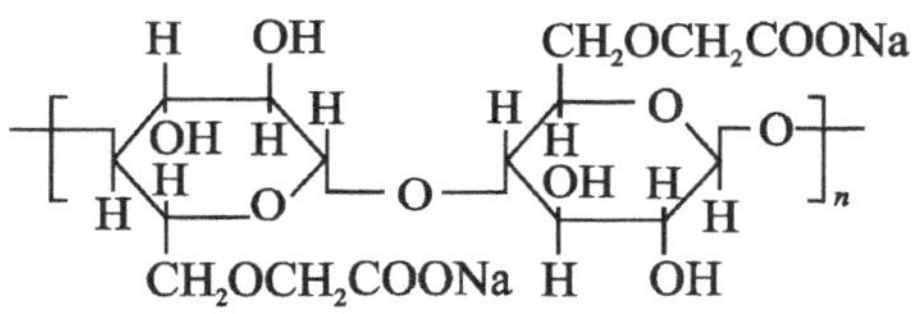

(Sodium carboxymethyl cellulose)

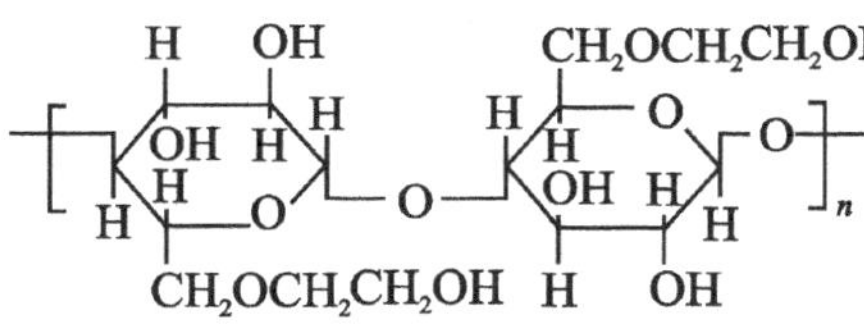

(Hydroxyethyl cellulose)

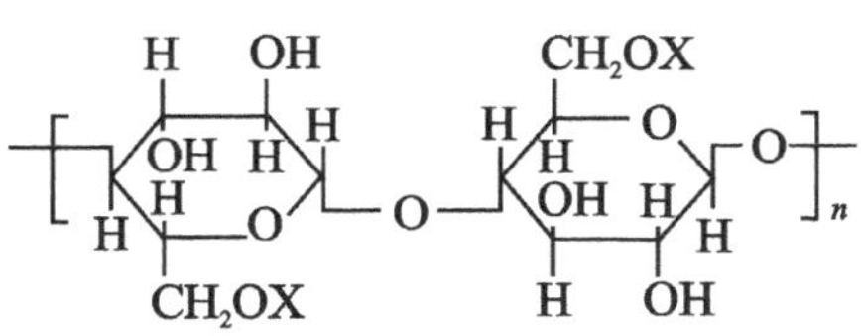

X: CH_2COONa or CH_2CH_2OH

(Sodium carboxymethyl hydroxyethyl cellulose)

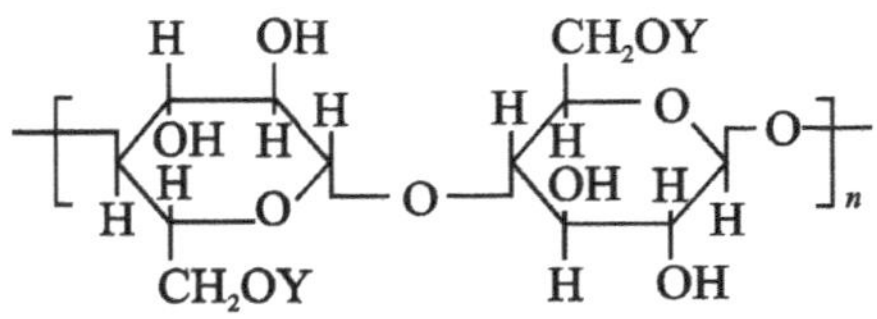

Y: CH_2COONa or CH_3CHCH_2OH

(Sodium carboxymethyl hydroxypropyl cellulose)

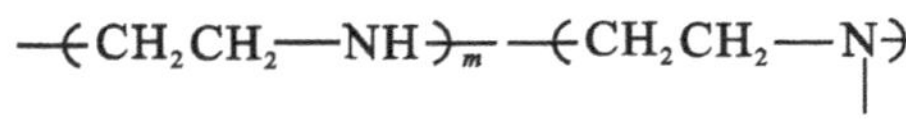

[Branched poly(1,2- ethylene imine)]

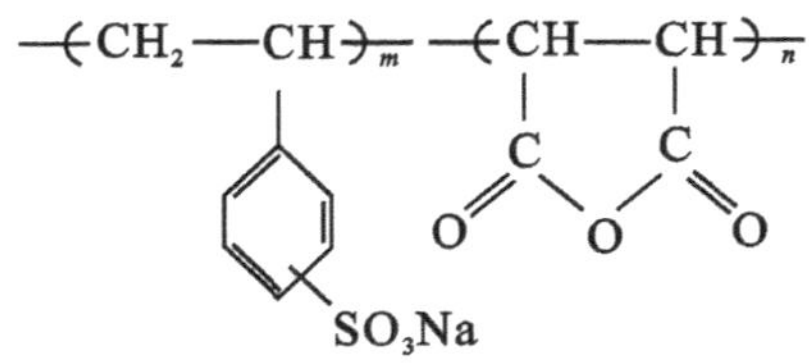

(Copolymer of sulfonated styrene and maleic anhydride)

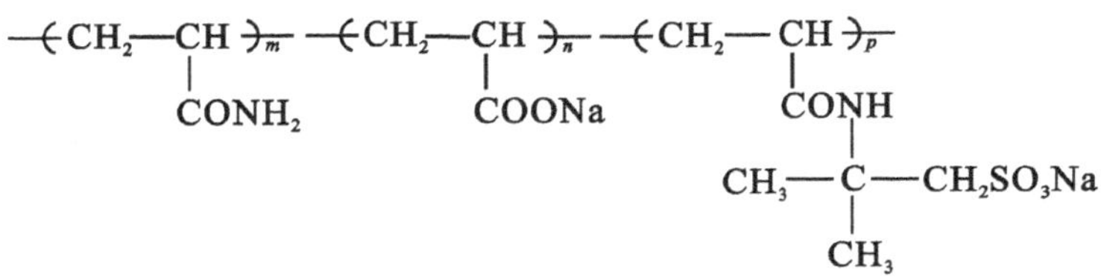

[Copolymer of acrylamide, sodium acrylate and (2-acrylamido-2-methyl) propyl sulfonate]

Water-soluble polymers can reduce filtrate loss through thickening mechanism, adsorption mechanism, trapping mechanism, and/or physical plugging mechanism. The mechanism for water-soluble polymers to reducing filtrate loss of cement slurry is the same as that of drilling fluid.

3.6 Gas Channeling and Its Control

Gas channeling is a common problem encountered in the cementing process. It refers to the phenomenon that the gas in the high-pressure layer enters the low-pressure layer or uplifts to the ground along the gaps between cement stone and

Table 3.8 Shrinkage ratios of system volumes of different compositions in cement slurry after hydration

The composition in cement slurry	Shrinkage ratio of system volume after hydration
$3CaO{\cdot}SiO_2$	5.3
$2CaO{\cdot}SiO_2$	2.0
$3CaO{\cdot}Al_2O_3$	23.8
$4CaO{\cdot}Al_2O_3{\cdot}Fe_2O_3$	10.0

well wall and/or between cement stone and casing (Bannister et al. 1983; Sutton and Sabins 1984; Cheung and Beirute 1985; Wilkings and Free 1989).

The reason why gaps exist between cement stone and well wall and/or between the cement stone and casing is that the cement slurry shrinks in curing stage and hardening stage (Chenevert and Shrestha 1987). The shrinkage ratios of system volumes (the volume of each composition plus water volume) of different compositions in cement slurry after hydration are shown in Table 3.8.

Obviously, the volume shrinkage of cement slurry in curing stage and hardening stage is the comprehensive consequence of the system volume shrinkage of each composition in cement after hydration.

To reduce the volume shrinkage of cement slurry in curing stage and hardening stage, expansive agent for cement slurry (also called gas channeling inhibitor) can be used.

The common expansive agents for cement slurry are listed below (Yao and Peng 1992; Ghofrani and Plack 1993; Wang et al. 1996).

3.6.1 Hemihydrate Gypsum

After added into cement slurry, the hemihydrate gypsum produces dihydrate gypsum by hydration first and then reacts with tricalcium aluminate hydrate to form ettringite.

$$\underset{\text{(Hemihydrate gypsum)}}{CaSO_4 \cdot 1/2H_2O} + 3/2H_2O \rightarrow CaSO_4 \cdot 2H_2O$$

$$\underset{\text{(Tricalcium aluminate)}}{3CaO \cdot Al_2O_3} + 6H_2O \rightarrow 3CaO \cdot Al_2O_3 \cdot 6H_2O$$

$$3(CaSO_4 \cdot 2H_2O) + 3CaO \cdot Al_2O_3 \cdot 6H_2O + 20H_2O \rightarrow \underset{\text{(Ettringite)}}{3CaO \cdot Al_2O_3 \cdot 3CaSO_4 \cdot 32H_2O}$$

The ettringite generated by the reaction contains lots of crystal water which results in the expansion of volume, thus inhibiting the volume shrinkage of cement slurry.

Table 3.9 Volume expansion ratio of cement slurry caused by aluminum powder

w(Al) (%)	The volume expansion ratio of cement slurry (%)	
	0.10 MPa	27.7 MPa
0.05	11.8	0.71
0.10	17.9	0.91
0.25	24.0	1.64
0.50	56.5	2.64
1.00	57.2	5.17

3.6.2 Aluminum Powder

After added into cement slurry, the aluminum powder can react with calcium hydroxide to generate hydrogen.

$$2Al + Ca(OH)_2 + 2H_2O \rightarrow Ca(AlO_2)_2 + 3H_2 \uparrow$$

The hydrogen generated by the reaction is dispersed in cement slurry, which results in the volume expansion of cement slurry, thus inhibiting the volume shrinkage of cement slurry. The volume expansion ratio caused by aluminum powder is shown in Table 3.9.

3.6.3 Magnesium Oxide

After added into cement slurry, magnesium oxide can react with water to form magnesium hydroxide.

$$MgO + H_2O \rightarrow Mg(OH)_2$$

Since the solid phase density of magnesium oxide is 3.58 g/cm^3 and that of magnesium hydroxide is 2.36 g/cm^3, the volume increases after the reaction of magnesium oxide with water, inhibiting the volume shrinkage of cement slurry.

Figure 3.4 shows the changes of volume expansion ratio of cement slurry caused by magnesium oxide with time and temperature.

Since the volume expansion ratio of cement slurry caused by magnesium oxide increases with temperature, magnesium oxide is applicable to high-temperature well cementation.

In order to decrease the permeability of cement stone and prevent the leakage of gas, water-soluble polymer, water-soluble surfactant, and latex (Cheung and

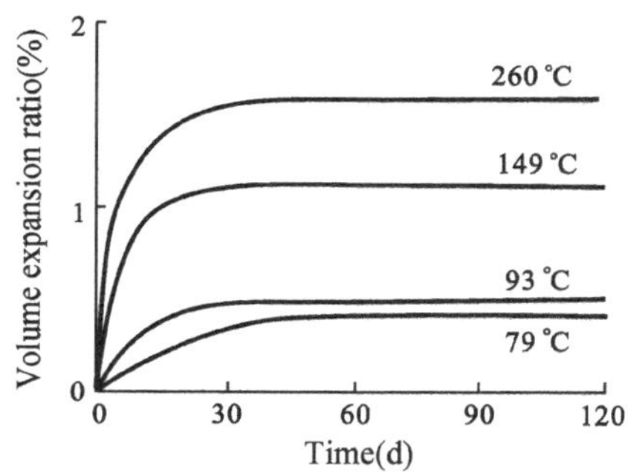

Fig. 3.4 Volume expansion ratio of cement slurry caused by magnesium oxide

Myrich 1983; Wang 1991) can be added into cement slurry. Water-soluble polymer decreases the permeability of cement stone by increasing the water phase viscosity or through physical plugging. Water-soluble surfactant decreases the permeability of cement stone through superimposed Jamin effect caused by the foams generated by gas permeating into the cement stone pore. While latex decreases the permeability of cement stone through superimposed, Jamin effect caused by the viscous polymer oil droplets in the cement stone pore and/or the film-forming function.

3.7 The Leakage of Cement Slurry and Its Treatment

Leakage should be controlled during the drilling process, which means during the circulation of drilling fluid, the leakage formation should be blocked. However, since the density of cement slurry is higher than that of the corresponding drilling fluid, the leakage of cement slurry may also occur during the cement injection.

There are two methods to treat the leakage of cement slurry. One is to reduce the density and/or the flow pressure drop of cement slurry as much as possible, ensuring that the downhole pressure at cement injection is lower than the maximum downhole pressure when corresponding drilling fluid is circulating. The other method is injecting spacer fluid (added with plugging material) before cement injection and adding plugging material into cement slurry. These plugging materials are mainly fibrous materials (such as cotton linter and asbestos fibers) or particulate materials (such as the particles of walnut shells, peanut shells, corn cob, clay, diatomite, expanded perlite, and limestone).

If the plugging materials are surface-inert materials (such as walnut shells), the addition of them will not affect the thickening time of cement slurry and the strength of cement stone. However, if the plugging materials are surface-active materials (such as clay), the addition of them would affect the thickening time of cement slurry and the strength of cement stone.

3.8 Cement Slurry System

The cement slurry system refers to all kinds of cement slurries used for well cementing in general and special strata. Figure 3.5 shows different kinds of cement slurries in the cement slurry system, which will be described in detail below.

3.8.1 *Conventional Cement Slurry*

Conventional cement slurry is prepared by cement (including nine kinds of oil well cements in API standard and four kinds of cements in SY standard), freshwater, cement slurry additive, and admixture, which is applicable to general stratum. The preparation and construction of this kind of cement slurry are all simple.

3.8.2 *Special Cement Slurry*

Special cement slurry is applicable to special stratum.

Some important special cement slurries are listed below.

Salt Cement Slurry
This kind of cement slurry uses inorganic salt (such as sodium chloride and potassium chloride) as additive. It is applicable to well cementing in rock salt layer and shale layer.

Latex Cement Slurry
This kind of cement slurry uses latex (such as polyvinyl acetate latex, styrene and methyl methacrylate copolymer latex) as the main additive. The latex in cement slurry can improve the bond strength between the cement stone and well wall, and between the cement stone and casing, reducing the filtrate volume of cement slurry and the permeability of cement stone; therefore, latex cement slurry has good performance of gas channeling prevention (Skaller and Sveen 1991; Li et al. 1997).

Fiber Cement Slurry
This kind of cement slurry uses fiber as additive. The common fiber additives include carbon fiber, polyamide fiber, and polyester fiber. This kind of cement slurry is applicable to well cementing in leakage zone. The fiber can also improve the toughness of cement (Bedel et al. 2003; Bu et al. 2005; Luo et al. 2006).

Foam Cement Slurry
Foam cement slurry is prepared by water, cement, gas, blowing agent, and foam stabilizer (Benge and Spang 1982; Rozieres and Ferriere 1990; Qu et al. 1994; Chatterji et al. 2002).

The available gas is nitrogen or air.

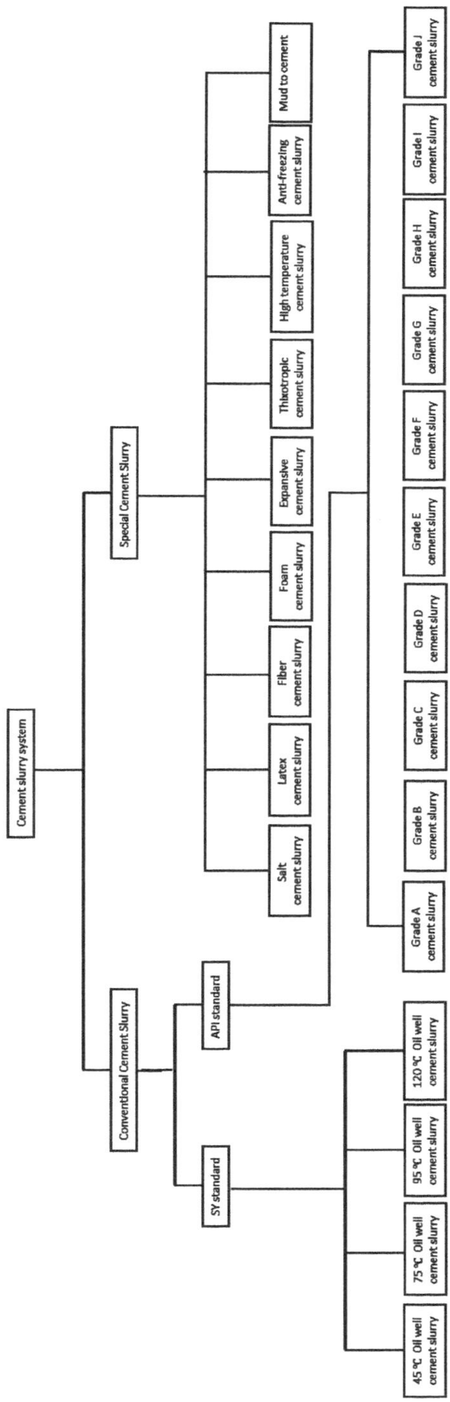

Fig. 3.5 Cement slurry system

The available blowing agent is water-soluble surfactant (such as alkyl benzene sulfonate, alkyl sulfate sodium salt, and polyoxyethylene alkyl phenol ether).

The available foam stabilizer is water-soluble polymer (such as sodium carboxymethyl cellulose, and hydroxyethyl cellulose).

The most prominent advantage of the foam cement slurry is the low density, which is suitable to the well cementing in high-permeability layer, crack layer, and cave layer.

Expansive Cement Slurry

This kind of cement slurry uses expansive agent as main additive. Expansive cement slurry can generate mild volume expansion when curing, which will solve the problem of volume shrinkage during the curing of conventional cement slurry, and improve the connection between the cement stone and well wall or between the cement stone and casing, thus preventing gas channeling (Seidel and Greene 1985).

Common expansive agents include hemihydrate gypsum, aluminum powder, and magnesia. The cement slurries added with these expansive agents are called hemihydrate gypsum cement slurry, aluminum powder cement slurry, and magnesia cement slurry, respectively.

Thixotropic Cement Slurry

This kind of cement slurry uses thixotropic material as additive. Hemihydrate gypsum is the most commonly used thixotropic material. Thixotropic cement slurry can be prepared by adding hemihydrate gypsum (8–12% of cement mass) into cement slurry. Hemihydrate gypsum hydrate can react with tricalcium aluminate hydrate to form ettringite, which is needle-like crystal that can deposit between cement particles to form gel structure. However, the gel structure is easily destroyed by shear force, restoring the mobility of cement slurry. If the shear force is removed, the gel structure is re-established. Therefore, the addition of hemihydrate gypsum can make the cement slurry thixotropic.

Thixotropic cement slurry is mainly used for the cementation in leakage layer. When thixotropic cement slurry comes into the leakage layer, its edge velocity gradually slows down (radial flow) and forms gel structure. The flow resistance is increased until the cement slurry no longer permeates into the leakage layer. After solidification of the cement slurry, the leakage layer can be effectively blocked.

The volume expansion during the formation of ettringite compensates for the volume shrinkage during the solidification of the cement slurry. Therefore, the thixotropic cement slurry prepared by hemihydrate gypsum can also be applied for well cementing in formations where gas channeling occurs easily.

High-Temperature Cement Slurry

This kind of cement slurry uses active silica as additive and is applicable to the well cementation in high-temperature stratum (more than 110 °C).

Conventional cement slurry cannot be applied in high-temperature stratum because the $2CaO \cdot SiO_2 \cdot H_2O$ in cement hydrate will transform from β crystalline phase to α crystalline phase under high temperature, which will cause the shrinkage of volume and the destruction of the integrity of cement stone, resulting in the decrease of compressive strength and the increase of permeability of the cement stone.

Table 3.10 Composition of silica fume

Composition	*w*(composition) (%)	Composition	*w*(composition) (%)
SiO_2	85–98	Al	0.1–1.0
C	0.5–2.5	Fe	0.1–2.5
K	0.2–3.5	Ti	0.001–0.03
Na	0.1–1.5	P	0.005–0.1
Mg	0.1–2.5	S	0.002–0.5
Ca	0.1–0.5		

In order to make the conventional cement slurry applicable to high temperature, the active silica (35% of the mass of cement slurry) can be added into cement slurry to reduce the molar ratio of calcium oxide to silica and inhibit the transformation of $2CaO{\cdot}SiO_2{\cdot}H_2O$ from β crystalline phase to α crystalline phase. A series of temperature-tolerance, low-permeability, and high-strength hydrates can be generated. For example, $5CaO{\cdot}6SiO_2{\cdot}5H_2O$ (tobermorite) is generated in 110 °C and $6CaO{\cdot}6SiO_2{\cdot}H_2O$ (xonotlite) is generated in 150 °C.

Silica fume is the main source of active silica. Silica fume is a byproduct in the refining of silicon alloy whose main component is silica (Table 3.10). Silica fume has large specific surface area (15–20 m^2/g, 50–60 times of that of cement) and is an ideal active silica material. Cement slurry using silica fume as additive is called silica fume cement slurry, which is applicable to well cementing in high-temperature stratum.

Antifreezing Cement Slurry

This kind of cement slurry uses antifreezer and coagulant as main additive. The antifreezer is added to make cement slurry maintain good mobility under low temperature (below −3 °C). While the coagulant is added so that the thickening time of cement slurry can satisfy the construction requirement and high-strength cement stone can be generated.

The most commonly used antifreezer is inorganic salts (such as sodium chloride and potassium chloride) and low molecular alcohol (such as ethanol and glycol). The most commonly used low-temperature coagulants are calcium aluminate and plaster.

Mud to Cement (MTC) (Wang 1996; Gu and Wang 1996; Yu and Jia 2005)

The blast furnace slag (short for slag), the activator which can increase pH, and other additives such as drag reducer and retarder can be added into drilling fluid to prepare cement slurry. This kind of cement slurry is called mud to cement, MTC.

Slag is the waste product in the steelmaking process whose main components are CaO, SiO_2, and Al_2O_3. The composition of a typical slag is shown in Table 3.11.

Table 3.11 Composition of a typical slag

Composition	*w*(composition) (%)	Composition	*w*(composition) (%)
CaO	37.62	MgO	10.95
SiO_2	34.39	Fe_2O_3	3.72
Al_2O_3	11.43		

From Table 3.11, it can be obtained that the composition of slag is similar with that of cement slurry. Different from cement slurry, the slag must be used under the condition of high pH (pH > 12). That is because under this alkaline condition, the main components of slag dissolve and hydrate, before forming network structure to solidify the system.

MTC is prepared by the solidification of slag under alkaline condition. Slag is added into the drilling fluid so that filter cake containing slag can form on the well wall surface during drilling. Then by adding activator which can increase the pH of the system (such as sodium hydroxide, potassium hydroxide, and sodium carbonate) and other additives during well cementing, the drilling fluid can be transformed into cement slurry for well cementation. The slag in filter cake can also solidify under the action of activator to improve the quality of well cementation.

This kind of mud to cement can decrease the usage of additive for cement slurry and alleviate the contamination of waste drilling fluid to environment.

References

Allouche M, Maroy P (2003) Additives for cement compositions. WO Patent 016239A2, 27 Feb 2003

Bannister CE, Shuster GE, Wooldriage LA et al (1983) Critical design parameters to prevent gas invasion during cementing operations. SPE 11982

Bedel JP, Card R, Vidick B (2003) Low density, fiber-reinforced cement composition. EP Patent 1284248A1, 19 Jan 2003

Benge OG, Spang LB (1982) Foamed cement—solving old problems with a new technique. SPE 11204

Brothers LE, Rao SP (1985) Oil field cementing methods and compositions. US Patent 4500357, 19 Feb 1985

Brothers LE, Braden JA, Dao B et al (2003) Well cement compositions. EP Patent 1316540A2, 4 June 2003

Bu Y, Wang R, Cheng R (2005) Performance of fiber cement slurry for well cementation. Oil Drill Prod Technol 27(2):25–27

Chatterji J, Cromwell RS, Brenneis CR et al (2002) Foamed well cement slurries, additives and methods. US Patent 096090A1, 25 Jul 2002

Chenevert ME, Shrestha B (1987) Shrinkage properties of cement. SPE 16654

Cook C, Commingham W (1977) Filtrate control: a key in successful cementing practices. JPT 29(8):951–1012

Cheung PR, Myrich BD (1983) Field evaluation of an impermeable cement system for controlling gas migration. SPE 11983

Cheung PR, Beirute RM (1985) Gas flow in cements. JPT 37(7):1041–1048

Eoff LS, Buster D (1995) High temperature synthetic cement retarder. SPE 28957

Ghofrani R, Plack H (1993) CaO and/or MgO-swelling cements: a key for providing a better annular sealing? SPE 25697

Gulf Publishing Company (1996) Cementing products and additives. World Oil 217(3):c3–c19

Gu J, Wang X (1996) Present status of mud-to-cement conversion in China. Oilfield Chem 13(3):269–272

Lebo SE, Jr., Resch SL (2002) Set retarders for foamed cements. US Patent 056405A1, 16 May 2002

Li H (1987) Concrete with naphthalene superplasticizer. Fuel Chem Process 18(1):95–97

Liu X (1988) Well completion engineering. Petroleum Industry Press, Beijing, p 98
Li W, Yao X, Wang T (1997) Foreign latex cement slurry techniques. Pet Drill Tech 25(2):34–36
Li H, Wang Q (1998) A study on the water-soluble polymers modified cement: I. Effect of water-solution polymers on the fluidity of cement. J Funct Polym 11(1):16–21
Luo Y, Fan D, Jiang C (2006) Applied research on leak protection and toughness-reinforcement fiber cement slurry. Drill Fluid Complet Fluid 23(4):4–6
Mckenzie LF, Mcelfresh PM, Reese DW (1986) Non retarding fluid loss additives for well cementing compositions. US Patent 4,602,685, 29 July 1986
Nelson EB (1990) Well cementing. Elsevier, Amsterdam. Chinese edition: Nelson EB (1994) Well cementing (trans: Liu D, Tian X, Liao R). Liaoning Science and Technology Publishing House, Shenyang, pp 14–16
Qu J, Du H, Huang B (1994) Study and application of foam cement. Drill Fluid Complet Fluid 11(5):1–7
Rao SP, Burkhalter JF (1985) Hydrolytically stable polymers for use in oil field cementing methods and compositions. US Patent 4,555,269, 26 Nov 1985
Rozieres J, Ferriere RF (1990) Foamed cement characterization under downhole conditions and its impact on job design. SPE 19936
Sabins FL, Autton DL (1982) The relationship of thickening time, gel strength, and compressive strengths of oil well cements. SPE 11205
Seidel FA, Greene TG (1985) Use of expanding cement improves bonding and aids in eliminating annular gas migration in Hobbs Grayburg-San Andres wells. SPE 14434
Skaller P, Sveen J (1991) Emulsion cement. SPE 23075
Sutton DL, Sabins FL (1984) Annular gas flow theory and prevention methods described. Oil Gas J 82(50):84–91
Wang Q (1991) Development and application of slurries additives for oil well cementing. Oilfield Chem 8(1):66–73
Wang W, Jiang S, Song M (1996) The study and initial application of slag MTC. Pet Drill Tech 24(2):29–30
Wang W, Liu F, Quan Y et al (1996) The development and applications of CX-18 anti-channeling agent. J Xi'an Shiyou Inst 11(5):50–51
Wilkings RD, Free D (1989) A new approach to the prediction of gas flow after cementing. SPE 18622
Xu H (1988) Handbook of drilling technology (III) well cementation. Petroleum Industry Press, Beijing, p 237
Yao X, Peng X (1992) MC type expansive agents for oil well cement laboratory evaluation. Oilfield Chem 9(4):297–301
Yu Z, Jia Z (2005) Laboratory test on solidification of waste drilling fluid. Environ Prot Oil Gas Fields 15(3):40–42

Part II
Oil Production Chemistry

Oil production chemistry is a part of oilfield chemistry.

Oil production chemistry is a frontier science between oilfield production engineering and chemistry.

Oil production chemistry is used to solve problems in the process of oil recovery with chemical methods.

The most important thing during oil production is to enhance oil recovery (EOR). EOR methods that have relationship with chemistry, such as chemical flooding, miscible flooding, profile control in water well, and water shutoff in oil well, are major contents of this part.

The heavy oil resources are rich, but it is difficult to explore and transport due to its high viscosity. According to the property of heavy oil, technology of viscosity reduction of heavy oil is introduced in this part.

Acidizing and fracturing are critical technology for stimulation of water injection well and improving oil well production, during which the injected working fluid (acidizing fluid and fracturing fluid) and additives are closely related with chemistry. In this part, acidizing fluid, fracturing fluid, and additives are introduced in detail.

Sand production from oil and water wells and wax deposition in oil wells will affect the normal oil production process. Essential introduction for sand control of oil and water wells, wax control, and removal of oil wells are conducted in this part.

In conclusion, the main content in this part includes chemical flooding and miscible flooding, profile control in water well and water shutoff in oil well, viscosity reduction of heavy oil, acids and additives used in acidizing, fracturing fluid and additives, sand control in oil and water wells, wax control, and removal in oil wells (totally seven chapters).

Chapter 4
Chemical Flooding and Miscible Flooding

The recovery of crude oil is quite low (Smith 1988). Generally, the oil recovery for water injection method (water flooding) only reaches 30–40% and most of the oil is left unrecovered underground. The low oil recovery is attributed to the heterogeneity of the reservoir, which makes oil displacement agent break into oil wells through the high permeable layers, leaving low permeable layers unswept. Concept of sweep efficiency is involved here. Sweep efficiency refers to the ratio between swept reservoir volume and the whole reservoir volume. However, even in the swept area, the oil cannot be completely recovered because of the wettability of formation surface and drag effect of capillary (Jamin effect). Then, concept of displacement efficiency is introduced. Displacement efficiency is the ratio between the amount of produced oil and oil reserve in the swept area. The following conclusion can be drawn based on the concept of sweep efficiency and displacement efficiency.

Oil recovery = sweep efficiency × displacement efficiency

Therefore, there are two ways to enhance oil recovery. One is to improve the sweep efficiency, and the other one is to improve the displacement efficiency.

The main method to enhance sweep efficiency is to change mobility of the displacement agents and (or) oil. Mobility measures the capability of fluid to flow through porous media. Its definition formula is given as follows, where λ is the fluid mobility, k is the effective permeability for fluid through porous media, and μ is the fluid viscosity.

$$\lambda = k/\mu \tag{4.1}$$

Because the mobility of displacement agents is far above that of oil, it is very easy for displacement agents to break into oil wells through high permeable layers during oil displacement. To enhance sweep efficiency of displacement agents, mobility of displacement must be reduced and (or) oil mobility must be increased. The main method to enhance oil recovery is to change wettability of rock surface and reduce the adverse impact of capillary effect.

C. Dai and F. Zhao, *Oilfield Chemistry*,
https://doi.org/10.1007/978-981-13-2950-0_4

Four methods have been developed to enhance oil recovery so far.

Chemical Displacement Method (Chemical Flooding)
It can be divided into polymer displacement method (polymer flooding), surfactant displacement method (surfactant flooding), alkali displacement method (alkali flooding), and their combination (combined flooding).

Miscible Displacement Method (Miscible Flooding)
It can be divided into hydrocarbon displacement method (hydrocarbon miscible flooding) and non-hydrocarbon displacement method (non-hydrocarbon miscible flooding).

Thermal Recovery Method (Thermal Recovery)
It can be divided into hot water displacement method (hot water flooding), steam displacement method (steam flooding), and in situ combustion method (in situ combustion).

Microbial Displacement Method (Microbial Flooding)
It can be divided into recovery method which activates indigenous microbial in reservoir and recovery method which injects suitable microbial, respectively.

As chemical flooding and miscible flooding are closely related to chemistry, only these two displacement methods are introduced here.

4.1 Polymer Flooding

4.1.1 Definition of Polymer Flooding

Polymer flooding is an oil displacement method using the polymer solution.

Polymer flooding can also be described as polymer solution flooding, polymer-enhanced water flooding, thickened water flooding, and tackifying water flooding.

4.1.2 Polymers Used in Polymer Flooding

The polymers used in the polymer flooding can be divided into two major categories, partially hydrolyzed polyacrylamide (HPAM) (Chen 1993) and xanthan gum (XC) (Wernau 1981).

Though the XC can tolerate temperature as high as 93 °C, the HPAM is mainly used for oil displacement because of the poor biological stability of XC. For better mobility controllability, the average molecular weight of HPAM used is above 2.5×10^7, and the hydrolysis degree is up to 35%. For better heat resistance (over 90 °C) and salt tolerance (higher than 3×10^4 mg/L), copolymers with acrylamide as main chain, while with ring structures, strong hydrophilic groups and associating hydrocarbon chains (Liu et al. 2003) have been designed. Some examples are given below.

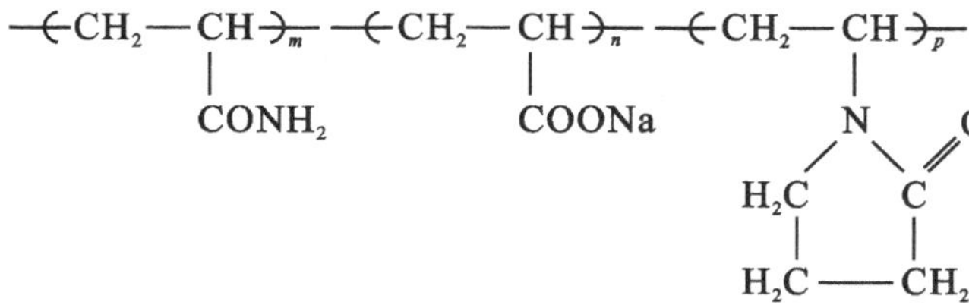

(Copolymer of acrylamide, sodium acrylate and *N*-vinyl pyrrolidone.)

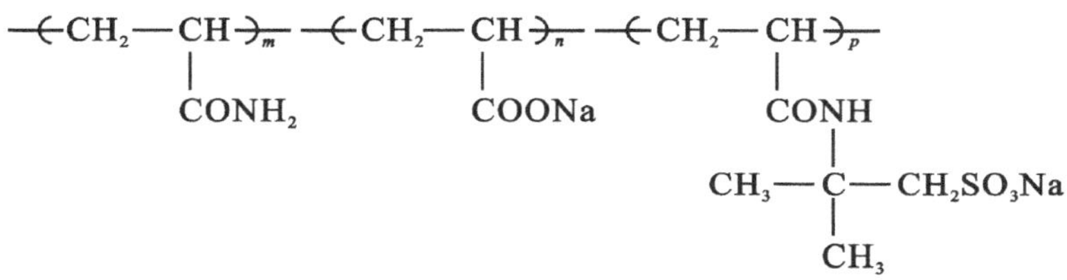

[(Copolymer of acrylamide, sodium acrylate and (2-Propenamide-2-methyl) propyl sulfonate)]

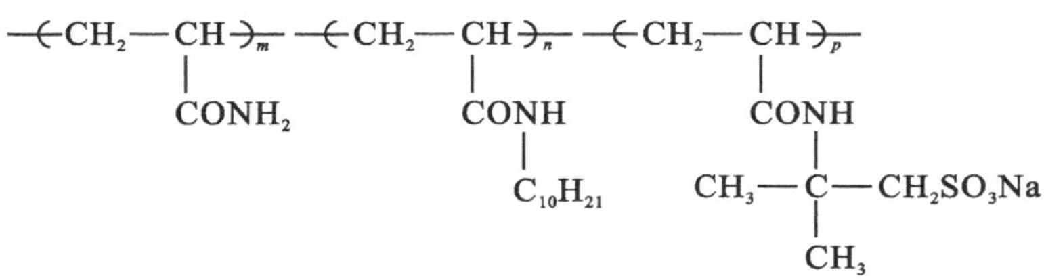

[(Copolymer of acrylamide, *N*-decyl acrylamide and (2-Propenamide-2-methyl) propyl sulfonate)]

4.1.3 Thickening Ability of Polymer to Water

The viscosity–concentration curves of polymer solutions are shown in Fig. 4.1. As shown in Fig. 4.1, both HPAM and XC exhibit excellent thickening abilities.

The thickening effects of polymers to water can be attributed to the following reasons.

- When the polymer concentration is higher than certain value, the polymer molecules would tangle with each other to form structure, generating the structural viscosity.
- The hydrophilic groups in polymer chains hydrate in the solution.

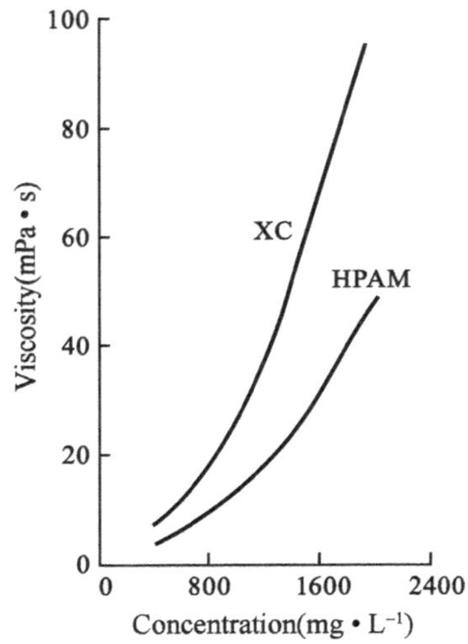

Fig. 4.1 Thickening effects of HPAM and XC to water (the viscosity is measured using the NaCl solution with a mass concentration of 1%, under 23 °C with 6 rpm)

- If the polymers are ionic polymers, they can ionize in water to form diffused double layers and generate lots of chain segments (composed of several chain elements, the chain segments are the minimum units that can move independently) with the same charges. This will make the polymer molecules disperse in water to form loose irregular coils, which leads to good thickening ability of polymers.

4.1.4 Viscoelasticity of Polymer Solution

The polymer solution is a kind of viscoelastic material, which has viscosity (the property of liquids) as well as elasticity (the property of solids).

The viscosity of polymer solution has been discussed above, and its elastic characteristic can be observed by the stretching effect when the polymer solution passes through the pore throats of a core, as shown in Fig. 4.2.

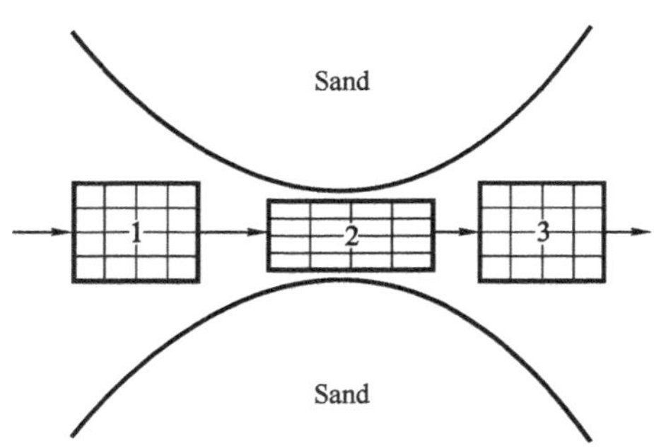

Fig. 4.2 Stretching effect when the polymer solution passes through the pore throats

1 is used to show the state of polymer molecule before stretching.
2 is used to show the state of polymer molecule under stretching.
3 is used to show the state of polymer molecule after stretching.

The reason for the elasticity of polymer solutions is that the polymer molecules show extended conformation under stretching, while adopting curly conformation when the stretching disappears.

The viscoelasticity of polymer solution can be measured by a rheometer (such as Haake RS150 rheometer). During the measurement, the sinusoidal shear is applied to the polymer solution by means of small amplitude oscillation. The corresponding variations of stress and strain with time are measured, based on which the values of storage modulus (G') and loss modulus (G'') are calculated. The former one is related with the elasticity of polymer solution, while the latter one is related with the viscosity of polymer solution. By comparing the values of storage modulus and loss modulus, whether the elasticity or the viscosity of polymer solution plays a dominant role can be determined under that specific condition.

4.1.5 Retention of Polymer Molecules in Porous Media

The polymer molecules can remain in porous media by means of adsorption or trapping.

4.1.5.1 Adsorption

Adsorption is a phenomenon that the polymer molecules gather on the surface of rock pore structures by dispersion force, hydrogen bond, etc.

4.1.5.2 Trapping

This is a phenomenon that the polymer molecules, whose diameters are smaller than those of the pore throats, are trapped in the pores by means of bridging action, as shown in Fig. 4.3.

Due to the trapping of polymer molecules in the pore structures, the flow resistance of fluid in the pores increases, therefore reducing the effective rock permeability to water.

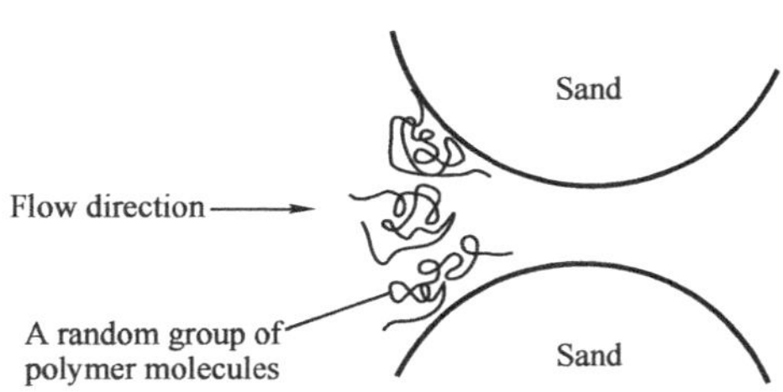

Fig. 4.3 Trapping phenomenon of polymer molecules in a pore

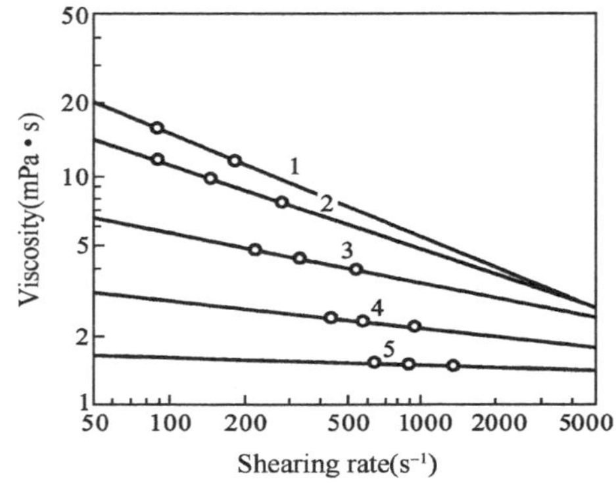

Fig. 4.4 Salt sensitivity effect of HPAM. w(NaCl): 1—0; 2—0.01%; 3—0.1%; 4—0.5%; 5—10%

4.1.6 Salt Sensitivity Effect of Polymer

The salt sensitivity effect of polymer refers to the special effect of salt to the viscosity of polymer solution, which can be observed in Fig. 4.4. As shown in Fig. 4.4, the salt concentration has remarkable effects on the viscosity of HPAM solution.

The salt sensitivity effect of HPAM is attributed to the fact that the salt compresses the diffusive double layers formed by carboxyl groups and sodium ions around the HPAM molecules. Before the addition of salt, the chain segments of HPAM are negatively charged and repellent with each other due to the existence of diffusive double layers. Therefore, the HPAM molecules form loose irregular coils, enhancing their water-thickening capability. However, the addition of salt will compress the diffusive double layers, decreasing the electronegativity of the chain segments and making the HPAM molecules form compact irregular coils, which will lead to obvious decline of thickening ability of HPAM to water (Mei 1982; Kang et al. 2007).

4.1.7 Enhanced Oil Recovery Mechanisms of Polymer Flooding

Polymer flooding can increase the oil recovery by two mechanisms.

4.1.7.1 The Mechanism of Decreasing Water–Oil Mobility Ratio

The definition formula of water–oil mobility ratio is presented as follows according to the concept of fluid mobility, where M_{wo} is the water–oil mobility ratio, λ_w and λ_o are, respectively, the mobilities of water and oil, k_w and k_o are, respectively, the

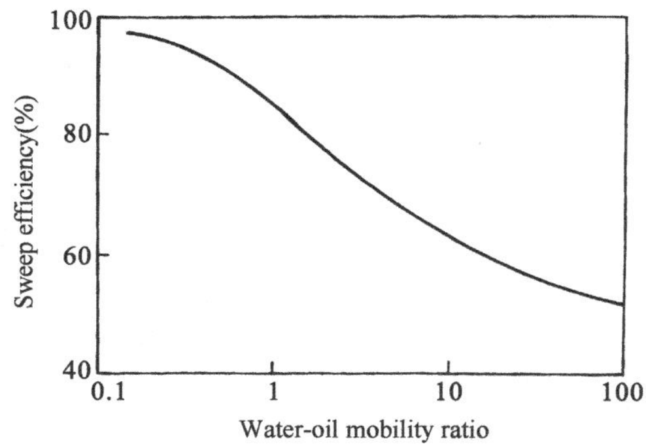

Fig. 4.5 Impact of water–oil mobility ratio on sweep efficiency

effective permeabilities of water and oil, and μ_w and μ_o are, respectively, the viscosities of water and oil.

$$M_{wo} = \frac{\lambda_w}{\lambda_o} = \frac{k_w/\mu_w}{k_o/\mu_o} = \frac{k_w\mu_o}{k_o\mu_w} \tag{4.2}$$

According to Eq. 4.2, polymer can increase water viscosity by thickening effect. Also, polymer can decrease the effective permeability of porous media to water by adsorption and trapping in porous media. These will all decrease the water–oil mobility ratio, increase the sweep efficiency, and thus enhance the oil recovery.

Figure 4.5 shows the influence of water–oil mobility ratio on sweep efficiency. As shown in Fig. 4.5, the sweep efficiency increases with the decrease in water–oil mobility ratio.

Also, as shown in the results of oil displacement tests of plate model (Fig. 4.6), polymer flooding has higher sweep efficiency than that of water flooding, which means higher oil recovery.

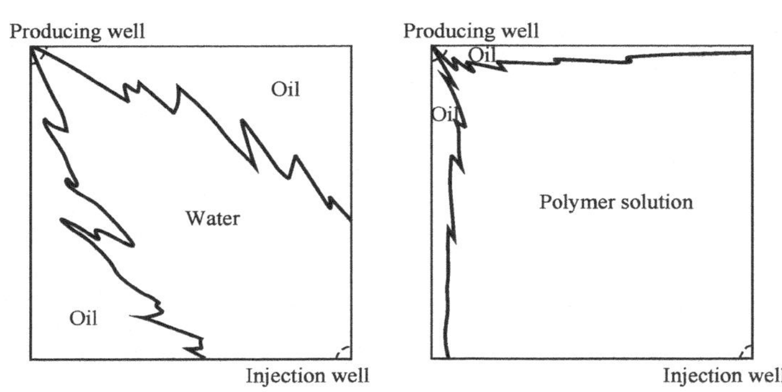

Fig. 4.6 Sweep efficiencies of water flooding and polymer flooding

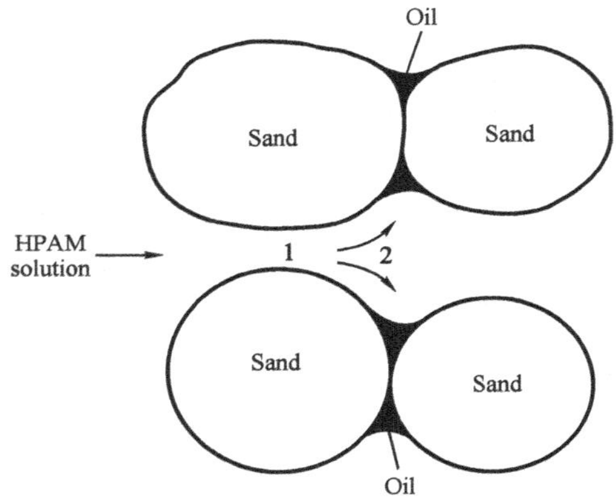

Fig. 4.7 Oil displacement by viscoelasticity of polymer solution. 1 is the position where the polymer solution is stretched, while 2 is the position where the viscoelasticity of polymer solution functions and the oil in between the sands can be displaced

4.1.7.2 The Mechanism of Oil Displacement by Viscoelasticity of Polymer Solution

The mechanism is shown in Fig. 4.7. As shown in Fig. 4.7, the polymer molecules in the solution show extended conformation under tension when the polymer solution flows through the pore throats. When the polymer solution leaves pore throats, the tension disappears and polymer molecules adopt curly conformation, making the polymer solution expand along the normal direction of the flow direction (exhibiting elasticity). Therefore, polymer solution can displace the remaining oil in between the sands which cannot be displaced by water flooding, thus increasing oil displacement efficiency.

4.1.8 Field Tests of Polymer Flooding

The polymers used in the field tests are mainly HPAM at present, with the mass concentration ranging from 800 to 2000 mg/L, and the injection dosage ranging from 0.30 to 0.60 pore volume (PV). There have been many successful field tests of polymer flooding. Figure 4.8 shows a production curve of a field test of polymer flooding (Lake 1989). As shown in Fig. 4.8, the polymer flooding has demonstrated excellent performance in enhanced oil recovery.

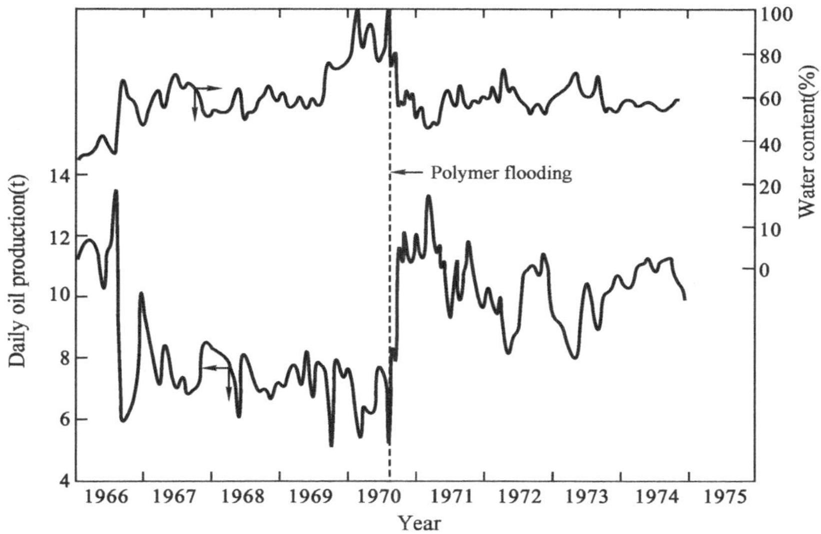

Fig. 4.8 A production curve of a field test of polymer flooding

4.2 Surfactant Flooding

4.2.1 Definition of Surfactant Flooding

Surfactant flooding is an oil displacement method using the surfactant systems as the oil displacement agents.

The surfactant systems used for oil displacement include dilute surfactant system and concentrated surfactant system. The former one includes active water and micellar solution, while the latter one includes water-external microemulsion, oil-external microemulsion, and middle-phase microemulsion (all called microemulsion).

Foam flooding and emulsion flooding are also included in surfactant flooding because foams and emulsions are oil displacement agents stabilized by surfactants.

4.2.2 Surfactants Used in Surfactant Flooding

The surfactants used in surfactant flooding mainly include the following six kinds.

4.2.2.1 Sulfonate Surfactants

Some examples are given as follows (Kamath et al. 1981; Hsieh and Koepke 1985; Guo et al. 2003; Ren 2001).

$R-SO_3M$
(Alkane sulfonate)

$RAr-SO_3M$
(Petroleum sulfonate)

$R-CH=CHSO_3M$
(α-olefin sulfonates)
(Fan et al. 2005; Levitt et al. 2006)

$C_nH_{2n-1}-CH_2SO_3M$
(Internal olefinsulfonate)

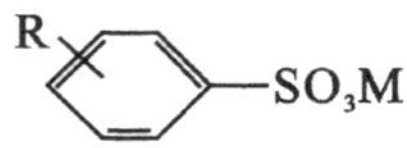
(Alkylbenzene sulfonate)

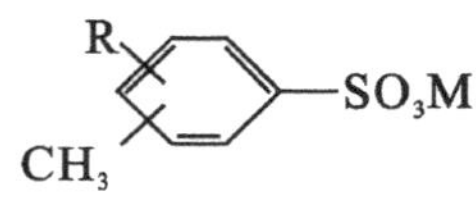
(Alkyl toluene sulfonate)

4.2.2.2 Carboxylate Surfactants

Some examples are given as follows (Huang et al. 1991).

$R-COOM$
(Fatty acid salt)

$RAr-COOM$
(Petroleum carboxylate)

4.2.2.3 Polyether Surfactants

Some examples are given as follows.

$R-O-[CH_2CH_2O]_n-H$
(Peregal surfactant)

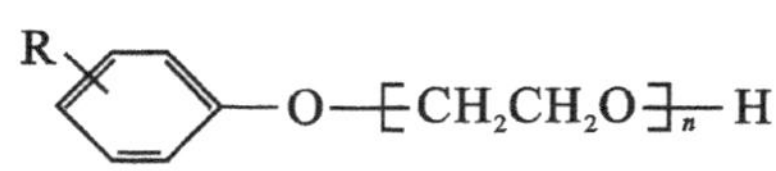
(OP surfactant)

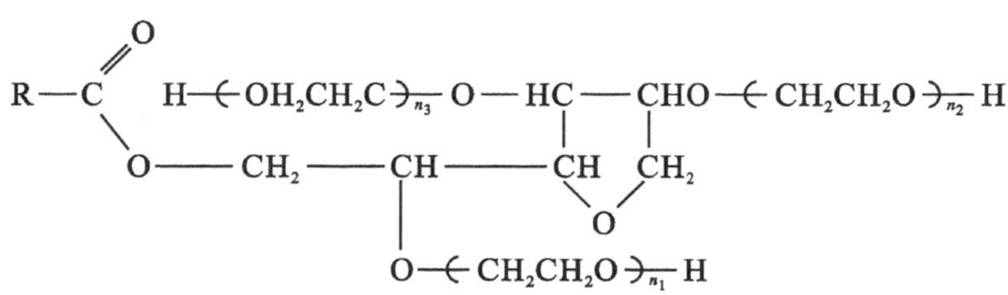
(Tween surfactant)

4.2.2.4 Nonionic–Anionic Surfactants

Some examples are given as follows, where R is alkyl, Ar is aryl, M is Na, K, NH_4, etc.

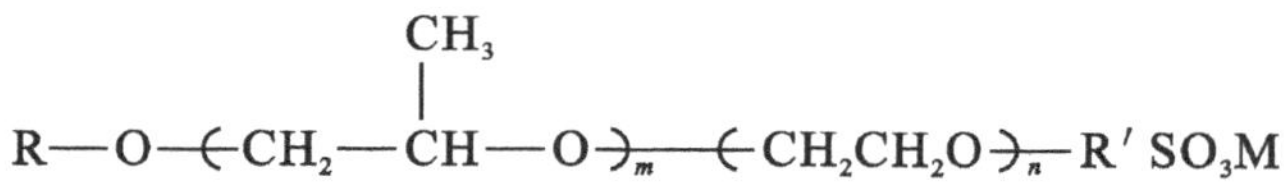

(Polyoxyethylene polyoxypropylenealkyl alcohol ether sulfonates)
(Mccoy 1984; Wang and Zhao 1996)

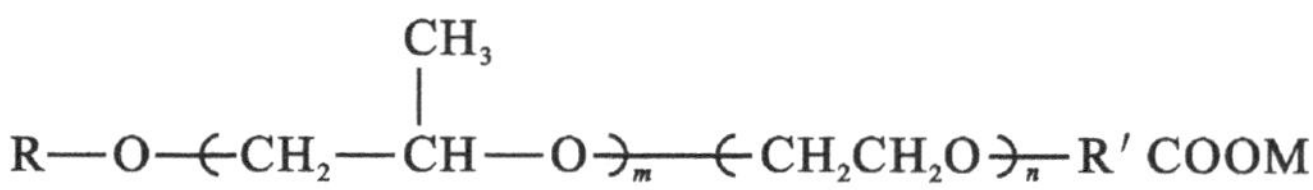

(Polyoxyethylene polyoxypropylenealkyl alcohol ether carboxylates)
(Zhao et al. 1985; Li et al. 2003; Wang and Zhao 2006; Kang et al. 2006)

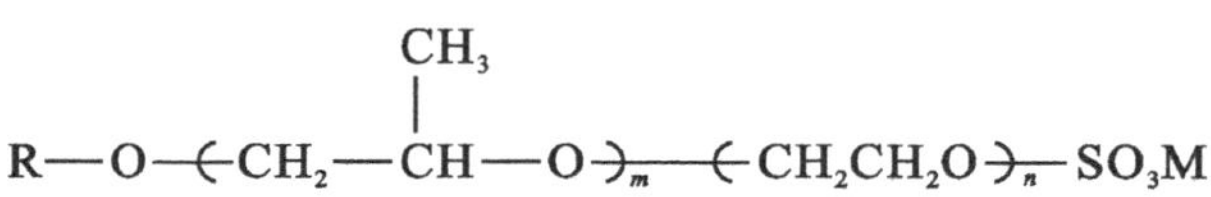

(Polyoxyethylene polyoxypropylenealkyl alcohol ether sulphates)
(Schievelbein 1981; Talley 1988)

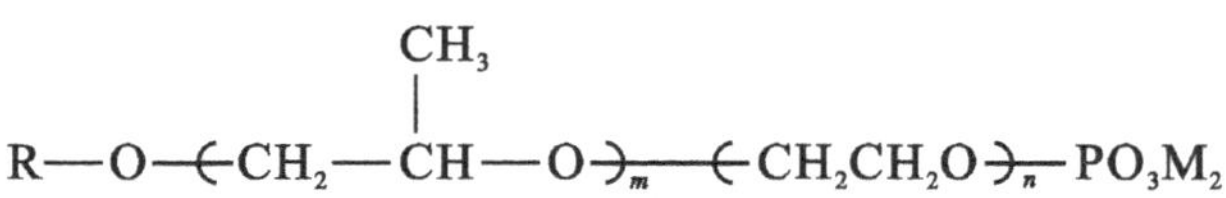

(Polyoxyethylene polyoxypropylenealkyl alcohol ether phosphates)
(Zhao et al. 1985)

4.2.2.5 Gemini Surfactants

Gemini surfactants are formed by two or more same surfactant molecules connecting with each other at or near hydrophilic groups through linking groups. Some examples are given as follows.

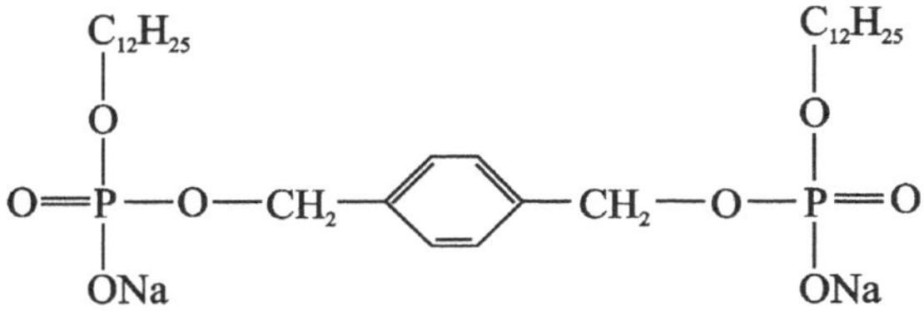

[Xylene double(sodium dodecyl phosphate)]

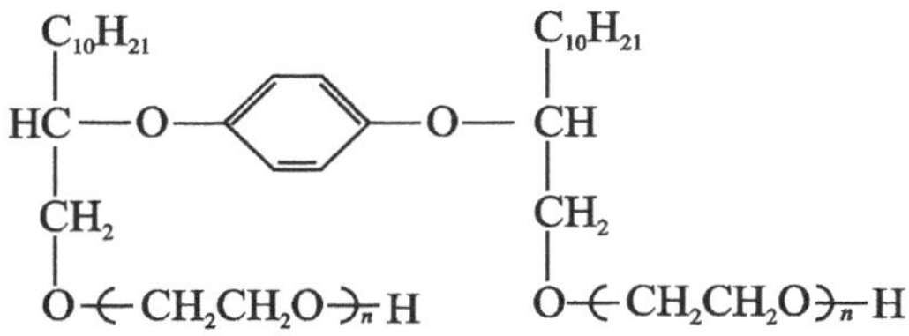

Gemini surfactant is a kind of newly developed surfactant with high surface activity, which is complementary with other surfactant structures to create synergy effect (Rosen 1993; Zhao 1997).

4.2.2.6 Biosurfactants

Biosurfactants are the compounds produced by microbial with characteristics of surfactants. Some examples are given below.

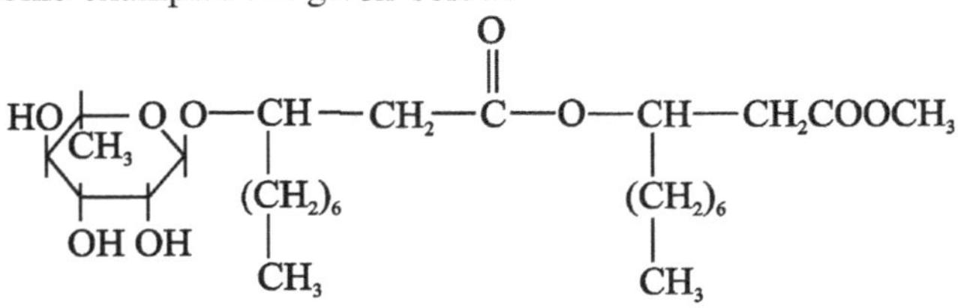

(Rhamnose ester)

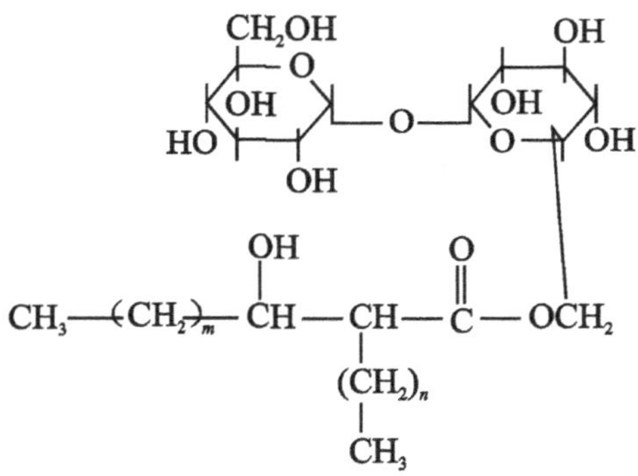

m : 20–21; n : 9–11

(Trehalsoe ester)

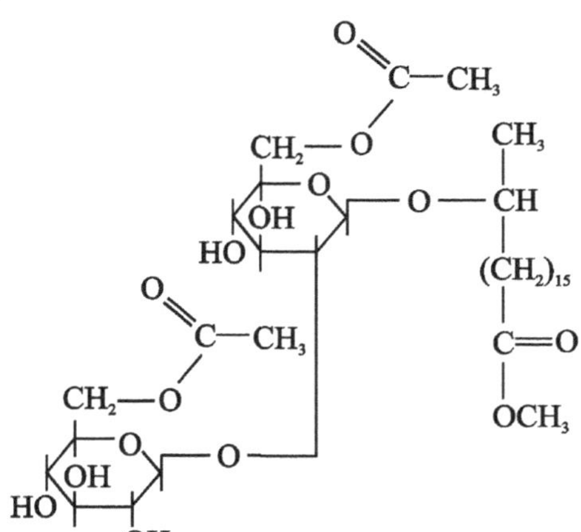

(Sophorose ester)

Due to the environmental friendliness, biosurfactant is a kind of newly developed surfactant which receives particular attention (Wang et al. 2005).

The surfaces of formation rocks are generally negatively charged. Therefore, cationic surfactants and nonionic–cationic surfactants are seldom used in oil displacement to reduce losses of surfactants.

4.2.3 Active Water Flooding

The active water belongs to the dilute surfactant system, in which the concentration of surfactant is lower than the critical micelle concentration.

The oil displacement method using active water as the displacement agent is called active water flooding, which is the simplest surfactant flooding.

The active water flooding can increase the oil recovery by the mechanisms described as follows.

4.2.3.1 Mechanism of Low Interfacial Tension

The oil–water interfacial tension can be reduced due to the adsorption of surfactant molecules on the oil–water interface. It can be observed from the following adhesive work formula that the decrease in oil–water interfacial tension will result in the reduction of adhesive work, which means oil can be washed off from the rock surface more easily and the oil-washing capability is improved.

$$W = \sigma(1 + \cos\theta) \tag{4.3}$$

In Eq. 4.3, W is the adhesive work, σ is the oil–water interfacial tension, and θ is the wetting angle of oil on rock surface.

4.2.3.2 Mechanism of Wettability Reversal

The hydrophilicity of the surfactants used for oil displacement is stronger than the hydrophobicity. The lipophilic rock surface (formed by the adsorption of natural surface-active substance (Buckley 1997) will convert to hydrophilic surface by the

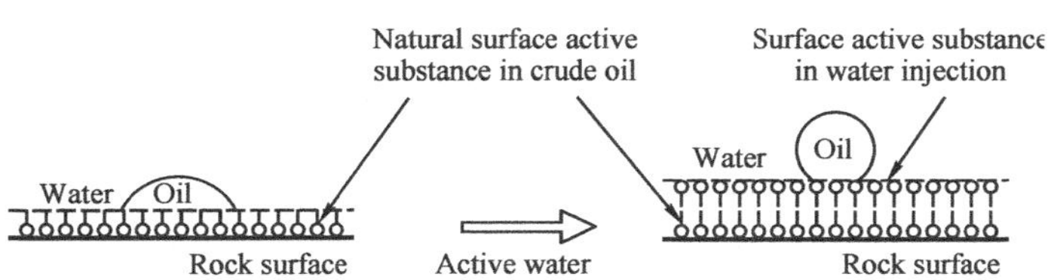

Fig. 4.9 Wettability reversal of rock surface due to surfactants

adsorption of surfactants on rock surface, and the wetting angle of oil on rock surface increases (as shown in Fig. 4.9). It can be seen from Eq. 4.3 that the increase in wetting angle of oil on rock surface will decrease the adhesive work, indicting enhanced oil-washing capability.

4.2.3.3 Mechanism of Emulsion

The HLB value of surfactants used for oil displacement generally ranges from 7 to 18, and the adsorption of surfactants on oil–water interface will stabilize the oil-in-water emulsion. The emulsified oil is not easy to adhere to rock surface again in the process of moving forward, so the oil displacement efficiency is improved. Moreover, the overlaid Jamin effect caused by the emulsified oil in high-permeability layers can make the water proceed more uniformly in the formation, which increases the sweep efficiency.

4.2.3.4 Mechanism of Increasing Surface Charge Density

When the surfactants used for oil displacement are anionic or nonionic–anionic surfactants, their adsorption on the surfaces of oil drops and rock will increase the surface charge density (Fig. 4.10), which will increase the electrostatic repulsion between oil drops and rock surface, making the oil drops displaced by displacing medium more easily and increasing the oil displacement efficiency.

4.2.3.5 Mechanism of Formation of Oil Zone by Coalescence

When more and more oil are displaced from the formation surface, they will collide with each other during the process of moving forward. The dispersed oil will gather

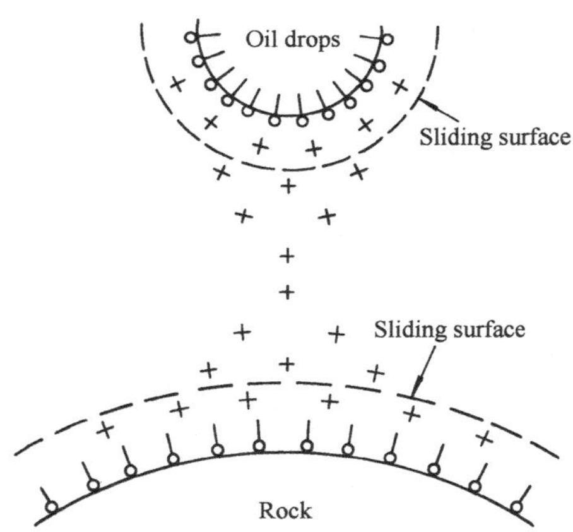

Fig. 4.10 Adsorption of anionic surfactants increases the surface charge density of oil drops and rock

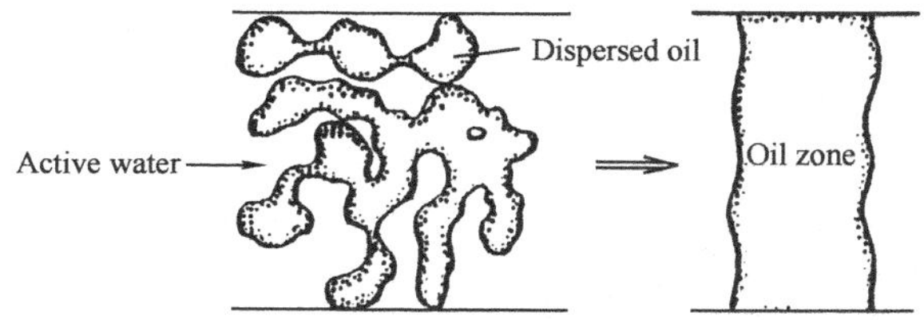

Fig. 4.11 Formation of oil zone by coalescence of displaced oil

together when the collision energy overcomes the electrostatic repulsion between them. Oil zone forms by the coalescence of dispersed oil (Fig. 4.11). During the proceeding of oil zone, the dispersed oil continues to merge into the oil zone to enlarge it (Fig. 4.12) until finally produced from the oil well.

On account of the low surfactant concentration in active water and the loss of surfactant due to its adsorption on rock surface, a large slug is needed to make the active water have effect on the formation.

4.2.4 Micellar Solution Flooding

Micellar solution also belongs to dilute surfactant system. Its surfactant concentration is higher than the critical micelle concentration, while its mass fraction is less than 2%.

The displacement method which uses micellar solution as displacement agents is called micellar solution flooding. It is one type of surfactant flooding which falls in between active water flooding and microemulsion flooding that will be mentioned later.

To reduce the interfacial tension between micellar solution and oil, alcohol (such as isopropanol and n-butyl alcohol) and (or) salt (such as sodium chloride) are also added in micellar solution besides surfactants.

Compared with active water, micellar solution has two characteristics. One is that the surfactant concentration in micellar solution surpasses the critical micelle concentration, so micelles exist in solution. The other is that besides surfactants, additives such as alcohol and (or) salt have also been added in micellar solution. All function mechanisms for active water flooding apply to the micellar solution flooding. The difference is that for micellar solution flooding, solubilizing mechanism caused by the existence of micelles also applies. Because micelles can increase the solubility of oil, the displacement efficiency of micellar solution is enhanced. Besides, the addition of accessory ingredients such as alcohol and salt helps with the adjustment of the polarity of oil and water phase, making the lipophilicity and hydrophilicity of surfactants fully balanced. Therefore, the surfactants can maximally adsorb onto the oil–water interface to generate ultra-low interfacial tension (lower than 10^{-2} mN/m), strengthening the mechanism of low interfacial tension for micellar solution flooding.

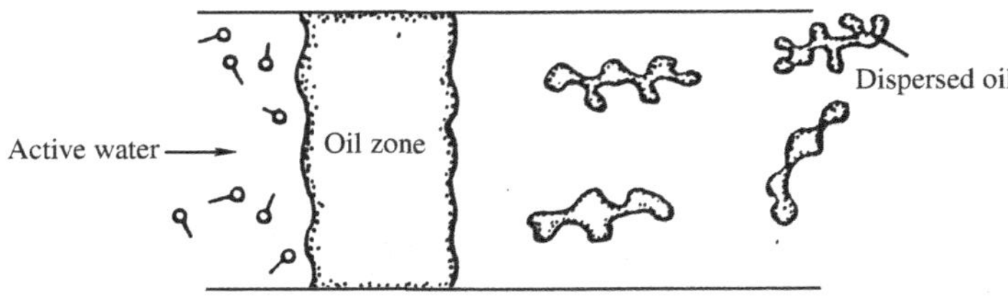

Fig. 4.12 Oil zone continues to expand during the process of moving forward

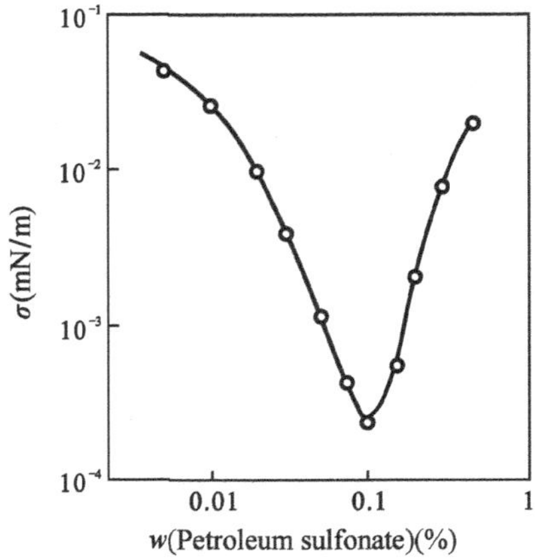

Fig. 4.13 Typical relationship between oil–water interfacial tension and surfactant concentrations. Conditions: Surfactant: TRS10-410 (a kind of petroleum sulfonate); alcohol: IBA (isobutyl alcohol). Oil phase: dodecane; water phase: TRS10-410 + IBA + NaCl; w(NaCl) = 1.5%; m(TRS10-410)/m(IBA) = 5/3

Figure 4.13 shows typical relationship between oil–water interfacial tension and surfactant concentrations. It can be seen from Fig. 4.13 that when w (petroleum sulfonate) reaches 0.1% and interfacial tension is as low as 2.6×10^{-4} mN/m, micelles appear in the system and micellar solution is formed. When w (petroleum sulfonate) ranges between 0.02% and 0.30%, the interfacial tension between surfactant system and oil all reaches ultra-low value, exhibiting great oil displacement capability. After the formation of micelles in system, the recovery of interfacial tension is caused by the solubilization of the highly active ingredients in petroleum sulfonate (a kind of mixed surfactants) by micelles.

4.2.5 Microemulsion Flooding

Microemulsion belongs to the concentrated surfactant system. It has two basic types and one transitional type. The former types are water-external microemulsion and oil-external microemulsion, while the latter type is middle-phase microemulsion.

The water-external microemulsion is prepared using water-soluble surfactants. It is a dispersion system formed by surfactant micelles dissolved with oil dispersing in water.

The oil-external microemulsion is prepared using oil-soluble surfactants, and it is a dispersion system formed by surfactant micelles dissolved with water dispersing in oil.

The hydrophilicity and hydrophobicity of surfactants are decided not only by the hydrophilic and oleophilic parts, but also by the temperature, oil property (carbon number of hydrocarbon), electrolyte in water (types and concentrations), cosurfactant in the systems (types and concentration), and so on. Therefore, mutual transformation between the basic types of microemulsion would happen under the effects of factors described above (Zhao et al. 1995). For example, the water-external microemulsion prepared with petroleum sulfonate could transform into oil-external microemulsion with the addition of salts (such as NaCl); reverse transformation would happen with the removal of salts. During the transformation of basic types of microemulsion, the transitional type, middle-phase microemulsion, would generally appear (as shown in Fig. 4.14).

There are differences between microemulsion and emulsion. The microemulsion is stable dispersion system because oil and water are solubilized in the surfactant micelles. While emulsion is unstable dispersion system as there are interfaces between oil and water. Though microemulsion is different from emulsion, mutual transformation can happen between them under certain conditions (as shown in Fig. 4.15).

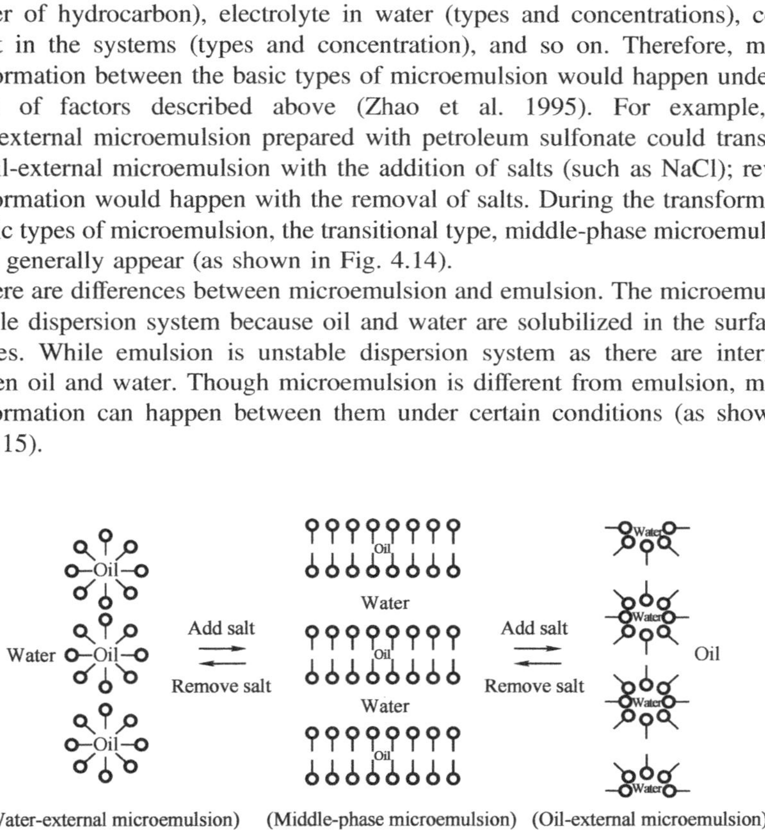

Fig. 4.14 Mutual transformation between the basic types of microemulsion

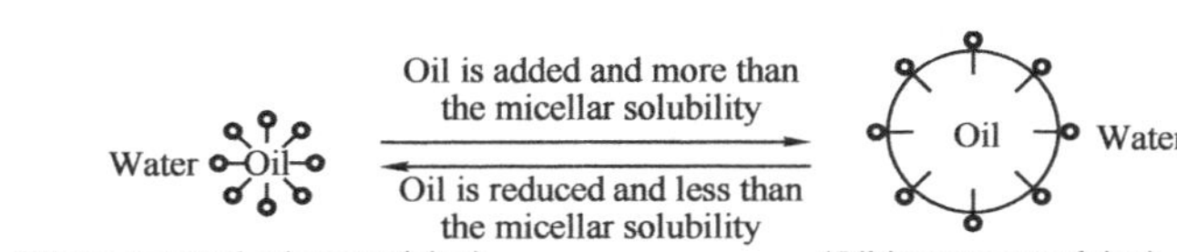

Fig. 4.15 Mutual transformation between microemulsion and emulsion

Three main components and two assistant components are needed to prepare microemulsion. The three main components are oil, water, and surfactants, while the two assistant components are cosurfactants and electrolytes.

The oil used to prepare microemulsion can be crude oil or crude oil distillates (such as gasoline, kerosene, and diesel).

The water used to prepare microemulsion can be freshwater or brine.

The surfactants used to prepare microemulsion can be anionic surfactants, nonionic surfactants, and nonionic–anionic surfactants, but it is better to use petroleum sulfonate (sodium salt or ammonium salt) to prepare microemulsion.

The cosurfactants used to prepare microemulsion would better be alcohol or phenol. Cosurfactants could not only adjust the polarities of water and oil (water-soluble alcohol can decrease the polarity of water, and oil-soluble alcohol can increase the polarity of oil), but also take part in the formation of micelles, increase the spaces of micelles, and increase the solubility of micelles to oil or water.

The electrolytes used to prepare microemulsion can be inorganic acids, inorganic bases, and inorganic salts, but inorganic salts such as sodium chloride, potassium chloride, and ammonium chloride work the best. Electrolytes can facilitate the formation of micelles in lower surfactant concentrations by decreasing the solvation degree of the polar parts of surfactants and cosurfactants, which will also generate ultra-low interfacial tension between microemulsion and oil or water.

Microemulsion can be used for oil displacement.

Microemulsion flooding is the oil displacement method using microemulsion as the oil displacement agent.

Microemulsion flooding can improve oil recovery by different mechanisms.

Assume that the oil displacement agent is the water-external microemulsion.

When the microemulsion contacts with oil layer, it is miscible with water (homogeneous phase) because the microemulsion is water-external. And because its micelles can dissolve oil, the microemulsion is also miscible with oil (homogeneous phase). Therefore, the oil displacement belongs to miscible microemulsion flooding when the water-external microemulsion initially contacts with oil layer. This kind of oil displacement has two characteristics. Firstly, microemulsion has no interfaces with oil and water; that is to say the interfacial tension is zero and no capillary resistance exists. Thus, the sweep efficiency of microemulsion flooding is higher than that of water flooding, active water flooding, and micellar solution flooding. Secondly, the microemulsion is completely miscible with oil, so its oil displacement efficiency is much higher than that of water flooding, active water flooding, and micellar solution flooding.

When microemulsion enters oil layer and the micelle is saturated with displaced oil by solubilization, interfaces are generated between microemulsion and displaced oil. At this time, miscible microemulsion flooding changes to immiscible microemulsion flooding.

When microemulsion further enters oil layer and more displaced oil enters micelle, the previous micelle changes to oil droplet and the water-external microemulsion changes to oil-in-water emulsion. Emulsion is also an oil displacement agent, whose EOR mechanisms are the same with those of the foam flooding which will be introduced later.

Thus, the oil displacement mechanism of microemulsion is complicated, which is due to the phase transitions caused by water and oil entering microemulsion.

Oil-external microemulsion and middle-phase microemulsion have similar cases.

4.2.6 Foam Flooding

Foam flooding is the oil displacement method using foam as the oil displacement agent.

Foam is composed of water, gas, and foaming agent. To generate foam, the foaming agents and gas can be alternately injected into the oil layer or simultaneously injected into the formation from tubing and casing.

The water used to generate foams can be freshwater or brine.

The gas used to generate foams can be nitrogen, carbon dioxide, natural gas, refinery gas, or flue gas.

The foaming agents used to generate foams are mainly surfactants such as alkyl sulfonate, alkylbenzene sulfonate, polyoxyethylene alkyl alcohol ether-15, polyoxyethylene alkyl phenol ether-10, polyoxyethylene alkyl alcohol ether sulfate, polyoxyethylene alkyl alcohol ether carboxylate. Certain amount of polymers (such as partially hydrolyzed polyacrylamide and sodium carboxymethyl cellulose) can be added to the foaming agents to increase the water viscosity, thus improving the foam stability.

The foam characteristic value is a key physical quantity to describe the properties of generated foam. The foam characteristic value is defined as the ratio of the gas volume in the foam to the total foam volume. Generally, the foam characteristic value ranges from 0.52 to 0.99. The foams with foam characteristic value smaller than 0.52 are called gas emulsion. The foams with foam characteristic value larger than 0.99 are easily to become fog by phase reversal. And the bubbles in the foams will turn into polyhedrons when the foam characteristic value is larger than 0.74.

Laboratory experiments indicate that for oil displacements, the oil recoveries are different by using foams with different foam characteristic values (Fig. 4.16).

Foam flooding can improve the oil recovery by mechanisms described as follows.

4.2.6.1 Mechanism of Superimposed Jamin Effect

For foams, Jamin effect refers to the resistance of bubbles to the fluid flowing through the pore throats, which is generated when the diameters of bubbles in the foams passing through the pore throats are larger than those of the pore throats. When the foams flow through heterogeneous formation, they will enter the high-permeability layer first. Because Jamin effect is superposable, the flow

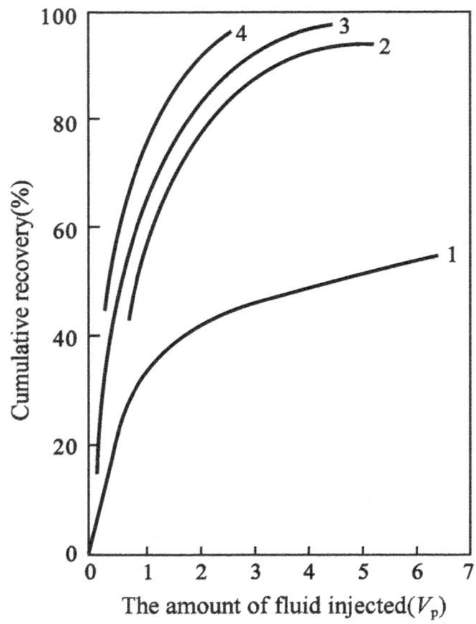

Fig. 4.16 Oil displacement results using foams. Foam characteristic values: 1—0, 2—0.72, 3—0.85, 4—0.91

resistance of the foams will increase gradually. Therefore, with the increase in injection pressure, the foams can successively enter the medium- and low-permeability layers in which originally the foams cannot enter due to high flow resistance, thus increasing the sweep efficiency.

4.2.6.2 Mechanism of Viscosity Increase

The viscosity change of foam related to the foam characteristic value is shown in Fig. 4.17. As shown in Fig. 4.17, the viscosity of foam is larger than that of water. This is because the viscosity of water comes only from the internal friction of fluid layers with relative movement, while the viscosity of foam derives not only from the internal friction of fluid (dispersion medium) layers with relative movement, but also from the intercollision of dispersion phases. The viscosity of foam increases sharply when the foam characteristic value surpasses certain value (0.74), which could be attributed to the deformation of bubbles caused by the intersqueezing of dispersion phases. The intercollision of dispersion phases is a primary origin of the flow resistance of foams.

The viscosity of foam can be calculated by the following experimental formula. When the foam characteristic value is smaller than 0.74,

$$\mu_f = \mu_0(1.0 + 4.5\varphi) \tag{4.4}$$

In the formula, μ_f is the viscosity of foam, μ_0 is the viscosity of dispersion medium, and φ is the foam characteristic value.

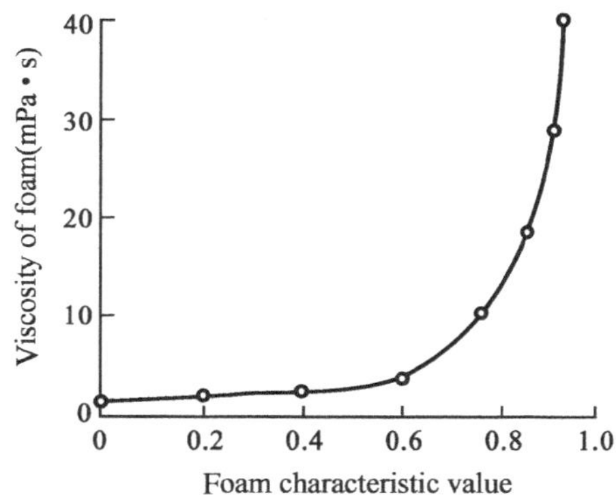

Fig. 4.17 Relationship between viscosity of foam and foam characteristic value

When the foam characteristic value is larger than 0.74,

$$\mu_f = \mu_0 \cdot \frac{1}{1 - \sqrt[3]{\varphi}} \tag{4.5}$$

It can be calculated from Eq. 4.5 that when the foam characteristic value is 0.90, viscosity of the foam is about 29 times of that of the dispersion medium (water).

Compared with water, foam has greater sweep efficiency due to its larger viscosity; thus, the foam flooding can achieve higher oil recovery than water flooding does.

4.2.6.3 Mechanism of Oil Displacement by Dilute Surfactant System

The dispersion medium of foams is surfactant solution. According to the surfactant concentration in foam system, the foam flooding should have properties of dilute surfactant system (such as active water, micellar solution). Thus, the foam flooding has same oil displacement mechanisms with the dilute surfactant system.

4.3 Alkaline Flooding

4.3.1 Definition of Alkaline Flooding

Alkaline flooding is the oil displacement method using alkaline solution as oil displacement agent.

Alkaline flooding is also called alkaline solution flooding or alkali-enhanced water flooding.

4.3.2 Alkalis Used in Alkaline Flooding

The alkalis used in alkaline flooding include not only alkali (such as NaOH, KOH, and NH_4OH) but also salts (such as $NaBO_2$, Na_2CO_3, Na_2SiO_3, Na_4SiO_4, and Na_3PO_4) (Lorenz and Peru 1989; Hurk 1983; Flaaten et al. 2008). The salts are called potential alkalis because they can generate OH^- in water by the following reactions.

Sodium carbonate and sodium hydrogen carbonate are a set of buffer substances which can buffer the pH value of the systems through the following reaction. Therefore, sodium carbonate and sodium hydrogen carbonate can be compounded to prepare alkaline systems with buffer function (Rudin and Wasan 1993).

Similarly, Na_3PO_4 and Na_2HPO_4 are a set of alkaline systems with buffer function. The alkalis used in alkaline drive can also be organic bases, such as

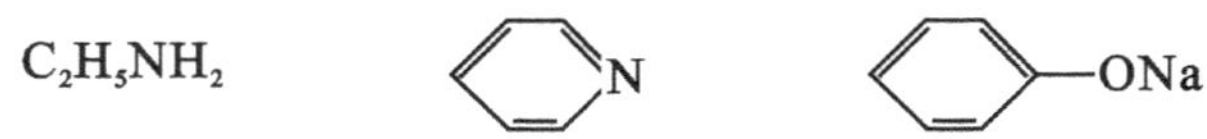

(Ethylamine) (Pyridine) (Sodium phenolate)

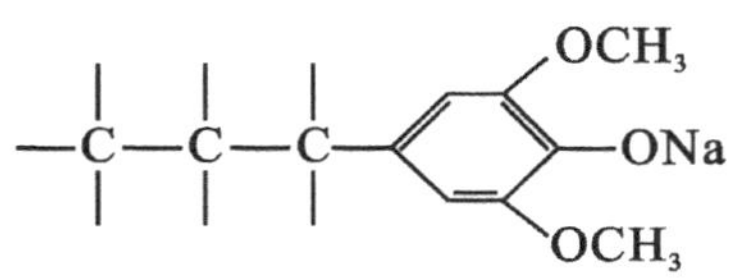

(The main structure of the alkali lignin in the black liquor in papermaking)

4.3.3 Reaction of Petroleum Acids and Alkalis

The petroleum acids in crude oil such as fatty acids, naphthenic acids, glial acids, and asphaltene acids can react with alkalis (such as sodium hydroxide) to form the corresponding petroleum acid salts.

$$R\text{—}COOH + NaOH \longrightarrow R\text{—}COONa + H_2O$$

(Fatty acid)

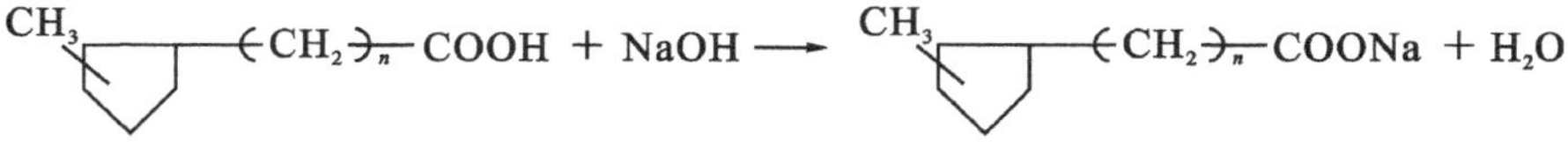

(Naphthenic acid)

$$\boxed{\text{Colloid}}\text{—}COOH + NaOH \longrightarrow \boxed{\text{Colloid}}\text{—}COONa + H_2O$$

(Colloid acid)

$$\boxed{\text{Asphaltene}}\text{—}COOH + NaOH \longrightarrow \boxed{\text{Asphaltene}}\text{—}COONa + H_2O$$

(Asphaltenic acid)

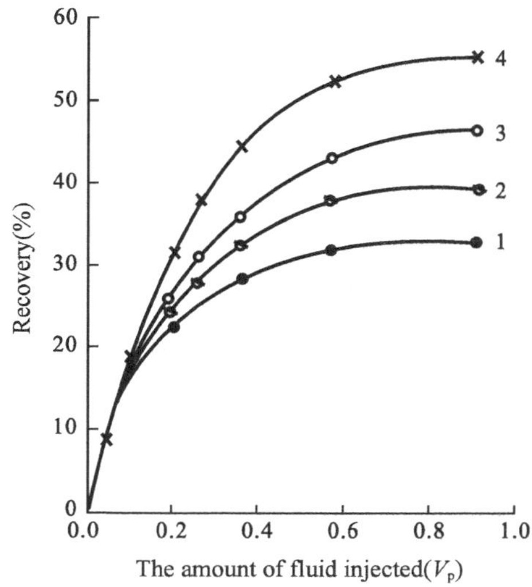

Fig. 4.18 Comparison of function effects of alkaline flooding and water flooding. *w*(NaOH): 1—0 (water flooding), 2—0.005%, 3—0.01%, 4—0.05%

The formed petroleum acid salts whose hydrophilicity and hydrophobicity are basically balanced are usable surfactants which can decrease the oil–water interfacial tension.

Certain amounts of salt (such as NaCl) need to be added into the alkali solution, to make the surfactants formed by reaction of alkalis and petroleum acids have the required hydrophilic–lipophilic balance.

4.3.4 Comparison of Alkaline Flooding and Water Flooding

Figure 4.18 shows the comparison of function effects of alkaline flooding and water flooding (Neil et al. 1983). As shown in Fig. 4.18, alkaline flooding can achieve higher oil recovery than the water flooding.

4.3.5 The Enhanced Oil Recovery Mechanism of Alkaline Flooding

The mechanism of alkaline flooding is very complex. Some mechanisms have been proposed to explain the role of alkaline flooding in enhancing oil recovery (Mayer et al. 1982; Castor et al. 1981).

4.3.5.1 Mechanism of Low Interfacial Tension

Under the conditions of low alkali content and optimal salt content, alkali can react with petroleum acid to generate surfactant which can decrease the interfacial tension between oil and water to values below 1×10^{-2} mN/m (Fig. 4.19). In this way, the alkali flooding can have the same effect as surfactant flooding. The concept of optimal salinity can be referenced from Fig. 4.20.

All the other mechanisms described below are based on this mechanism.

Fig. 4.19 Relationship between the interfacial tension of one kind of crude oil and the mass fraction of NaOH

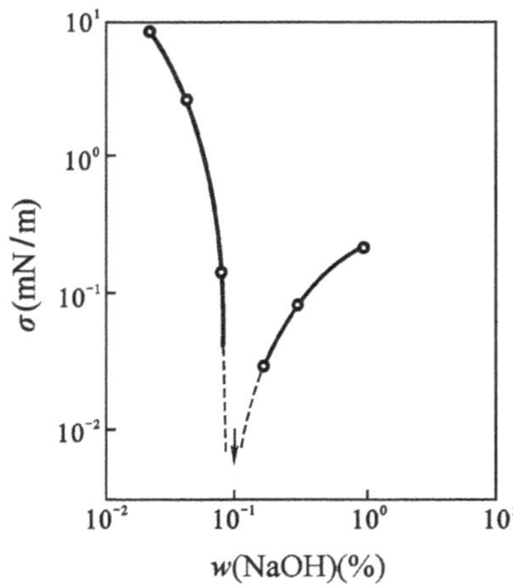

Fig. 4.20 Concept of optimal salinity in alkali solution/crude oil system. w(NaCl): 1—0; 2—0.1%; 3—0.75%; 4—2.0%; 5—5.0%; 6—10.0%; 7—30.0%

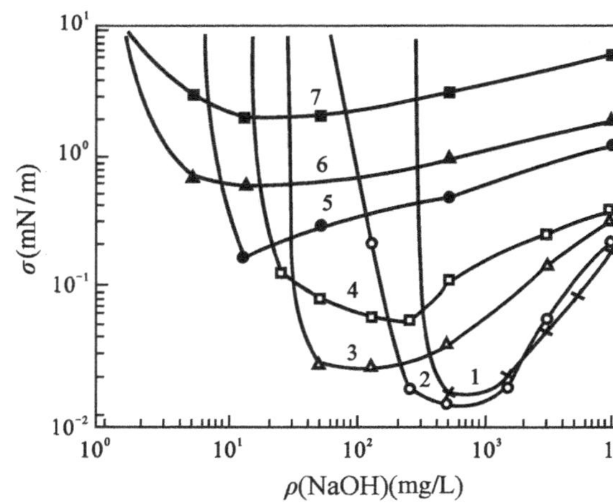

4.3.5.2 Emulsifying–Carrying Mechanism

Under the conditions of low alkali content and low salinity, the surfactant generated by alkali and petroleum acid can emulsify the remaining oil in the formation, which can be carried by alkali solution to pass through the formation layer. According to this mechanism, alkali flooding has the following characteristics.

- The emulsion with very small oil droplets can be formed.
- The oil displacement efficiency of alkali flooding can be increased through emulsification.
- The petroleum production cannot increase before the breakthrough of alkali solution from the production well.
- The coalescence of oil droplets has adverse effect on the process.

4.3.5.3 Emulsifying–Entrapping Mechanism

Under the conditions of low alkali content and low salinity, the oil can be emulsified into the alkali solution due to the low interfacial tension. The diameter of oil droplets is very large. Therefore, when the oil droplets move forward, they get trapped, increasing the flow resistance of water. In this way, the mobility of water is decreased and mobility ratio is improved, which can increase the sweep efficiency and enhance oil recovery. According to this mechanism, alkali flooding has the following characteristics.

- The emulsion with large oil droplets can be formed.
- The dispersed oil droplets are trapped in the smaller channel.
- The petroleum production can increase before the breakthrough of alkali solution from the production well.
- The coalescence of oil droplets has beneficial effect on the process.

4.3.5.4 Oil-Wet Converting to Water-Wet Mechanism

Under the conditions of high alkali content and low salinity, the alkali can change the solubility of oil-soluble surfactant adsorbed on the rock surface in water, which results in the desorption of surfactant and restoration of water-wet rock surface. This can make the rock surface convert from oil-wet to water-wet, thus increasing the displacement efficiency and enhancing oil recovery.

4.3.5.5 Water-Wet Converting to Oil-Wet Mechanism

Under the conditions of high alkali content and high salinity, the surfactants generated by alkali and petroleum acid mainly locate in the oil phase and adsorb on rock surface, making the rock surface convert from water-wet to oil-wet. In this way, the discontinued remaining oil can form continued oil phase on the rock

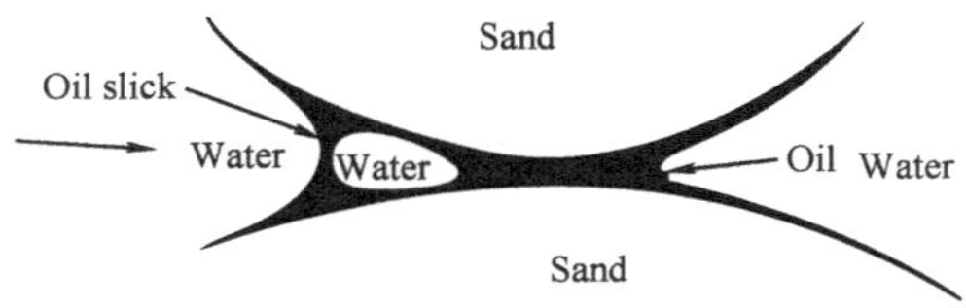

Fig. 4.21 Enhanced oil recovery through water-wet converting to oil-wet mechanism

Table 4.1 Realization conditions of alkaline flooding

Mechanism	*w*(chemical agents) (%)	
	NaOH	NaCl
Low interfacial tension	Low, <1	Low, 1–2
Emulsifying–carrying	Low, <1	Low, 0.5–1.5
Emulsifying–entrapping	Low, <1	Low, <0.5
Oil-wet converting to water-wet	High, 1–5	Low, <5
Water-wet converting to oil-wet	High, 1–5	High, 5–15

surface, which can provide flowing channels for oil. Meanwhile, the lipophilicity of surfactants generated by alkali flooding and the low surface tensions lead to the formation of water-in-oil emulsion. The water droplets in the emulsion can block the flow channel, which increases the injection pressure. High injection pressure forces the oil out from the continuous oil phase between emulsified water droplets and rock surface. The high-water-content emulsions are remained, and the purpose of enhanced oil recovery is achieved. Figure 4.21 illustrates this mechanism.

Table 4.1 lists the realization conditions of alkaline flooding (Symonds et al. 1991).

The precondition of conducting alkaline flooding is that there are petroleum acids which can generate surfactant in the crude oil; therefore, sufficiently high acid value is required for crude oil reservoir subject to alkali flooding operations. The acid value is the mass of KOH consumed by the neutralization of 1 g oil, with a unit of mg/g. When the acid value of oil is smaller than 0.2 mg/g, the reservoir is not applicable for alkaline flooding. Certain acid value is the necessary condition, but not sufficient condition, of conducting alkaline flooding. The sufficient condition of conducting alkaline flooding is that the reaction product of petroleum acid in the oil and alkali is surfactant. For example, xylenol in the oil belongs to the petroleum acid. However, after reaction with alkali, its reaction product xylenol salt is not surfactant.

4.4 Combination Flooding

4.4.1 Concept of Combination Flooding

Combination flooding is an oil displacement method using two or more combined oil displacement agents (Baviere et al. 1995). Here, the oil displacement agents

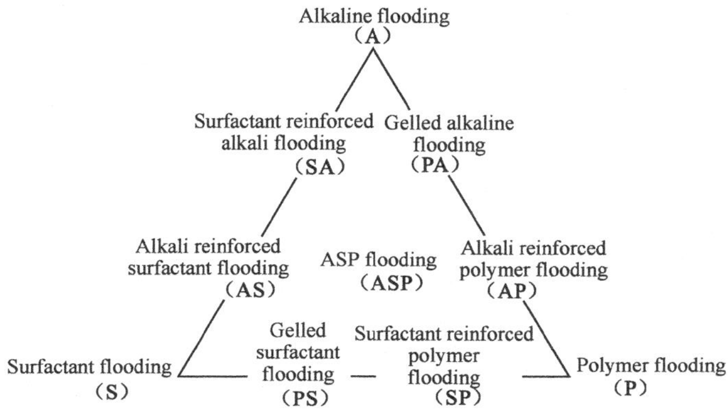

Fig. 4.22 Combination of various flooding ways in chemical flooding

refer to the main agents in chemical flooding such as polymer, alkali, and surfactant, which can be combined in different ways to form various combination floodings. For example, the flooding composed of alkali and polymer is called gelled alkaline flooding or alkali-reinforced polymer flooding (Doll 1986); the flooding composed of surfactant and polymer is called gelled surfactant flooding or surfactant-reinforced polymer flooding (Bock et al. 1992); the flooding composed of alkali and surfactant is called alkali-reinforced surfactant flooding or surfactant-reinforced alkali flooding (Surkalo 1990; Taylor et al. 1990) the flooding composed of alkali (A), surfactant (S), and polymer (P) is called ASP flooding (Wyatt et al. 1995; Shuler et al. 1989; Clark and Pitts 1988; Wang et al. 2009). We can use the pseudo-tricomponent phase diagram to indicate the various combination ways in chemical flooding (Fig. 4.22) (Labrid 1991).

In Fig. 4.22, the floodings at three vertices in the tricomponent phase diagram belong to the single-component flooding; the floodings at any point of the three sides of the phase diagram belong to the binary combination flooding, while the floodings at any point inside the phase diagram belong to the ASP combination flooding.

4.4.2 Comparison of the Oil Displacement Effects of Some Floodings

4.4.2.1 Combination Flooding Vs. Single-Component Flooding

Compared with single-component flooding, combination flooding usually has higher oil recovery. Figure 4.23 shows the comparison among the alkali + polymer flooding, pure alkaline flooding, pure polymer flooding, alkaline flooding first and polymer flooding next, polymer flooding first and alkaline flooding next

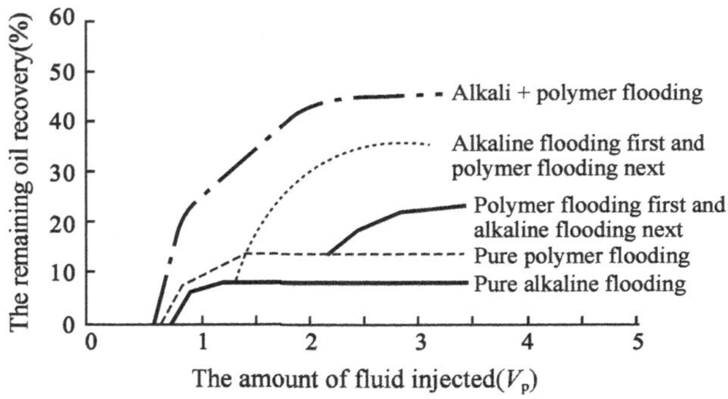

Fig. 4.23 Comparison of the oil recovery of different floodings (The viscosity of crude oil is 180 mPa · s. The polymer is polyacrylamide and the alkali is sodium orthosilicate)

Table 4.2 Comparison of the effects of different combination floodings

The combination flooding after water flooding①	AS	AP	ASP
Original oil saturation, S_{oi} (%)	81.9	86.1	76.9
Residual oil saturation after water flooding, S_{or} (%)	50.6	50.9	49.7
Residual oil saturation after combination flooding, S_{orc} (%)	39.2	39.1	22.5
The oil-phase permeability under the condition of S_{or}, k_o (10^{-3} μm^2)	754	1,124	690
The water-phase permeability under the condition of 1-S_{oi}, k_w (10^{-3} μm^2)	29.0	57.7	26.3
The recovery of water flooding (%)	38.2	40.9	35.4
The recovery of combination flooding (%)	22.5	23.2	54.7

Note ①The recipe of combination flooding is as follows
AS: $w(Na_2CO_3)$ = 1%, w(R–O—[CH_2CH_2O]$_3$—SO_3Na) = 0.1%
AP: $w(Na_2CO_3)$ = 1%, w(HPAM) = 0.1%
ASP: $w(Na_2CO_3)$ = 1%, w(R–O—[CH_2CH_2O]$_3$—SO_3Na) = 0.1%, w(HPAM) = 0.1%

(Krumrine and Falcone 1983). As shown in Fig. 4.23, the remaining oil recovery of alkali + polymer flooding is five times of that of the pure alkaline flooding, and three times of that of the pure polymer flooding.

4.4.2.2 Binary Combination Flooding and ASP Flooding

Table 4.2 gives the results of flooding tests using, respectively, alkali + surfactant (AS), alkali + polymer (AP), and alkali + surfactant + polymer (ASP) (Hawkins et al. 1994). The crude oil used in the tests has a viscosity of 67.0 mPa · s, a density of 0.92 g/cm^3, and an acid value of 0.45 mg/g. As given in Table 4.2, the oil displacement effect of the ASP flooding is better than that of the AS and AP floodings.

4.4.3 Synergistic Effect Among the Displacement Components in Combination Flooding

The reason why ASP flooding has better flooding effect than single-component flooding and binary combination flooding is that the polymer, surfactant, and alkali in combination flooding have synergistic effects. They play their own roles in the synergistic effects.

4.4.3.1 Role of the Polymer

- Polymer can improve the mobility ratio of surfactant and (or) alkali solution to oil.
- The thickening effect of polymer to displacement medium can reduce the diffusion rate of surfactant and alkali, thus reducing their consumption.
- Polymer can react with calcium and magnesium ions. So it can protect surfactant from forming the low surface-active calcium and magnesium salts.
- Polymer can improve the stability of oil-in-water emulsion formed by alkali and surfactant, which can greatly improve the sweep efficiency (based on the emulsifying–trapping mechanism) and (or) oil displacement capacity (based on the emulsifying–carrying mechanism).

4.4.3.2 Role of the Surfactant

- Surfactant can reduce the interfacial tension between the polymer solution and the oil, so it has the ability to displace oil.
- Surfactant can emulsify oil to improve the viscosity of displacement medium. The more the amount of emulsified oil is, the higher the viscosity of the emulsion becomes.
- If the surfactant can form complex structure with polymer, the surfactant can increase the thickening ability of the polymer.
- Surfactant can cover the shortage of the amount of the surfactant generated from the reaction of alkali and petroleum acid.

4.4.3.3 Role of the Alkali

- The surfactant generated from the reaction of alkali and petroleum acid can emulsify the oil and improve the viscosity of the flooding medium, therefore enhancing the polymers' capability in controlling the mobility.
- The surfactant generated from the reaction of alkali and petroleum acid has a synergistic effect with the synthetic surfactant (Table 4.3).

Table 4.3 Synergistic effect between alkali and surfactant

Solution	w(Chemical agents) (%)	σ[①] (mN/m)
Polymer	0.1	18.2
Sodium hydroxide	0.8	2.1
Petroleum sulfonate	0.1	5.5
Sodium hydroxide	0.8	0.02
Petroleum sulfonate	0.1	

Note ① The density of crude oil is 0.900 g/cm^3, and the interfacial tension between oil and water is 18.2 mN/m

Table 4.4 Adsorption of R–O—$[CH_2CH_2O]_n$—SO_3Na on Berea cores (71 °C)

Adsorption type	The adsorption amount on 100 g Berea cores (mmol)		
	pH = 7.0	pH = 12.7 (Na_2O SiO_2)	pH = 12.7 (NaOH)
Static adsorption	0.064	0.005	0.015
Dynamic adsorption	0.026	0.006	0.017

- Alkali can react with calcium and magnesium ions or exchange ions with the clay, functioning as sacrificial agent to protect the polymer and surfactant.
- Alkali can increase the electronegativity of the surface of sandstone, reducing the adsorption amount of polymer and surfactant on the sandstone surface (Table 4.4).

Due to the interactions among all components, the oil displacement efficiency of the combination system is high. Therefore, combination flooding has reduced consumption in chemical agents and lower costs.

4.4.4 Field Test of ASP Flooding

At present, the polymer used in the field test of ASP flooding is partially hydrolyzed polyacrylamide, the surfactant used is sulfonate surfactant, and the alkali used is sodium hydroxide (or sodium carbonate). Their mass concentrations are given as follows.

Polymer 1.5×10^3–2.5×10^3 mg/L
Alkali 1.0×10^4–2.0×10^4 mg/L
Surfactant 2.0×10^3–6.0×10^3 mg/L

The injection amount of ASP flooding ranges between 0.25 and 0.50 times of the pore volume. There have been successful field tests of the ASP flooding (Wyatt et al. 1997; Wang et al. 1996; Wang et al. 1998). Figure 4.24 shows the oil production curve of a field test of ASP flooding. As seen in Fig. 4.24, the effect of the ASP flooding on enhancing oil recovery is obvious.

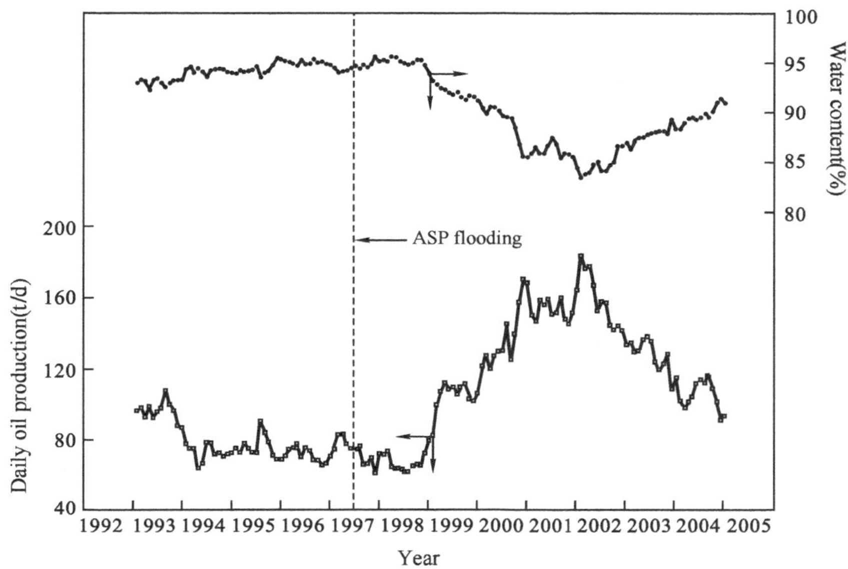

Fig. 4.24 Oil production curve of a field test of ASP flooding

4.5 Miscible Flooding

4.5.1 Concept of Miscible Flooding and Miscible Phase Injectant

Miscible phase means the disappearance of interfaces between phases.

Miscible flooding is an oil displacement method using miscible phase injectant as the displacing agent.

Miscible phase injectant is a kind of material which can form miscible phase with reservoir oil, when injected into the stratum under certain conditions.

There are two kinds of miscible phase injectants.

4.5.1.1 Hydrocarbon Miscible Phase Injectants

Based on the content[1] (mole fraction) of C_2–C_6, the hydrocarbon miscible phase injectants can be divided into liquefied petroleum gas (LPG, with C_2–C_6 content higher than 50%), rich gas (with C_2–C_6 content in the range of 30–50%), and lean

[1]In this chapter, the content refers to mole fraction. If the gas applies to the ideal gas law, the content also refers to volume fraction.

Table 4.5 Compositions of flue gas

Compositions	Contents (%)
CO_2	16.5
N_2	64.6
O_2	5.6
H_2O	13.3

gas (with C_2–C_6 content less than 30%). In lean gas, the gas with C_1 content higher than 98% is called dry gas.

The hydrocarbon gas of C_2–C_6 is called enriched agent, which can facilitate the formation of miscible phase. The generally mentioned gas enrichment refers to the increase in the C_2–C_6 content in gas.

4.5.1.2 Non-hydrocarbon Miscible Phase Injectants

This kind of injectants refers to injectants like CO_2 and N_2.

Flue gas is the industrial waste gas which belongs to the non-hydrocarbon miscible phase injectant. The compositions of flue gas are given in Table 4.5.

In the flue gas, the content of CO_2 ranges between 5 and 20%, which is mainly generated from the coal-fired power stations.

According to the properties of miscible phase injectants, the miscible flooding can be divided into the hydrocarbon miscible flooding and non-hydrocarbon miscible flooding. The former one includes LPG flooding, rich gas flooding, and high-pressure dry gas flooding, while the latter one includes CO_2 flooding and N_2 flooding.

As the LPG flooding and CO_2 flooding are the most representative miscible floodings, only these two floodings are introduced here.

4.5.2 LPG Flooding

The LPG flooding can be illustrated using C_1–C_4–C_{10} tricomponent phase diagram (Fig. 4.25). In the three components, C_4 represents enrichment agent C_2–C_6 and C_{10} represents C_7+ (oil). The content of enrichment agent (C_4) in the LPG is larger than 50%; therefore, in the LPG region of Fig. 4.25, miscible phase can be formed immediately when LPG contacts with the oil (C_{10}). This kind of miscible phase is called first-contact miscibility (FCM). The FCM indicates that no immiscible phase stage exists in the whole process; therefore, it has the highest efficiency.

The dry gas, N_2, and flue gas can be used as driving gases after the LPG, which can form miscible phase with LPG after exceeding certain pressure.

Under the condition of 70 °C and 17.2 MPa, using C_1 as driving gas and 75% C_4 as LPG to displace oil (C_{10}), the whole process of LPG flooding can be shown

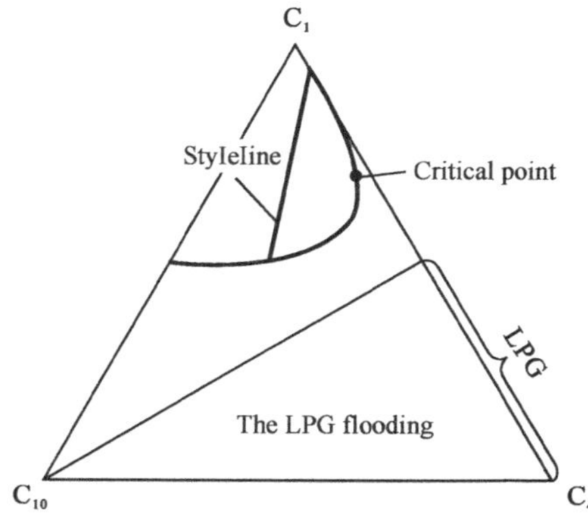

Fig. 4.25 Tricomponent phase diagram of C_1–C_4–C_{10} (70 °C, 17.2 MPa)

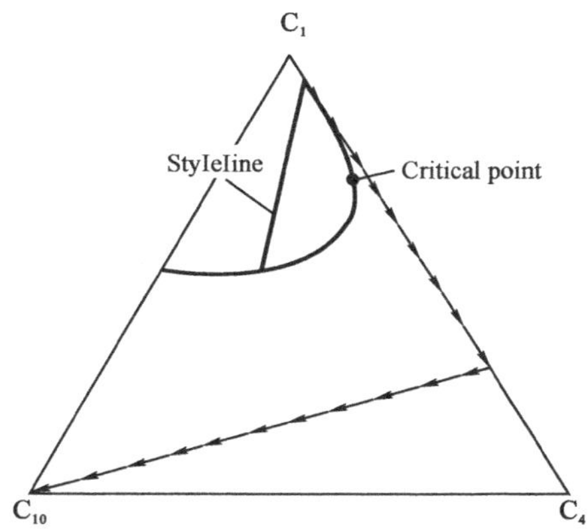

Fig. 4.26 Whole process of LPG flooding (70 °C, 17.2 MPa)

with arrows in Fig. 4.26. From Fig. 4.26, LPG flooding can be used to displace the crude oil without enrichment agents.

LPG can enhance oil recovery through the following mechanisms.

4.5.2.1 Low Interfacial Tension Mechanism

LPG and oil follow the FCM process, which means there are no interfaces in the miscible phases. So the interfacial tension between LPG and oil is zero, indicating LPG has very high oil displacement efficiency.

4.5.2.2 Viscosity Reduction Mechanism

The viscosity of LPG is low. After mixing with oil, it can help in reducing the viscosity and increasing the mobility of oil. The mobility ratio of displacing agent to oil is modified, which is conductive to increase sweep efficiency.

4.5.3 CO_2 Flooding

The miscible process of CO_2 flooding can be analyzed using pseudo-tricomponent phase diagram. The pseudo-tricomponent means that the system is not composed solely of three components. In the process of analyzing CO_2 flooding, CO_2 is regarded as one component, while the reservoir oil is regarded as two components (light component and heavy component). Due to the partial mutual dissolution of CO_2 and heavy component, there exists a two-phase region in the pseudo-tricomponent phase diagram (Fig. 4.27). The system in the two-phase region (e.g., I) can be divided into two phases (1 and 1′) according to the connecting line. Phase 1 is rich in CO_2, while also contains some light components and heavy components. Phase 1′ is rich in heavy components, while also contains light components and CO_2. If all the connecting lines have a concurrent *G* and a tangent line of the two-phase boundary is drawn through *G*, the point of tangency *k* is defined as the critical point.

Figure 4.28 shows the whole generation process of CO_2 miscible phase. In Fig. 4.28, the point O represents the composition of reservoir oil. When CO_2 just contacts with reservoir oil, point I is obtained according to the connection line rule and lever rule, and the quantity ratio of CO_2 to reservoir oil. As point I lies in the two-phase region, it can be divided into two phases (1 and 1′) according to the connecting line rule. The phase 1 continues to move forward and contact with reservoir oil. According to the connection line rule and lever rule, point II is obtained through the quantity ratio of phase 1 to reservoir oil and is separated into two phases (2 and 2′). During the further moving forward and oil contacting of

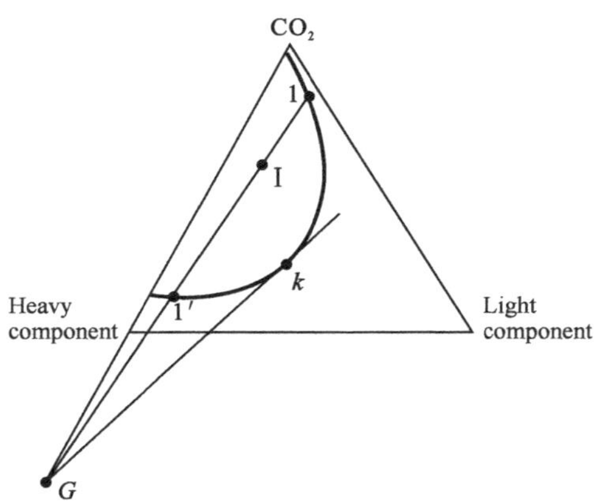

Fig. 4.27 Pseudo-tricomponent phase diagram for analyzing CO_2 flooding

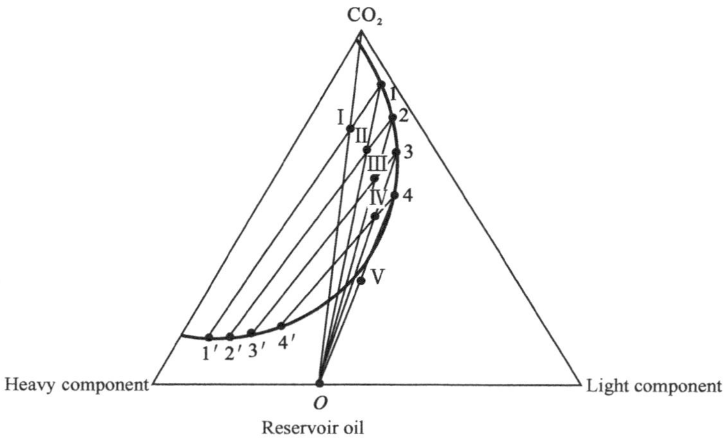

Fig. 4.28 Whole process of CO_2 flooding

phase 2, compositions III and IV are sequentially generated. When the system reaches a composition of point V located in the homogeneous zone (miscible zone), it can function as miscible flooding. In summary, CO_2 flooding can achieve miscible phase after multiple contacts with the reservoir oil. This kind of miscible phase is called multiple contact miscibility (MCM), which is different from the FCM using miscible phase injectant like LPG.

CO_2 can enhance oil recovery through the following mechanisms (Klins and Bardon 1991; Martin and Taber 1992; Mungan 1992).

4.5.3.1 Low Interfacial Tension Mechanism

The process of CO_2 flooding is a continuous enrichment process of CO_2. The enrichment of CO_2 is obtained by its extraction of C_2–C_6 components from oil. The richer the CO_2 is, the lower the interfacial tension between CO_2 and oil, and the higher the oil displacement efficiency can be.

4.5.3.2 Viscosity Reduction Mechanism

CO_2 can dissolve in oil to reduce the viscosity and increase the mobility of oil (Fig. 4.29), which can help in increasing the sweep efficiency of displacement agent.

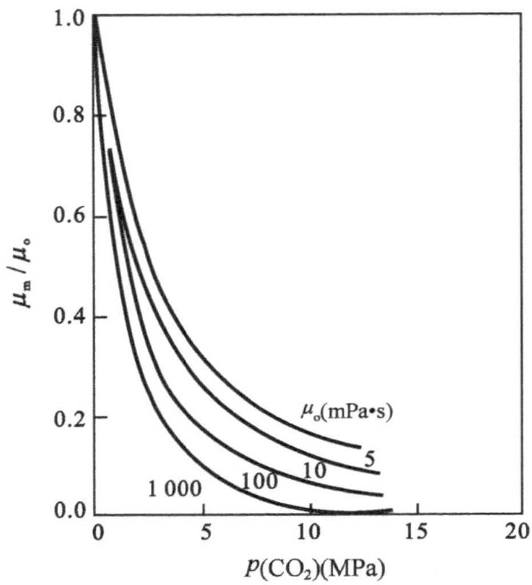

Fig. 4.29 Relationship between viscosity reduction of oil and CO_2 saturation pressure (50 °C). μ_o—the density of crude oil; μ_m—the density of crude oil containing dissolved CO_2

4.5.3.3 Oil Expansion Mechanism

After the solubilization of CO_2 in oil, the volume of oil expands. The expanded oil is easier to be displaced by the displacement media. The expansion degree of oil can be expressed using expansion coefficient. The expansion coefficient is the volume ratio of oil under certain temperature and CO_2 saturation pressure to the oil under the same temperature and 0.1 MPa. Figure 4.30 shows the relation of expansion coefficient of oil to the molar fraction of CO_2 (Lake 1989). It can be observed from Fig. 4.30 that higher molar fraction of CO_2 in oil will lead to larger density, smaller relative molecular weight, and larger expansion coefficient of oil.

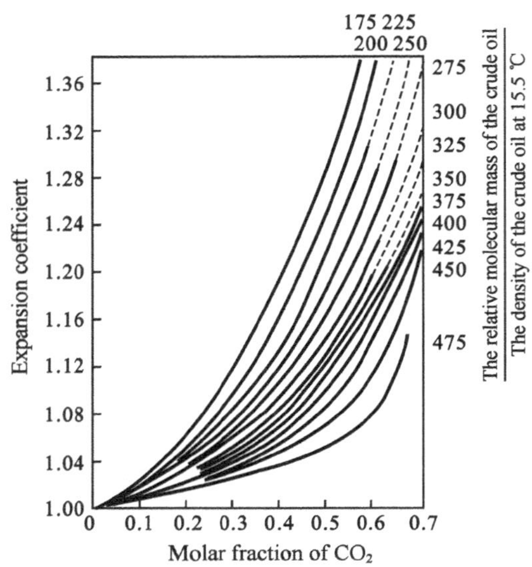

Fig. 4.30 Variation of expansion coefficient of oil with the molar fraction of CO_2

4.5.3.4 Increasing Formation Permeability Mechanism

After the solubilization of CO_2 in water, carbonic acid is generated. The carbonic acid can react with limestone and dolomite in the formation to generate water-soluble bicarbonate, which can increase the formation permeability and the sweep efficiency of flooding media, contributing to enhanced oil recovery.

4.5.3.5 Solubilized Gas Flooding Mechanism

The flooding process from injection well to production well is a pressure-reduction process. With the decrease in pressure, CO_2 can be extracted from oil, which can generate gas driving in the oil layer and increase oil recovery.

In addition, part of CO_2 becomes irreducible gas, which is also conductive to enhancing oil recovery.

References

Baviere M, Glenat P, Plazanet V et al (1995) Improved EOR by use of chemicals in combination. SPE Reserv Eng 10(3):187–193

Bock J, Siano DB, Pace SJ (1992) Enhanced oil recovery with hydrophobically associating polymers. CA 1300362, 12 May 1992

Buckley JS, Liu Y, Monsterleet S (1997) Mechanisms of wetting alteration by crude oils. SPE 37230

Castor TP, Somerton WH, Kelly JF (1981) Recovery mechanisms of alkaline flooding//Shah DO. Surface phenomena in enhanced oil recovery. Plenum Press, New York, pp 249–291

Chen L (1993) Polyacrylamide used for oil displacement. Oilfield Chem 10(3):283–290

Clark SR, Pitts MJ (1988) Design and application of an alkaline-surfactant-polymer recovery system to the West Kiehl field. SPE 17538

Doll TE (1986) An update of polymer-augmented alkaline flood at the Isenhour Unit, Sublette County, Wyoming. SPE/DOE 14954

Fan S, Li Y, Song W et al (2005) The development of oil displacement active agent α-olefin sulfonate from α-olefin. J Daqing Pet Inst 29(3):19–22

Flaaten AK, Nguyen QP, Pope GA et al (2008) A systematic laboratory approach to low-cost, high-performance chemical flooding. SPE 113469

Guo L, Dai C, Wang Y et al (2003) Study on the oi displacement agent used in the profile controlling and flooding test in Ng 3^2 of Northwest Chengdong oilfield east zone. J China Univ Pet (Ed Nat Sci) 27(2):67–69

Hawkins BF, Taylor KC, Nasr-El-Din HA (1994) Mechanisms of surfactant and polymer enhanced alkaline flooding: application to David Lloydminster and Wainwright Sparky fields. JCPT 33(4):52–63

Hsieh WC, Koepke JW (1985) Process for enhanced oil recovery employing synthetic sulfonates and petroleum sulfonate blends. US Patent 4,530,400, 23 July 1985

Huang H, Yang Y, Tang J et al (1991) The development of oil displacement agent petroleum sulfonate. Oilfield Chem 8(3):235–239

Hurk JH (1983) Comparison of sodium carbonate, sodium hydroxide and sodium orthosilicate for EOR. SPE 12039

Jennings HY Jr. (1974) A caustic water flooding process for heavy oils. JPT 26(12):1344–1352

Kamath KI, Comberiati JR, Zammerilli AM et al (1981) Tertiary oil recovery by flooding with aqueous chemical solution. SPE 9929
Kang W, Meng L, Niu J et al (2007) Mechanisms of the influence of degree of mineralization on the viscosity of HPAM solution. Polym Mater Sci Eng 22(5):175–181
Kang W, Xu D, Tan J et al (2006) The synthesis and interfacial tension of fatty alcohol polyoxyethylene polyoxypropylene ether carboxylate. Pet Nat Gas Chem 35(4):304–306
Klins MA, Bardon CP (1991) Carbon dioxide flooding//Baviere M. Basic concepts in enhanced oil recovery processes. Elsevier Applied Science, London and New York, pp 216–217
Krumrine PH, Falcone JS Jr. (1983) Surfactant, polymer, and alkali interactions in chemical flooding processes. SPE 11778
Labrid J (1991) The use of alkaline agents in enhanced oil recovery processes//Baviere M. Basic concepts in enhanced oil recovery processes. Elsevier Applied Science, London and New York, pp 124–125
Lake LW (1989) Enhanced oil recovery. Prentice Hall, New Jersey, pp 257–259
Lake LW (1989) Enhanced oil recovery. Prentice Hall, New Jersey, pp 343–344
Levitt DB, Jackson AC, Heinson C et al (2006) Identification and evaluation of high-performance EOR surfactants. SPE 100089
Li Y, Zhao F, Wang Y (2003) The synthesis of alkyl phenol polyoxyethylene ether carboxylate using acetone as solvent. Acta Pet Sin (Pet Process) 19(2):33–38
Liu P, Yang N, Niu Y (2003) The synthesis of hydrophobic association AM/C10AM/AMPS terpolymer and its viscosity properties in saline solution. Oilfield Chem 20(1):40–42
Lorenz PB, Peru DA (1989) Guidelines help select reservoir for $NaHCO_3$ EOR. Oil Gas J 87 (37):53–57
Martin FD, Taber JJ (1992) Carbon dioxide flooding. JPT 44(4):396–400
Mayer EH, Weinbrandt RM, Irani MR (1982) Alkaline waterflooding: its theory, application and status. In: Transactions of 1982 European symposium on enhanced oil recovery, pp 191–202
Mccoy DR (1984) Oil recovery method utilizing an alkylarylpolyalkoxyalkylene sulfonate. US Patent 4,465,602, 14 Aug 1984
Mei Y (1982) The salt sensitivity effect of partially hydrolyzed polyacrymide. Acta Pet Sin 3 (1):41–49
Mungan N (1992) Carbon dioxide flooding as an enhanced oil recovery process. JCPT 31(9):13–15
Neil JD, Chang HL, Geffen TM (1983) Waterflooding and improved waterflooding//Poettman FH. Improved oil recovery. Interstate Oil Compact Commission, Oklahoma City, Oklahoma, pp 56–58
Ren M (2001) The optimum synthesis of petroleum sulfonate KPS. Oilfield Chem 18(4):330–333
Rosen MJ (1993) Gemini: a new generation of surfactants. Chem Technol 23(3):30–33
Rudin J, Wasan DT (1993) Surfactant enhanced alkaline flooding: buffering at intermediate alkaline pH. SPE Reserv Eng 8(4):275–280
Schievelbein VH (1981) Surfactant waterflooding oil recovery method. US Patent 4,307,782, 29 Oct 1981
Shuler PJ, Kuehne DL, Lerner RM (1989) Improving chemical flood efficiency with micellar/alkaline/polymer processes. JPT 41(1):80–88
Smith RV (1988) Enhanced oil recovery update, part 1, improvement of sweep efficiency. Pet Eng Int 60(11):29–33
Surkalo H (1990) Enhanced alkaline flooding. JPT 42(1):6–7
Symonds RWP, Farouq Ali SM, Thomas S (1991) A laboratory study of caustic flooding for two Alberta crude oils. JCPT 30(1):44–49
Talley LD (1988) Hydrolytic stability of alkylethoxy sulfates. SPE Reserv Eng 3(1):235–242
Taylor KC, Hawkins BF, Rafiq Islam M (1990) Dynamic interfacial tension in surfactant enhanced alkaline flooding. JCPT 29(1):50–55
Wang Y, Zhao F (1996) The study on the synthesis of sodium fatty alcohol polyoxyethylene ether sulfonate from AES. Fine Petrochem Technol 5:22–24

Wang Y, Zhao F (2006) The synthetic method of a kind of ether carboxylate for oil displacement. China Patent 00,129,383.4, 27 Sept 2006
Wang H, Yang Y, Zhang G et al (2005) Biosurfactant and its application. Petroleum Industry Press, Beijing, pp 72–217
Wang F, Wu X, Chen G et al (2009) Advances on the ternary composite flooding technology in Daqing oilfield. Pet Geol Oilfield Dev Daqing 28(5):154–162
Wang D, Zhang Z, Chang J et al (1996) Pilot test of alkaline surfactant polymer flooding in Daqing oil field. SPE 36748
Wang D, Cheng J, Wu J et al (1998) An alkaline/surfactant/polymer field test in a reservoir with a long-term 100% water cut. SPE 49018
Wernau WC (1981) Fermentation process for production of xanthan. US Patent 4,282,321, 4 Aug 1981
Wyatt K, Pitts MJ, Harry VS et al (1995) Alkaline-surfactant-polymer technology potential of the Minnelusa trend, powder river basin. SPE 29565
Wyatt K, Pitts M, Surkalo H (1997) ASP flooding produces incremental oil at moderate cost. World Oil 218(7):87, 90, 92
Zhao G (1997) Advances on the surfactant science. Acta Phys Chim Sin 13(8):760–768
Zhao F, Zhu T, Wang X (1985) The phosphate salinization and carboxymethylation of OP-10. J East China Pet Inst 9(1):60–75
Zhao F, Zhang Y, Xu Y (1995) Phase diagram of microemulsion with intermediate phase. J East China Pet Inst 9(4):39–51

Chapter 5
Profile Control and Water Shutoff

Profile control in water injection wells and water shutoff in production wells are important technologies for enhanced oil recovery. Also they are indispensable supplemental technologies for other enhanced oil recovery technologies such as chemical flooding, miscible flooding, thermal recovery, and microbial flooding. Therefore, a comprehensive and thorough understanding of profile control in water injection wells and water shutoff in oil wells is required.

5.1 Profile Control in Water Injection Well

5.1.1 Concept of Profile Control

Formation is heterogeneous in permeability. The injected water is often absorbed by the thin and high permeable zone, so the water injection profile is non-uniform (Fig. 5.1).

In order to perform the function of low and medium permeable zones, high permeable zone must be shut off to adjust the water injection profile. The usually called profile control is to adjust the water absorption profile of water injection formation in water injection wells.

5.1.2 Importance of Profile Control

The importance of profile control is that it can improve oil recovery, which can be explained by Figs. 5.2 and 5.3.

C. Dai and F. Zhao, *Oilfield Chemistry*,
https://doi.org/10.1007/978-981-13-2950-0_5

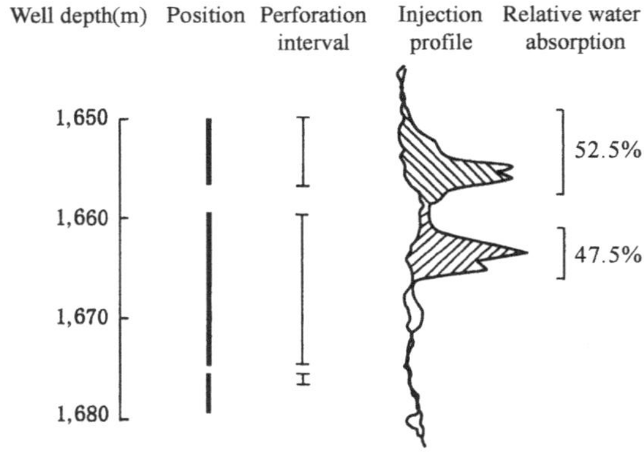

Fig. 5.1 Water injection profile of a water injection well

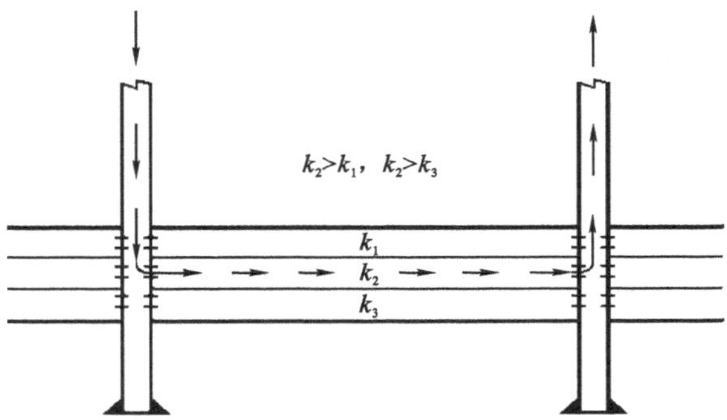

Fig. 5.2 Injected water flows into the oil well along the high-permeability zone

Fig. 5.3 Profile control in water injection wells

It is demonstrated in Fig. 5.2 that the injected water will inevitably flow into the production wells through high permeable zone, which will reduce the sweep efficiency of the injected water and decrease oil recovery efficiency.

Figure 5.3 also shows that the injected water can be forced to flow into the medium and low permeable zones by plugging high-permeability zone, which will improve the sweep efficiency of injected water and enhance oil recovery efficiency.

5.1.3 Profile Control Agent

Water injection well profile control agents (simplified as profile control agents) are the materials injected from injection wells into formations that are used to control water injection profile.

5.1.3.1 Classification of Profile Control Agents

The profile control agents can be classified according to different standards.

According to the injection technology, they can be classified into single-fluid profile control agent (e.g., chromium gel) and double-fluid profile control agent (e.g., water glass/$CaCl_2$ composite) (Fig. 5.4). Profile control agents are generally classified based on this standard.

According to the plugging distance, they can be classified into near wellbore profile control agent (e.g., silicic acid gel) and in-depth profile control agent (colloidal dispersion gel).

According to the application conditions, they can be classified into high-permeability zone profile control agent (e.g., clay/cement solid system), low-permeability zone profile control agent (e.g., ferrous sulfate), and high-temperature and high-salinity zone profile control agent (e.g., various inorganic profile control agents).

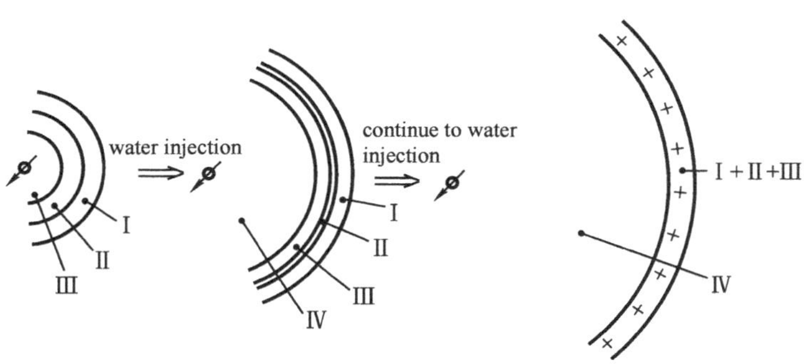

Fig. 5.4 Double-fluid profile control. I—first working fluid; II—spacer fluid; III—second working fluid; IV—the injected water; ⌀—the injection well, +—the plugging material generated at the contact of I and II

5.1.3.2 Some Important Single-Fluid Profile Control Agents

Sulfuric Acid

Sulfuric acid can produce profile control materials using calcium and magnesium sources from the formation. If concentrated sulfuric acid or chemical waste liquid containing concentrated sulfuric acid is injected into well, sulfuric acid will first react with the carbonate (carbonate contained in rock mass or cement) near wellbore to improve the water absorbing capacity of injection well. The calcium sulfate and magnesium sulfate generated from the reaction will enter the formation with acid and precipitate out to form pluggings in the appropriate position (e.g., throat of pore structure). Because most of sulfuric acids enter the high permeable zone to generate calcium and magnesium sulfate, the major plugging happens in high permeable zone.

Ferrous Sulfate

Ferrous sulfate can hydrolyze in water to generate ferrous hydroxide and sulfuric acid (Liao and Zhao 1994).

$$FeSO_4 + 2H_2O \rightarrow Fe(OH)_2\downarrow + H_2SO_4$$

As shown in this reaction, sulfuric acid can control profile as mentioned above, while ferrous hydroxide as a kind of precipitation can also control profile. As long as there is ferrous sulfate with the consumption of sulfuric acid in the formation, ferrous hydroxide precipitation would be constantly produced. As most ferrous sulfate enters the high permeable zone, the plugging effects of control agents mainly function in the high permeable zone.

Silicic Acid Gel

Silicic acid gel is generated by the reaction of water glass and activating agent. Water glass is also called sodium silicate, and its formula is $Na_2O{\cdot}mSiO_2$ where "m" is the molar ratio of SiO_2 to Na_2O. The molar ratio ranges between 1 and 4. The properties of sodium silicate change with the molar ratio. With smaller molar ratio, the water glass would obtain greater alkalinity and higher solubility. Activating agent refers to the kind of substance that can turn water glass first into sol and then into gel. There are two types of activating agents, which are inorganic activating agents such as hydrochloric acid, nitric acid, sulfuric acid, amino sulfonic acid, ammonium carbonate, ammonium bicarbonate, ammonium chloride, ammonium sulfate, and sodium dihydrogen phosphate, and organic activating agents such as formic acid, acetic acid, ammonium acetate, ethyl formate, ethyl acetate, chloroacetic acid, trichloroacetic acid, oxalic acid, citric acid, phenol, catechol, resorcinol and hydroquinone, phloroglucinol, formaldehyde, and urea.

Silica gel usually uses hydrochloric acid as activating agent. Its reaction with sodium silicate is shown as follows (Barbosa et al. 1987; Zhao et al. 1988; Lu et al. 2004; Nasr-El-Din and Taylor 2005).

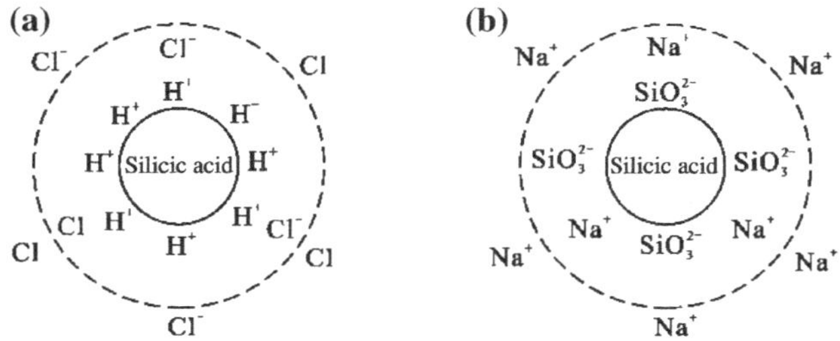

Fig. 5.5 Silica sol. **a** Acidic silica sol; **b** Alkaline silica sol

$$Na_2O \cdot mSiO_2(\text{Water glass}) + 2HCl \rightarrow H_2O \cdot mSiO_2(\text{Silicic acid}) + 2NaCl$$

Two kinds of silica gels, acidic silica gel and alkaline silicic acid gel, can be obtained based on different preparation methods. The former one is obtained by adding water glass into hydrochloric acid under the condition of excess H^+. According to Fajans rule, the structure shown in Fig. 5.5a is formed and the surface of colloidal particle is positively charged. The latter one is obtained by adding hydrochloric acid into water glass under the condition of excess silicate. If the molar ratio of water glass is 1 and the silicate is SiO_3^{2-}, according to Fajans rule, the structure shown in Fig. 5.5b is formed and the colloidal particle surface is negatively charged. Both of silicic acid gels can form silica sol first under certain temperature, pH, and silicate content, and then form silica gel under certain period of time. For example, silica sol with the pH value of 1.5 can be prepared by adding water glass with $w(Na_2O{\cdot}3.43\ SiO_2)$ of 4% to the hydrochloric acid with $w(HCl)$ of 10%. This sol can be turned into silica gel after 8 h at 70 °C to plug high permeable zone.

Aluminum Hydroxide Gel (Busolo et al. 2001; Davies et al. 2002)
Aluminum hydroxide gel is generated by injecting solution prepared by aluminum chloride and urea into formation. The urea decomposes under formation temperature to change the solution from acidic to alkaline. Aluminum hydroxide sol is then generated and turned into aluminum hydroxide gel afterward.

Zirconium Gel (Zhao and Li 1996)
Zirconium gel is generated by the crosslinking of multinuclear olation complex ions composed of Zr^{4+} with polymers containing $-COO^-$ (such as HPAM) in solution.

Zr^{4+} can come from $ZrOCl_2$ or $ZrCl_4$.

Zr^{4+} can generate multinuclear olation complex ions of zirconium through the following steps.

- Complexation

$$Zr^{4+} + 8\,H_2O \rightarrow \left[(H_2O)_8\,Zr\right]^{4+}$$

- Hydrolyzation

$$\left[(H_2O)_8Zr\right]^{4+} \rightarrow \left[(H_2O)_7Zr(OH)\right]^{3+} + H^+$$

- Olation

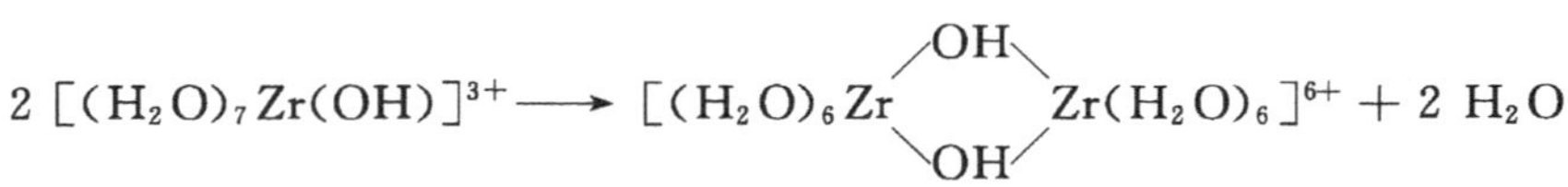

- Further hydrolysis and olation

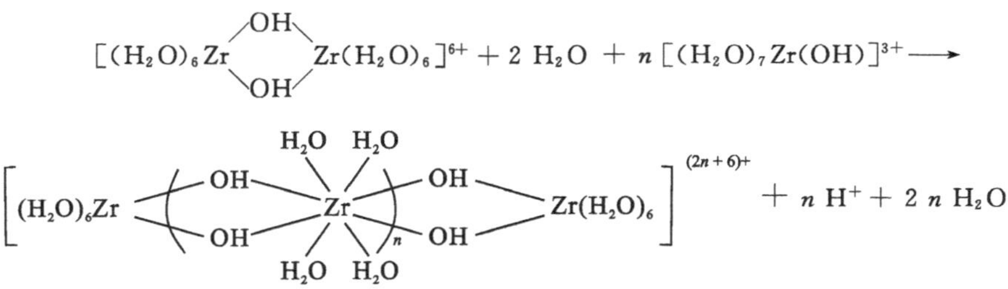

(Mutinuclearolation complex ions of zirconium)

Below are the crosslinks generated by the crosslinking of multinuclear olation complex ions of zirconium with polymers containing –COO⁻ (such as HPAM).

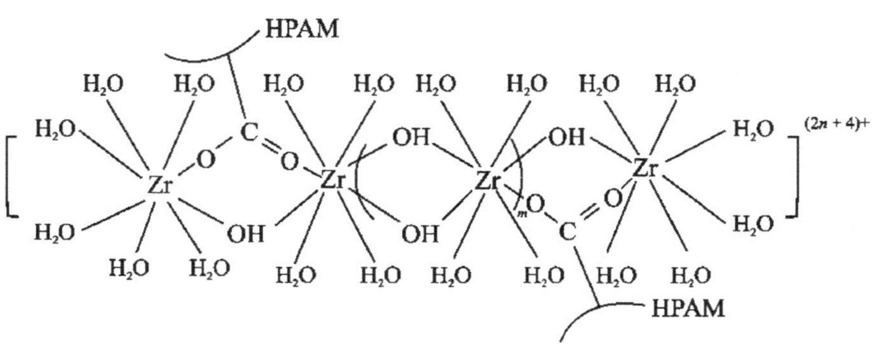

These crosslinks are called zirconium gel. For example, by mixing solutions with *w*(HPAM) of 0.75%, *w*($ZrOCl_2$) of 1.0%, and *w*(HCl) of 5.5% by volume ratio of 100:4:3, a kind of zirconium gel whose gelation time under 60 °C is 7 h can be prepared for plugging high permeable zone.

Chromium Gel (Mccool et al. 1991; Batarajan et al. 1998; Dolan et al. 1998)
Chromium gel is generated by the crosslinking of multinuclear olation complex ions composed of Cr^{3+} with polymers containing $-COO^-$ (such as HPAM) in solution.

Cr^{3+} can come from $KCr(SO_4)_2$, $CrCl_3$, $Cr(NO_3)_3$, and $Cr(CH_3COO)_3$, or be obtained from Cr^{6+} (such as $K_2Cr_2O_7$, $Na_2Cr_2O_7$) by reduction reaction using reducing agents (such as $Na_2S_2O_3$, Na_2SO_3, and $NaHSO_3$).

Cr^{3+} can also generate multinuclear olation complex ions of chromium through complexation, hydrolyzation, olation, and further hydrolysis and olation.

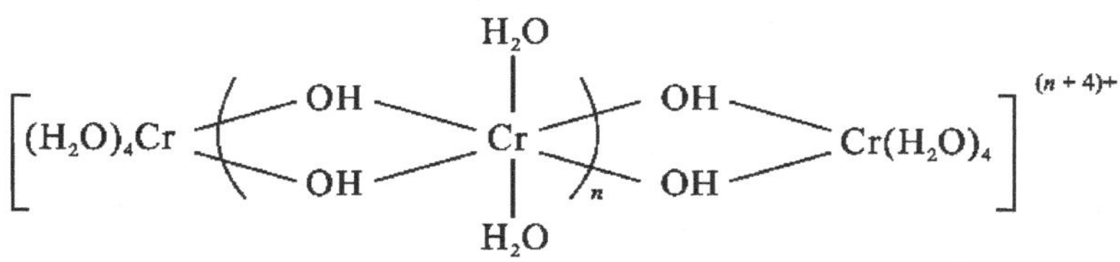

Below are the crosslinks generated by the crosslinking of multinuclear olation complex ions of chromium with polymers containing $-COO^-$ (such as HPAM).

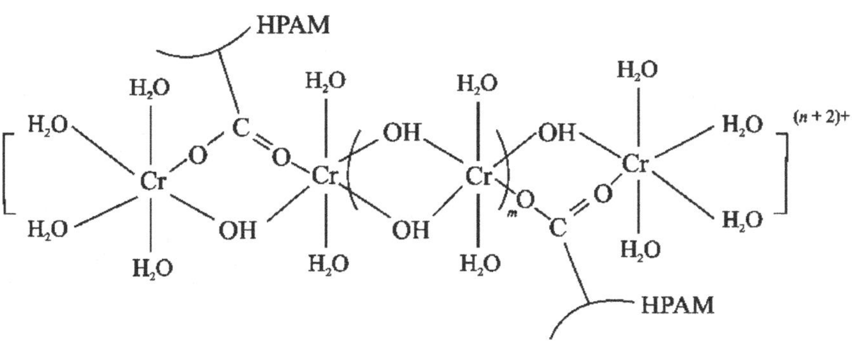

These crosslinks are called chromium gel. For example, when *w*(HPAM) is 0.4%, *w*($Na_2Cr_2O_7$) is 0.09%, and *w*(Na_2SO_3) is 0.16%, a kind of chromium gel whose gelation time under 60 °C is 2 h can be prepared for plugging high permeable zone.

Aluminum Gel

Aluminum gel is generated by the crosslinking of multinuclear olation complex ions composed of Al^{3+} with polymers (such as HPAM) containing $-COO^{-}$ in solution.

Al^{3+} can be from aluminum citrate.

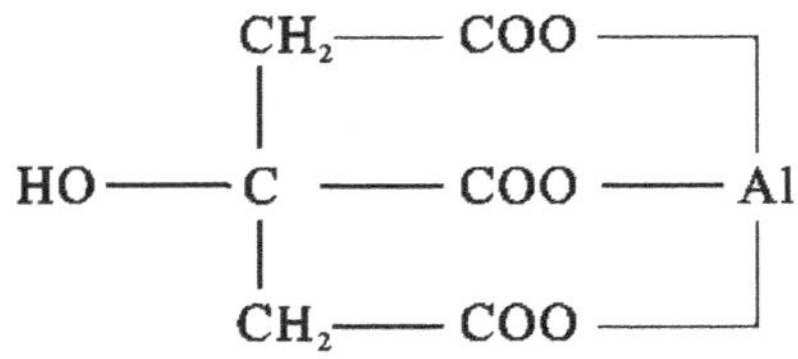

Due to the low strength of aluminum gel, it is usually prepared into colloidal dispersion gel (CDG) for application (Mack and Smith 1994; Fielding et al. 1994).

CDG is prepared with polymers of low mass concentration and cross-linkers of low mass concentration. The mass concentration of the polymer is in the range of 100–1,200 mg/L, and the mass concentration ratio of polymer to cross-linker is in the range of 20:1–100:1. Due to the low mass concentration, polymer and cross-linker cannot form a continuous network. They can only form the gel bundle slowly.

Gel bundles are formed by a small amount of polymer molecules which are cross-linked intramolecularly and/or intermolecularly by cross-linker; therefore, CDG is the dispersion of gel bundle. After the formation of the gel bundle, the flow resistance of CDG increases. If the differential pressure can overcome the flow resistance, CDG can flow, otherwise the CDG stops flowing to perform the plugging function. As CDG has low mass concentration, low cost, and can be used in large dosage, it is applicable for profile control in zones far from wellbore. The small differential pressure in zones far from wellbore is also beneficial for CDG to perform its plugging function.

Zirconium gel and chromium gel can also be prepared into CDG.

Phenolic Resin Gel

Phenolic resin gel is generated by the crosslinking of phenolic resin with polymers containing $-CONH_2$ (such as HPAM) in solution.

Phenolic resin is generated by condensation polymerization of formaldehyde and phenol under the catalysis of sodium hydroxide.

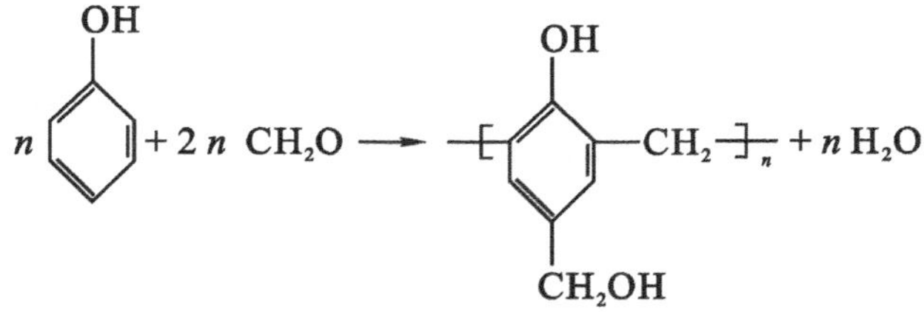

(Phenolic resin)

The $–CH_2OH$ of phenolic resin can cross-link with the $–CONH_2$ of polymer through dehydration reaction. The crosslinks generated by the crosslinking reaction are called phenolic resin gel. For example, when *w*(HPAM) is 0.4% and *w*(phenolic resin) is 0.8%, a kind of phenolic resin gel whose gelation time is 72 h under 80 °C can be prepared for plugging high permeable zone.

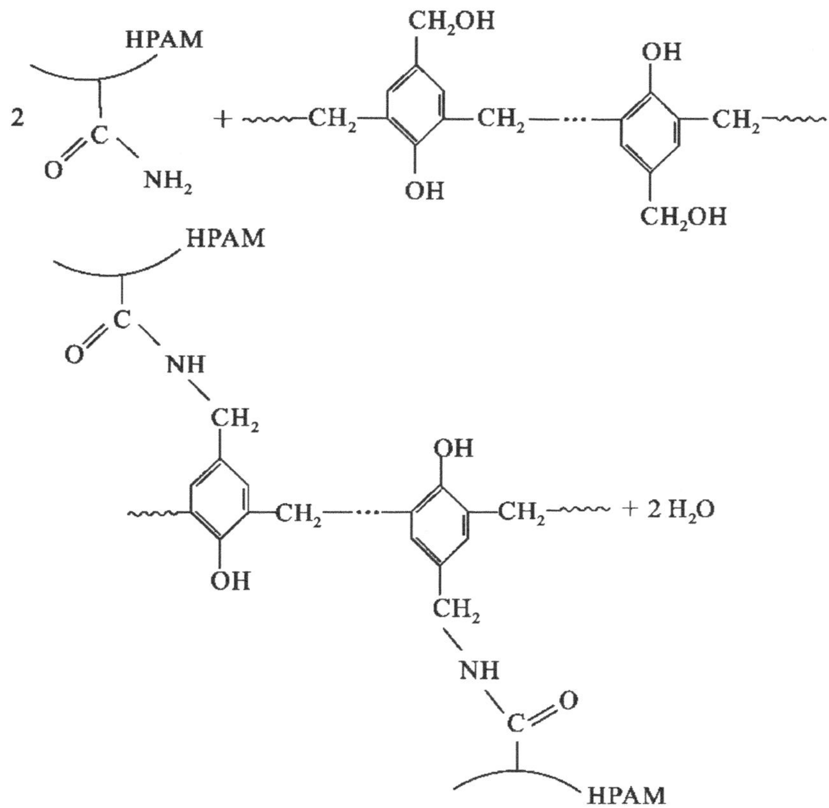

Polyethylene Imine Gel (Hardy et al. 1999; Eoff et al. 2007)
Polyethylene imine gel is generated by the crosslinking of polyethylene imine with polymers containing $–CONH_2$ (such as HPAM) in solution.

Polyethylene imine (PEI) is generated by the polymerization of ethylene imine.

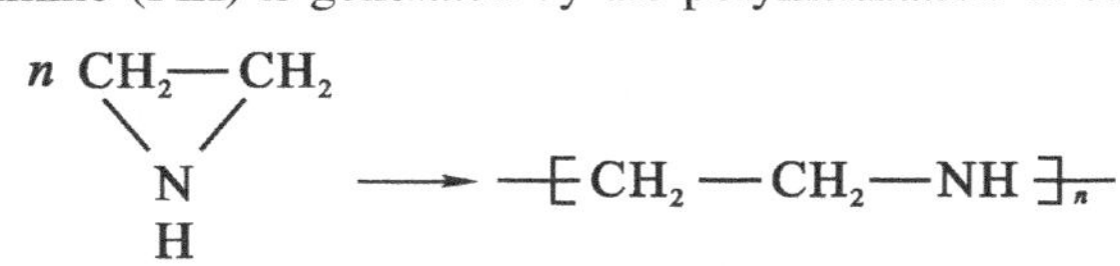

(Ethylene imine) (Polyethylene imine, PEI)

The imido (–NH–) of polyethylene imine can cross-link with the $-CONH_2$ of polymer by removing NH_3.

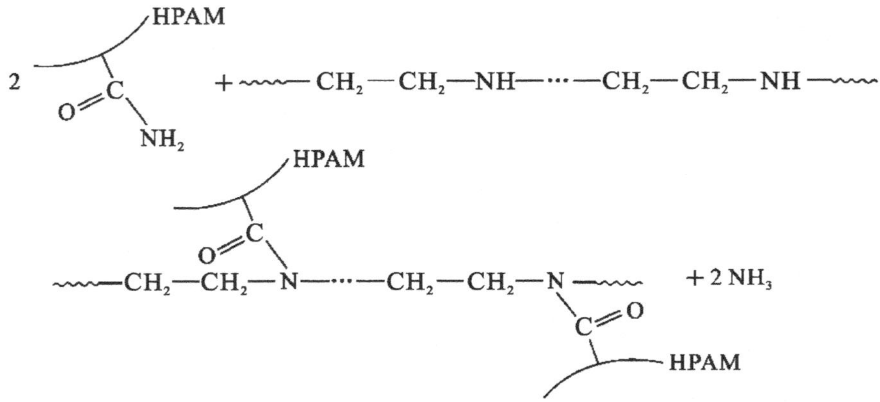

These crosslinks generated by the crosslinking reaction are polyethylene imine gel.

The copolymer of acrylamide and butyl acrylate (PA-*t*-BA) can also be used to prepare this kind of gel.

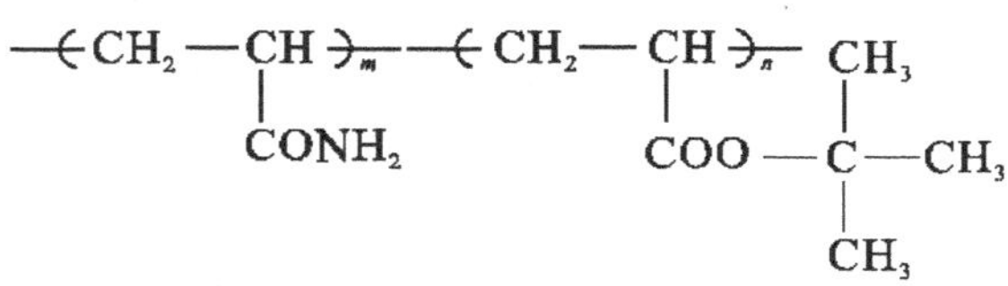

The following crosslinking reaction exists between the copolymer and polyethylene imine.

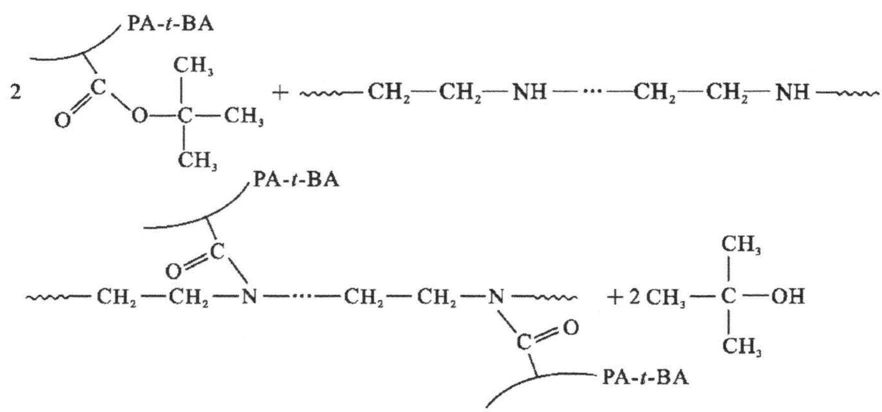

Fig. 5.6 Changes of the swelling ratio of water-swelling polyacrylamide in water as a function of time (30 °C)

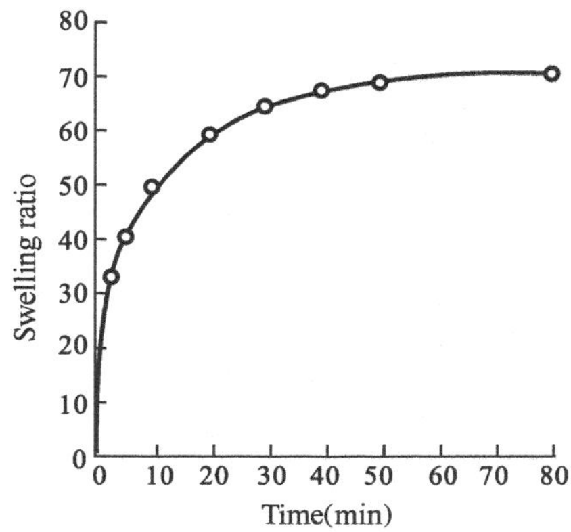

For example, when *w*(PA-*t*-BA) is 7.0% and *w*(PEI) is 0.33%, a kind of polyethylene imine gel whose gelation time is 12 hour under 95 °C can be prepared for plugging high permeable zone.

Water-Swelling Polymer (Dong 1987; Frampton 1997; Seright 1997; Bai et al. 2004)

Water-swelling polymer is a kind of properly crosslinked polymer which will expand rather than dissolve in water. For example, by adding small amount of crosslinker N, N′-methylene bisacrylamide during the polymerization of acrylamide and then drying and grinding, the water-swelling polyacrylamide is prepared. The swelling rate and swelling ratio of this kind of water-swelling polymer are both high (Fig. 5.6).

All the water-soluble polymers which are properly crosslinked can be used to prepare water-swelling polymers.

There are two methods to place the water-swelling polymers to deep formations. One method is to choose proper carrying media including kerosene, ethanol, and electrolyte solutions (such as sodium chloride solution and ammonium chloride solution) which can inhibit the swelling of the polymers. The other one is to coat a film (e.g., hydroxypropyl methylcellulose film) on the outer surface of the polymers. When the coated water-swelling polymers are carried into formation by water, they will not swell until the coatings dissolve in water. Coatings on the surface of the water-swelling polymers can be achieved by fluidized bed method (Wurster method (Wurster 1953)).

Gel Microspheres

Gel microspheres are dispersed gel with particle sizes reaching nanoscale. They can be prepared by microemulsion polymerization method (Sun et al. 2006). For example, the solution prepared by acrylamide or other alkenyl-containing monomers, *N, N′*-methylene-bis-acrylamide, and ammonium persulfate can be used to prepare oil-external microemulsion using mixed surfactant with high concentration (e.g., Span 80 + Tween 60). The aqueous solution solubilizes in the micelles of the microemulsion. The monomers copolymerize to form gel microspheres. In the process of utilization, the oil-external gel microspheres can be dispersed in water and injected into the formation by using phase-reversal agents (e.g., OP-10). In the formations, gel microspheres have certain swelling ratio, so they can conduct the profile control from the near to the distant by mechanisms of migrating, trapping, distorting, and remigrating, retrapping, redistorting.

Lime Cream

Lime cream is generated by dispersing calcium oxide in water. Since calcium oxide can react with water to form calcium hydroxide which has a low solubility in water (at 60 °C, 100 g water can dissolve 0.116 g calcium hydroxide), lime cream is the suspension of calcium hydroxide in water.

$$CaO + H_2O \rightarrow Ca(OH)_2$$

In the lime cream, $w(CaO)$ is general between 5% and 10%. This single-fluid profile control agent has the following characteristics.

- The particle size of calcium hydroxide is relatively large (around 62 μm); therefore, lime cream is particularly suitable for plugging the high permeable zone with cracks. Because calcium hydroxide particles cannot enter the medium- and low-permeability formations, lime cream can protect these formations.
- The solubility of calcium hydroxide decreases with increasing temperature (Table 5.1); therefore, it can be used to plug high-temperature formations.
- Calcium hydroxide can turn into water-soluble calcium chloride by reacting with hydrochloric acid. Therefore, the plugging can be removed with hydrochloric acid when it is not needed.

$$Ca(OH)_2 + 2HCl \rightarrow CaCl_2 + 2H_2O$$

Clay/Cement Dispersing System

Clay/cement dispersing system is generated by the suspending of clay and cement in water. This dispersing system is suitable for plugging extra-high-permeability zones. After the entrance of clay and cement into the formations, filter cake can be generated in the formation (mainly in the throat of pore structure). In the filter cake,

Table 5.1 Solubility of calcium hydroxide at different temperatures

Temperature (°C)	Solubility (g/kg)	Temperature (°C)	Solubility (g/kg)
30	1.53	70	1.06
40	1.41	80	0.94
50	1.28	90	1.85
60	1.16	100	1.77

the hydration reaction of cement facilitates the solidification of filter cake, which can generate effective plugging to extra-high-permeability zones.

In the clay/cement dispersion, both *w*(cement) and *w*(clay) are between 6% and 20%.

Similar to the calcium hydroxide in lime cream, clay and cement cannot enter the medium- and low-permeability formations, so they can protect medium- and low-permeability formations.

If needed, the plugging produced by clay/cement dispersions can be removed by conventional mud acid whose *w*(HCl) is 12% and *w*(HF) is 3%.

Besides clay/cement dispersions, calcium carbonate/cement dispersions and coal ash/cement dispersions can also be used to plug extra-high-permeability formation.

5.1.3.3 Some Important Double-Fluid Profile Control Agents

Precipitation-Typed Double-Fluid Profile Control Agents

This refers to the double-fluid profile control agent whose two working fluids after contacting with each other will generate precipitation to plug the high-permeability zone (Zhao et al. 1987; Zhao et al. 1991; Lakatos et al. 2000). Some examples are shown below.

Example 1:

First working fluid: solution with $w(Na_2CO_3)$ ranging between 5% and 20%.

Second working fluid: solution with $w(FeCl_3)$ ranging between 5% and 30%.

The reactions after they encounter are as follows.

$$3Na_2CO_3 + 2FeCl_3 \rightarrow Fe_2(CO_3)_3\downarrow + 6NaCl$$

$$Fe_2(CO_3)_3 + 3H_2O \rightarrow 2Fe(OH)_3\downarrow + 3CO_2\uparrow$$

Example 2:

First working fluid: solution with $w(Na_2O \cdot mSiO_2)$ ranging between 1 and 25%.

Second working fluid: solution with $w(FeSO_4)$ ranging between 5 and 13%.

The reaction after they encounter is as follows.

$$Na_2O \cdot mSiO_2 + FeSO_4 \rightarrow FeO \cdot mSiO_2\downarrow + Na_2SO_4$$

Example 3:

First working fluid: solution with $w(Na_2O \cdot mSiO_2)$ ranging between 1 and 25%.

Second working fluid: solution with $w(CaCl_2)$ ranging between 1 and 20%.

The reactions after they encounter are as follows.

$$Na_2O \cdot mSiO_2 + CaCl_2 \rightarrow CaO \cdot mSiO_2\downarrow + 2NaCl$$

Example 4:
First working fluid: solution with $w(Na_2O{\cdot}mSiO_2)$ ranging between 1 and 25%.
Second working fluid: solution with $w(MgCl_2)$ ranging between 1 and 15%.
The reaction after they encounter is as follows.

$$Na_2O \cdot mSiO_2 + MgCl_2 \rightarrow MgO \cdot mSiO_2 \downarrow + 2NaCl$$

Example 5:
First working fluid: solution with 10% $w(Na_2O{\cdot}mSiO_2)$.
Second working fluid: solution with 6% $w(HCl)$.
First, the hydrochloric acid in the second working fluid reacts with the calcium carbonate and magnesium carbonate in the formation to generate the calcium chloride and magnesium chloride, which react further with the sodium silicate in the first working fluid to generate calcium silicate and magnesium silicate precipitates to plug the high-permeability layers.

Gel-Typed Double-Fluid Profile Control Agents

This refers to the profile control agent which can generate gel to plug the high-permeability layers after the two kinds of working fluids contact with each other. For example, if sodium silicate and ammonium sulfate are injected into the formation alternately, separated by the spacer fluid (e.g., water), the two kinds of working fluids can have the following reactions after contacting with each other to generate gel to plug the high-permeability layers.

$$Na_2O \cdot mSiO_2 + (NH_4)_2SO_4 + 2\,H_2O \rightarrow$$
$$H_2O \cdot mSiO_2 \text{ (can change from sol to gel) } + Na_2SO_4 + 2NH_4OH$$

Gelly-Typed Double-Fluid Profile Control Agents

This refers to the profile control agent which can generate gel to plug high-permeability formation after the mixing of two kinds of working liquids. In the two working fluids, usually one is polymer solution and the other is cross-linker solution.

Some examples for gelly-typed double-fluid profile control agents are shown as follows.

Example 1:
First working fluid: HPAM solution or XC solution.
Second working fluid: Aluminum citrate solution.
Aluminum gel is generated after the mixing of these two kinds of working fluids.
Example 2 (Hudson and Shioyama 1987):
First working fluid: HPAM solution or XC solution.
Second working fluid: Propionic acid chromium solution.
Chromium gel is generated after the mixing of these two kinds of working fluids.
Example 3 (Zhao et al. 1989):
First working fluid: HPAM solution or XC solution containing Na_2SO_3.
Second working fluid: HPAM solution or XC solution containing $Na_2Cr_2O_7$.

Cr^{6+} in $Na_2Cr_2O_7$ is reduced to Cr^{3+} by Na_2SO_3 after the mixing of these two kinds of working fluids. Cr^{3+} can further generate multinuclear olation complex ions to crosslink polymers to form chromium gel.

Example 4 (Zhu et al. 1989):

First working fluid: HPAM solution.

Second working fluid: $ZrOCl_2$ solution.

Zirconium gel is generated after the mixing of these two kinds of working fluids.

Example 5 (Kang et al. 2002):

First working fluid: HPAM solution.

Second working fluid: Polyquaternium solution.

Polyquaternium gel is generated after the mixing of these two kinds of working fluids.

Foam-Typed Double-Fluid Profile Control Agents

If foaming agent solution and gas are injected into the formation alternatively, foam can be produced in the formation (mainly in high permeable formation) to work as profile control agent. Available foaming agents include non-ionic surfactant (e.g., polyoxyethylene alkyl phenol ether) and anionic surfactant (e.g., alkylaryl sulfonate). Available gases include nitrogen and carbon dioxide.

Flocculation-Typed Double-Fluid Profile Control Agents

If the clay suspension and HPAM solution are divided into several slugs separated by spacer fluids and injected into the formation alternately, flocs will be formed when they contact with each other. These flocs can effectively plug high permeable layer.

5.1.4 Selection of Profile Control Agents

5.1.4.1 High-Permeability Zone

For high-permeability zone, profile control agents such as zirconium gel, chromium gel, phenolic resin gel, water-swelling polymer, lime cream, clay/cement dispersion system, precipitation-typed double-fluid profile control agent, foam-typed double-fluid profile control agent, and flocculation-typed double-fluid profile control agent can be used.

5.1.4.2 Low-Permeability Zone

For low-permeability zone, profile control agents such as sulfuric acid, ferrous sulfate, gel microsphere, gel-typed double-fluid profile control agent, and precipitation-typed double-fluid profile control agent can be selected.

5.1.4.3 High-Temperature and High-Salinity Zone

For high-temperature and high-salinity zone, inorganic profile control agents such as sulfuric acid, ferrous sulfate, lime cream, clay/cement dispersion, and precipitation-typed double-fluid profile control agent are mainly used.

5.1.4.4 Near Borehole Zone

For near borehole zone, profile control agents such as silica gel, zirconium gel, chromium gel, water-swelling, lime cream, and clay/cement dispersion can be selected.

5.1.4.5 Far Borehole Zone

For far borehole zone, profile control agents such as colloidal dispersion gel, gel microsphere, gel-typed double-fluid profile control agent, and precipitation-typed double-fluid profile control agent can be selected.

5.1.5 Field Test of Profile Control in Water Injection Wells

Example 1 South test area of west Chengdong oilfield
The formation temperature of the test area is 70 °C. The formation water salinity is 5.14×10^3 mg/L. The formula of the used profile control agent is shown in Table 5.2. By injecting 11,267 m^3 profile control agents to the test area through 12 injection wells, the effects are shown in Fig. 5.7.

Example 2 South test area of Pucheng oilfield
The formation temperature of the test area is 86 °C. The formation water salinity is 2.40×10^4 mg/L. The profile control agents used are shown in Table 5.3. By injecting 19,300 m^3 profile control agents to this test area through 10 injection wells, the results are shown in Fig. 5.8.

Table 5.2 Profile control agents used in south test area of west Chengdong oilfield

Profile control agent		Formula
Gel-typed profile control agent	Single-fluid method	0.40% HPAM + 0.10% $Na_2Cr_2O_7$ + 0.20% Na_2SO_3
	Double-fluid method	0.40% HPAM + 0.20% $Na_2Cr_2O_7$ (the first working fluid) 0.40% HPAM + 0.40% Na_2SO_3 (the second working fluid)
Precipitation-typed profile control agent		20% $Na_2O \cdot mSiO_2$ + 15% $CaCl_2$

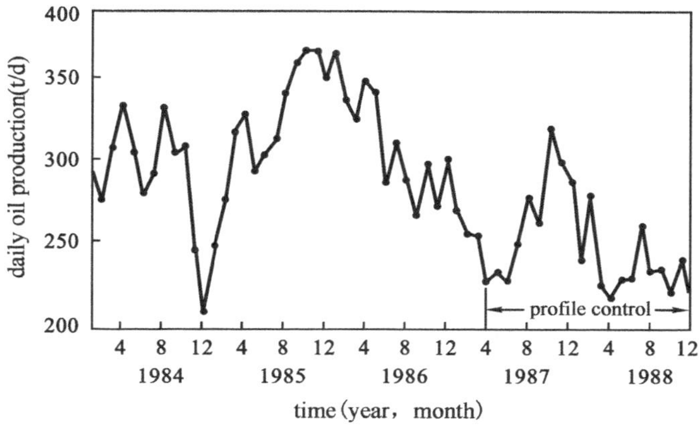

Fig. 5.7 Profile control effect in south test area of west Chengdong oilfield

Table 5.3 Profile control agents used in south test area of Pucheng oilfield

Profile control agent	Formula
Flocculation system	15–20% calcareous clay + 600–1000 mg/L HPAM 3–5% sodium bentonite + 600–1000 mg/L HPAM
Curing system	8% calcareous clay + 8% cement 12% calcareous clay + 12% cement

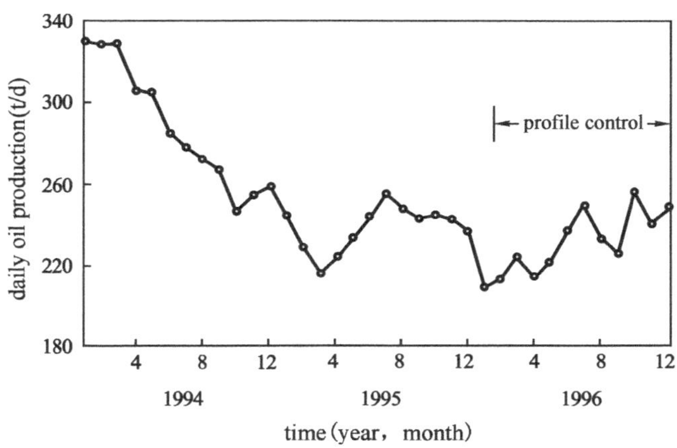

Fig. 5.8 Profile control effect in south test area of Pucheng oilfield

5.2 Water Shutoff in Oil Well

5.2.1 Concept of Water Shutoff in Oil Well

Edge water, bottom water, and injected water are the energy sources of oilfield development. Due to formation heterogeneity, these kinds of water often invade into production wells much too early through high permeable layers, resulting in high water cut and reduced oil production.

Water shutoff refers to the control of water production from the oil well. High-permeability layer must be blocked to control water production from the oil well. As shown in Fig. 5.9, water shutoff agent can be injected into production well to plug the high-permeability layer, controlling the production of injection water.

5.2.2 Importance of Water Shutoff

For reservoir developed with water flooding, the importance of water shutoff can be illustrated by the pressure drawdown curve between production wells and water injection wells, as shown in Fig. 5.10.

It can be seen from Fig. 5.10 that there is an inflection point on the pressure drawdown curve. The space from the inflection point to the injection well is named space for profile control, and the space between the inflection points to the production well is named space for water shutoff. Placement of water shutoff agent anywhere within these two spaces can divert the flow direction of the injection water, thus raise the sweep efficiency and enhance the oil recovery.

It can also be observed from Fig. 5.10 that the spaces for profile control and water shutoff are approximately of the same size. This indicates that profile control near the injection wells and water shutoff near the production wells are equally significant in the EOR process.

Fig. 5.9 Control of injection water from production well

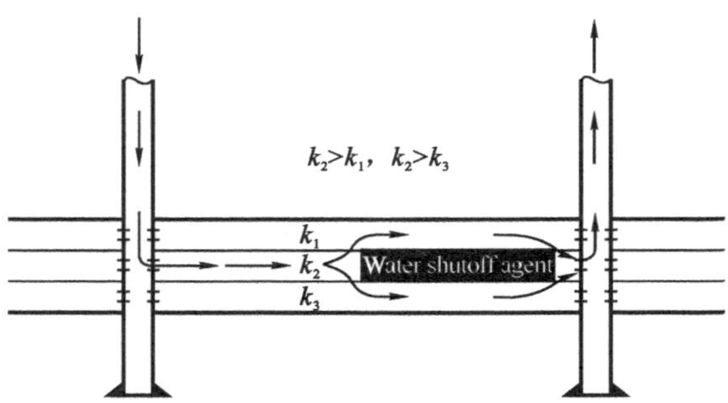

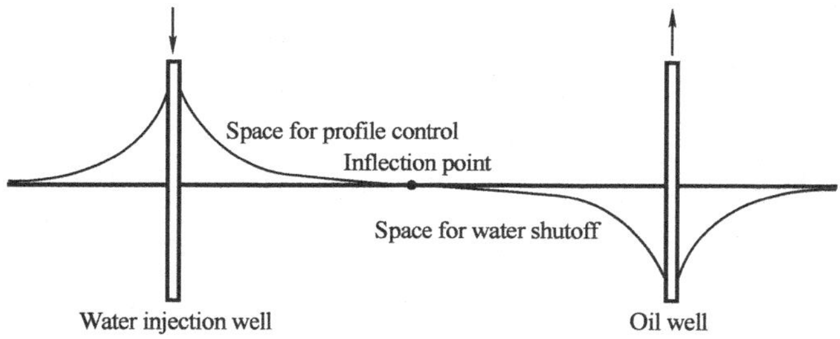

Fig. 5.10 Pressure drawdown curve between production well and water injection well

The significance of water shutoff lies in that it can enhance the oil recovery just as profile control does.

5.2.3 Water Shutoff Agent

Water shutoff agent refers to the matter that is injected into the formation from the production well to control the water production in the oil well.

5.2.3.1 Classification of Water Shutoff Agent

The classifications of water shutoff agents and profile control agents are similar. The difference is that water shutoff agents are often classified by their selectivity into selective and non-selective water shutoff agents. The former one refers to the agents that have different blocking effects to oil and water, or to oil-producing layers and water-producing layers, while the latter one refers to those that do not. Based on the solvent or dispersion media used in preparation, water shutoff agents are also classified into water-based agents, oil-based agents, and alcohol-based agents. In formations with high water saturation, water-based agents and alcohol-based agents are more permeable than oil-based ones.

Due to the crucial importance of selective water shutoff, this section will focus on selective shutoff agents. Besides, since high permeable layers (layers with high water saturation) have lower flow resistance, non-selective agents will preferentially enter these layers to act as selective agents; therefore, introduction of some non-selective agents will also be included.

5.2.3.2 Important Selective Water Shutoff Agents

HPAM (Water-Based)

HPAM has distinct selectivity to oil and water. It can decrease the rock permeability to oil by less than 10%, while decrease the rock permeability to water by more than 90%.

In production wells, the selectivity of HPAM can be demonstrated as follows.

- HPAM preferentially enters the layers with high water saturation.
- HPAM can absorb on rock surface (exposed after washing) with hydrogen bonds (Fig. 5.11).
- The part of HPAM molecules that are not absorbed can stretch in water, thus decrease the water phase permeability (Fig. 5.12a).
- HPAM can form a water film which decreases the flow resistance of oil (Fig. 5.12b).

HPAM with molecular weight between 3.0×10^6 and 1.2×10^7 and hydrolysis degree between 10% and 35% can be applicable for water shutoff.

To increase the adsorption of HPAM in the formation and thereby enhance the water shutoff ability, HPAM can be dissolved in brine before injected into the formation because salt can increase the adsorbing capacity of HPAM on rock surface (Zaitoun 1991). Another way is to pre-treat the formation with cross-linking agents such as aluminum sulfate or aluminum citrate to decrease the electronegativity of rock surface, and even alter the surface into electropositive, thus increasing the adsorbing capacity of HPAM. Or HPAM with low hydrolysis degree can be injected first (the adsorbing capacity is increased due to the non-ionic property of $-CONH_2$ in HPAM), followed by injection of alkaline to increase the hydrolysis

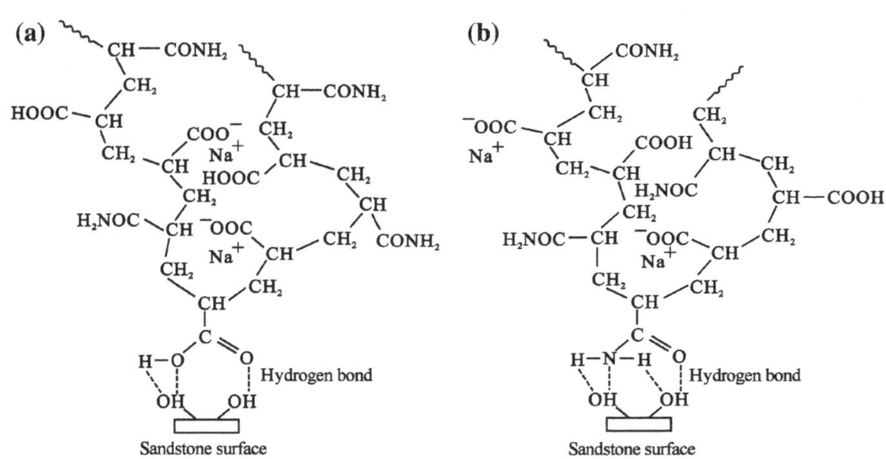

Fig. 5.11 Adsorption of HPAM on sandstone surface. **a** Hydrogen bond formed with $-COOH$; **b** hydrogen bond formed with $-CONH_2$

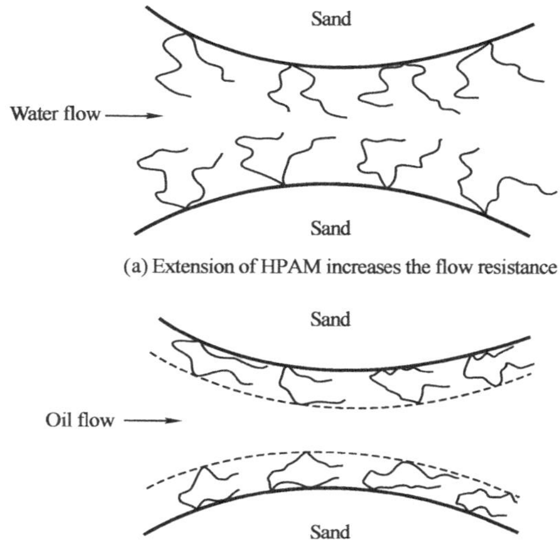

(b) HPAM builds flow resistance to oil while provides a water membrane to decrease flow resistance

Fig. 5.12 Selectivity of HPAM

degree of HPAM molecules that are not absorbed, so as to raise the water shutoff ability of HPAM.

HPAN (partially hydrolyzed polyacrylonitrile) has similar molecular structure as HPAM. Therefore, it has selective water shutoff properties as HPAM does.

Anionic–Cationic–Non-ionic Terpolymer (Water-Based)

One kind of anionic–cationic–non-ionic terpolymer could be obtained through copolymerization of acrylamide (AM) and 3-acrylamido-3-methyl butyl-trimethyl ammonium chloride (AMBTAC) and partial hydrolysis.

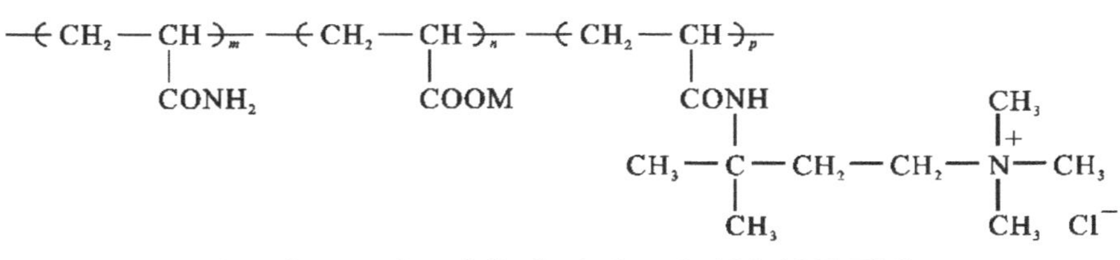

Copolymer of partially-hydrolyzed AM-AMBTAC

The molecule of this kind of water shutoff agent contains anionic, cationic, and non-ionic chains. When injected into formation, the cationic chains will absorb firmly on the electronegative rock surface while the anionic and non-ionic chains will extend into the water phase to increase the flow resistance, performing selective water shutoff function. Table 5.4 indicates that the copolymer has better plugging effect than HPAM.

Table 5.4 Plugging abilities of partially hydrolyzed copolymer AM-AMBTAC and HPAM

Polymer	Resistance factor[a]	Residual resistance factor[b]
Partially hydrolyzed AM-AMBTAC copolymer	7.229	3.739
HPAM	5.023	2.031

Note [a]The pressure ratio of polymer injection and brine injection under same injection rate
[b]The pressure ratio of brine injection before and after polymer treatment under same injection rate

Through copolymerization of acrylamide (AM) and diallyl dimethyl ammonium chloride (DADMC) and partial hydrolysis, another kind of anionic–cationic–non-ionic terpolymer can be obtained for selective water shutoff in the production well.

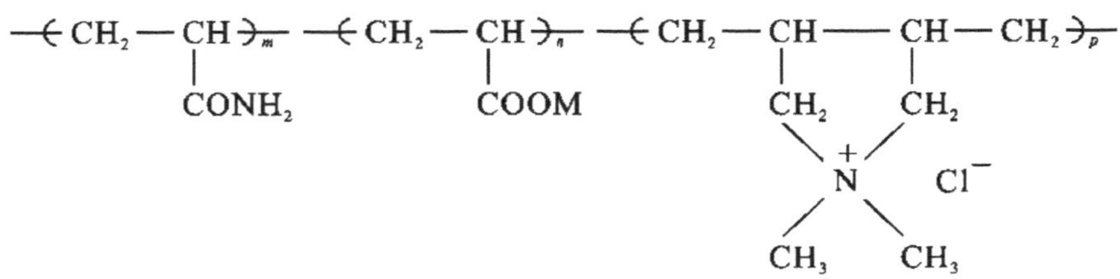

Copolymer of partially-hydrolyzed AM-DADMC

Cationic Polymer (Water-Based)
Cationic polymer refers to the kind of polymer which can dissociate in water to generate cationic chains. It is water-based and can preferentially enter the water-producing layers and absorb onto the electronegative rock surface exposed by washing (such as sandstone and some carbonate rocks). The unabsorbed chains can extend into water and perform water shutoff function.

Here are examples of cationic polymers that can be used as selective water shutoff agents.

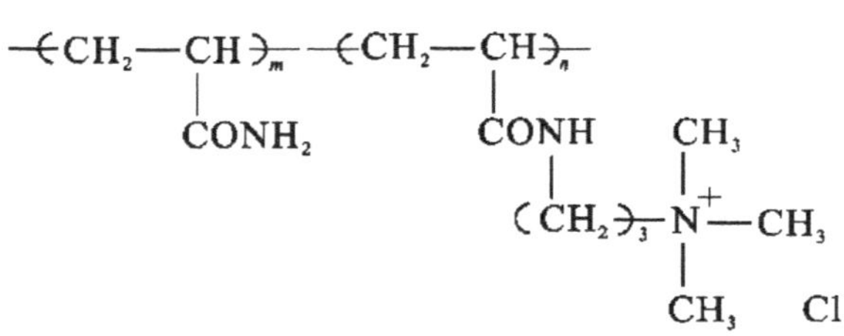

(Copolymer of acrylamide and acrylic acid-1, 2-ethylene trimethyl ammonium chloride)

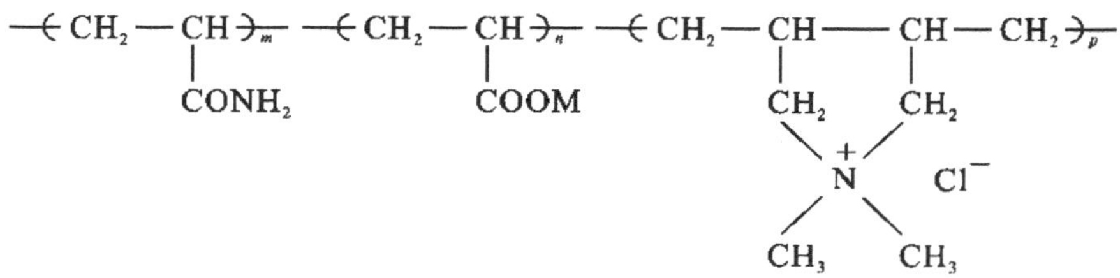

(Copolymer of acrylamide and acrylamide-propylidene trimethyl ammonium chloride)

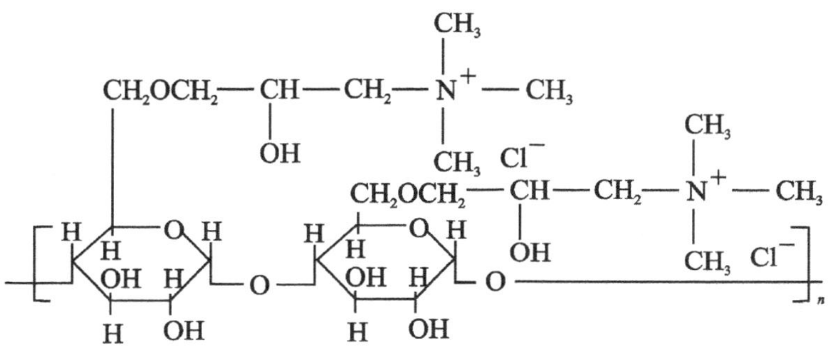

(Reaction product of glycidyl trimethyl ammonium chloride and starch)

Gel (Water-Based)

Gels (such as chromium gel, zirconium gel, phenolic resin gel, and polyethyleneimine gel) have distinct selectivity to oil and water; therefore, they can cause disproportionate permeability reduction (DPR) to oil and water (Liang et al. 1991). Figure 5.13 shows the DPR data from the literature.

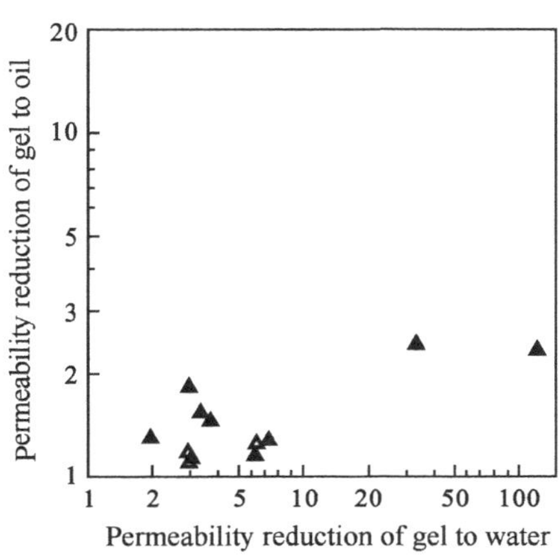

Fig. 5.13 Disproportionate permeability reduction of gel to oil and water (data from Δ Needham (1974), ▲ Zaitoun (1991))

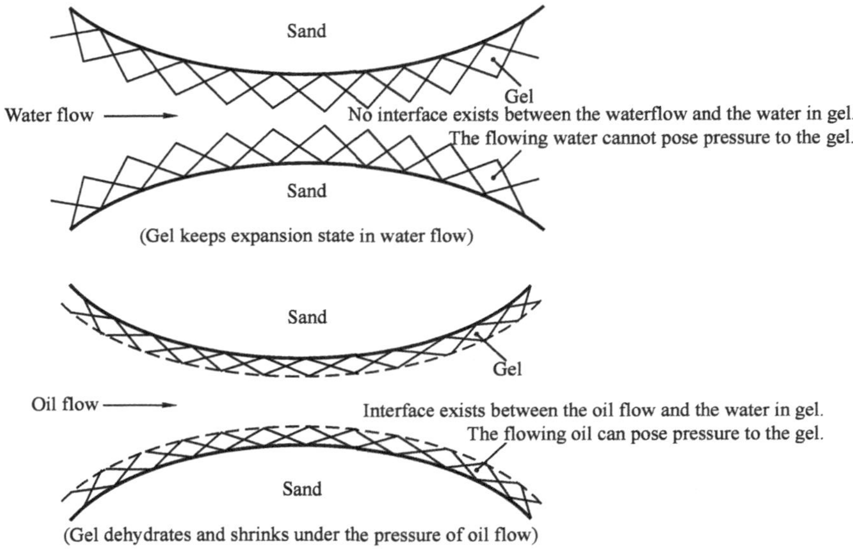

Fig. 5.14 Explanation of the selective expansion and shrinkage of gel

The selectivity of gel can be explained by expansion/shrinkage mechanism (Willhite et al. 2000) and oil/water separation mechanism (Nilsson 1998).

The expansion/shrinkage mechanism believes that the structure of cross-linked polymer is a network. The network is filled with water, which is of the same phase as flowing water. Therefore, when the water flows, the network maintains an expansion state, posing high flow resistance to water. As for oil which is of different phase from the water in the network structure, it can pose pressure onto the water phase when it flows. Therefore, the water in the network would be squeezed out and the gel network gets shrunk, reducing the flow resistance of oil. Figure 5.14 is an illustration of this mechanism.

The oil/water separation mechanism believes that oil and water have different flowing channels within porous media. Water flows through the water channel and oil through the oil channel. When gel is injected, it will flow through and occupy the water channel as it is water-based. After the gelation, it will plug water channel instead of oil channel. Therefore, gel has the characteristic of blocking water instead of oil. Figure 5.15 is an illustration of this mechanism.

Foam (Water-Based) (Zhdanov et al. 1996; Hirasaki and Miller 2003; Li et al. 2007)
With water as dispersion medium, foam can preferentially enter and remain stable in the water-producing layers. It can perform water shutoff function through superimposed Jamin effect. In the oil-producing layers, oil can be emulsified in the disperse medium of foam to form a three-phase foam. The oil droplet in the disperse medium can cause the foam failure through steps demonstrated in Fig. 5.16.

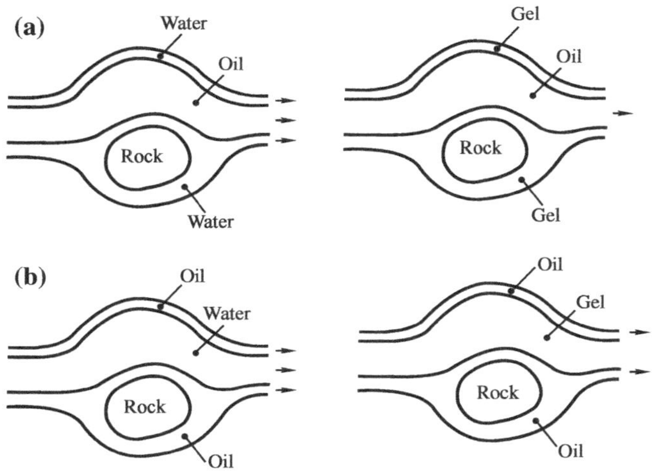

Fig. 5.15 Explanation of selective oil–water separation mechanism. **a** Water-wet surface; **b** oil-wet surface

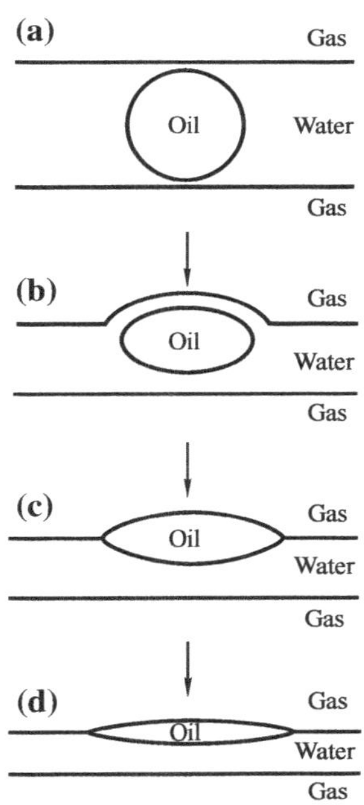

Fig. 5.16 Failure process of disperse medium film in three-phase foam. **a** Oil droplet in the foam dispersion medium; **b** oil droplet floats to form a bulged film; **c** the liquid in the bulged film is driven away, resulting in the broken of the film and the direct contact of oil and gas; **d** if $\sigma_{gas/water} > (\sigma_{gas/oil} + \sigma_{oil/water})$, the oil droplet spreads out and the water film breaks

Therefore, the foam that enters the oil-producing layers will not block the oil layers; it is also a kind of selective water shutoff agent.

The foaming agents are mainly sulfonate surfactants. To increase the stability of foam, thickeners such as sodium carboxymethyl cellulose, polyvinyl alcohol, and polyvinyl pyrrolidone can be added into the foaming agents.

The gas used to prepare foam can be nitrogen or CO_2, which can be converted from liquid state. Nitrogen can also be produced through chemical reaction. For example, NH_4NO_2 or other substance which can generate NH_4NO_2 such as $NH_4Cl + NaNO_2$ (Mitchell et al. 1984; Wu et al. 1995) and $NH_4NO_3 + KNO_2$ is injected into the formation first. Then the pH control system (e.g., $NaOH + CH_3COOCH_3$) is used to make the system change from alkaline to acidic. The system is alkaline at the beginning, so the generation of nitrogen is restrained. When the system enters the formation, the pH of the system changes into acidic so nitrogen can be generated through the following reaction, acting as foaming agent.

$$NH_4NO_2 \rightarrow N_2 \uparrow + 2H_2O$$

Water-Soluble Soap (Water-Based)

Water-soluble soap refers to the soluble high-carbon organic acid. Sodium abietate, sodium naphthenate, and sodium aliphatate are all water-soluble soaps.

Sodium abietate can be generated by the reaction between the rosin (where the mass fraction of abietic acid ranges from 80 to 90%) and sodium carbonate (or sodium hydrate).

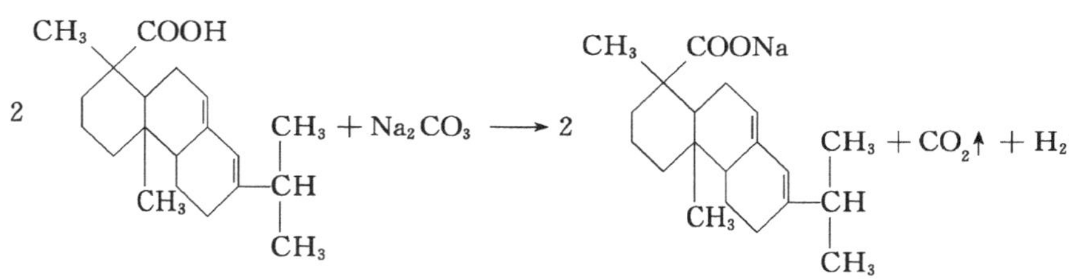

(Abietic acid) (Sodium abietate)

Because the sodium abietate can react with calcium and magnesium ions to form insoluble calcium/magnesium abietate precipitates, it can be applied for water shutoff in production well where the water contains high concentration of calcium and magnesium ions (e.g., above 1×10^3 mg/L). The oil in reservoir does not contain calcium and magnesium ions; therefore, the sodium abietate will not plug oil layer.

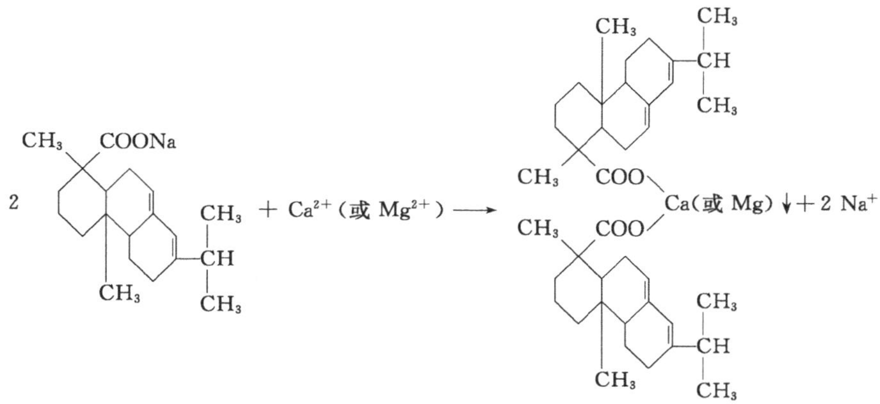

(Sodium abietate) (Calcium/ magnesium abietate)

The sodium naphthenate and the sodium aliphatate (such as sodium stearate and sodium oleate) can be generated from the reaction of naphthenic acid and aliphatic acid with sodium carbonate (or sodium hydrate), respectively. All these water-soluble soaps can selectively block the oil layers where the reservoir water contains high concentration of calcium and magnesium ions.

(Sodium naphthenate)

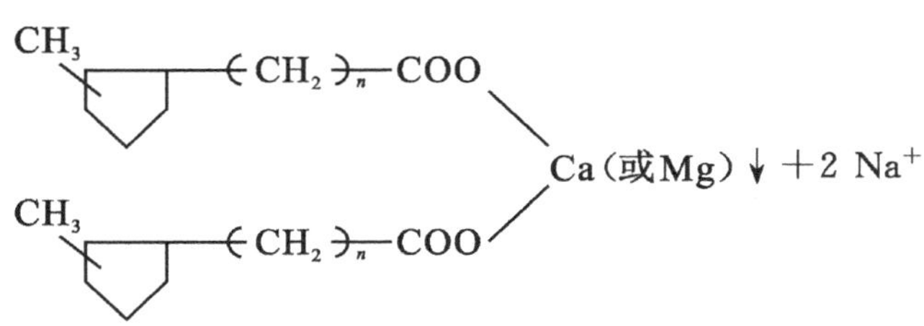

(Calcium/ magnesium naphthenate)

$$2\,R{-}COONa + Ca^{2+}\ (\text{或}Mg^{2+}) \longrightarrow \begin{matrix} R{-}COO \\ R{-}COO \end{matrix}\!\!>\!Ca\,(\text{或}Mg)\downarrow + 2\,Na^{+}$$

(Sodium aliphatate) (Calcium/ magnesium aliphatate)

Kalium Behenate (Water-Based)

Sodium behenate is water insoluble, but kalium behenate is water soluble. When kalium behenate solution is injected into the formation, it can react with the sodium ions in the formation water to form precipitates to plug the water-producing layer.

$$CH_3 - (CH_2)_{20} - COOK + Na^+ \longrightarrow CH_3 - (CH_2)_{20} - COONa \downarrow + K^+$$

(Kalium behenate) (Sodium behenate)

Hydrocarbyl Halogenosilicane (Oil-Based)

Hydrocarbyl halogenosilicane can be expressed using a general formula R_nSiX_{4-n}, where R is alkyl, X is halogen (fluorine, chlorine, bromine, or iodine), and n is an integer in 1–3.

The dimethyl dichlorosilane is a kind of hydrocarbyl halogenosilicane which is produced from the reaction of silicon powder with methane chloride.

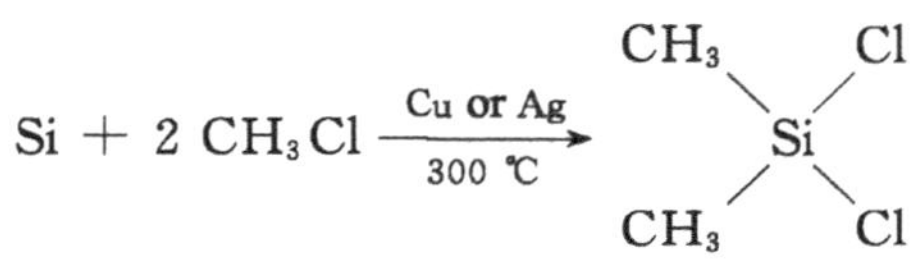

(Dimethyl dichlorosilane)

Hydrocarbyl halogenosilicane can react with water to generate the corresponding silanol. The polyol in the silanol can easily generate polysilanol precipitates through polycondensation, which can plug the water production layer. The reaction of dimethyl dichlorosilane with water is used as an example to demonstrate the generation of water-plugging precipitate.

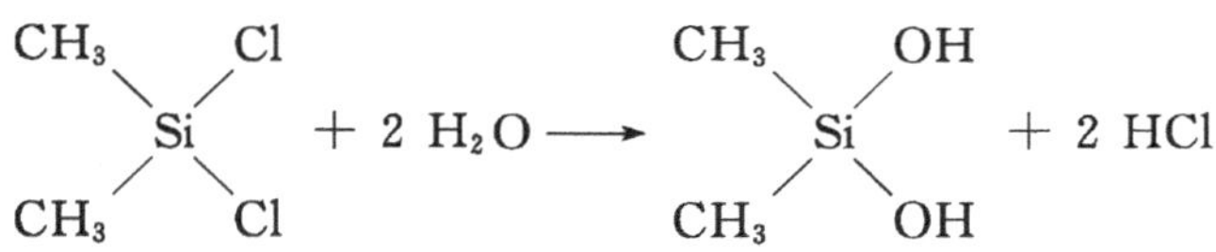

(Dimethyl silicone glycol)

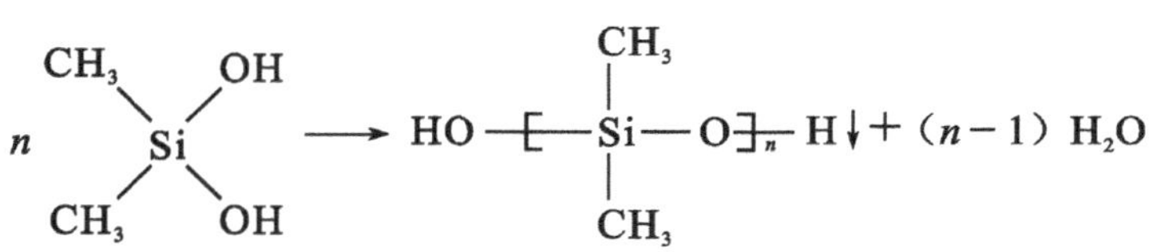

(Poly dimethyl silicone glycol)

As the hydrocarbyl halogenosilicane is oil soluble, it must be prepared into oil solution before usage.

Tetrahydroxy Orthosilicate (Oil-Based) (Karmakar et al. 2002; Grattoni et al. 2001)
The tetrahydroxy orthosilicate is dissolved in oil before injected into the formation. It can have the following reaction when it contacts with water in the formation.

- Hydrolysis

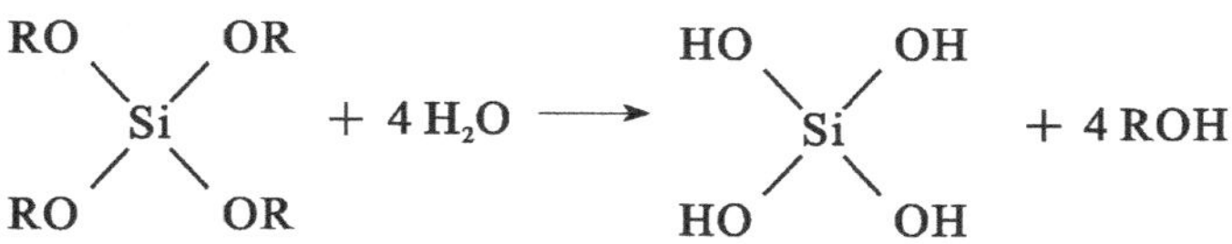

(Tetrahydroxy orthosilicate) (Orthosilicate)

The hydrolysate of the tetrahydroxy orthosilicate is water soluble.

- Condensation polymerization

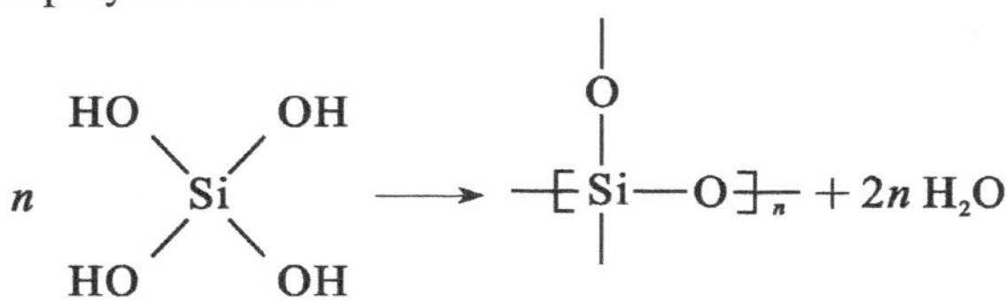

The hydrolysates in water can form network structure through the condensation polymerization, which can deprive the mobility of water and therefore perform the water shutoff function.

Available tetrahydroxy orthosilicates involves tetramethyl orthosilicate, tetraethyl orthosilicate, and tetrapropyl orthosilicate.

Polyurethane (Oil-Based)
Polyurethane is polymerized by polyhydroxy compound and polyisocyanate. If the number of the isocyan acid group is more than that of hydroxyl during the polymerization, the polyurethane which can work as selective water shutoff agent will be prepared. After encountering water, polyurethane can undergo a series of reactions to form amino group and CO_2.

$$-NCO + H_2O \rightarrow -NH_2 + CO_2 \uparrow$$

The amino group can react further with isocyan acid group to generate the urea bond.

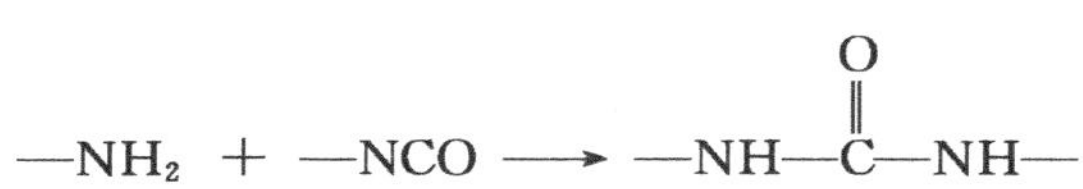

(Urea bond)

The active hydrogens in the urea bond can also react with the unreacted isocyan acid groups, changing the initial mobile linear polyurethane into the immobile three-dimensional polyurethane and blocking the water production layer. If the polyurethane encounters the oil, the above reactions will not take place and no plugging will be generated. Therefore, the polyurethane is a water shutoff agent with good selectivity and strong plugging capability.

Three kinds of other ingredients are also added into the polyurethane water shutoff agent.

- Diluent
  Diluent is used for diluting the polyurethane to enhance its flowability. Xylene, dichloroethane, petroleum fractions, and so on can be used as diluent.
- Sealant
  Sealant can react with all the isocyan acid groups of the polyurethane within certain time, inhibiting the water shutoff agent from changing into three-dimensional structure. Thus, even if the water shutoff agent stays in the oil production layer, it will not result in the bad effect. Alcohols of low molecular weight (C_1–C_8, such as ethanol and isopropanol) can be used as sealant.
- Catalyst
  It can change the reaction rate of the closed reaction. The following chemicals can be used as catalyst.

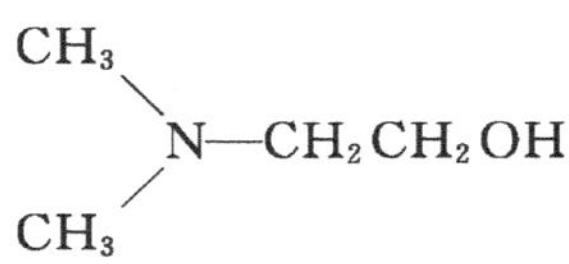

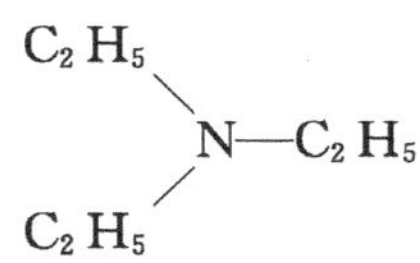

(Dimethylethanolamine) (Trimethylamine)

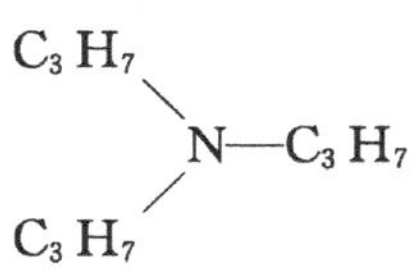

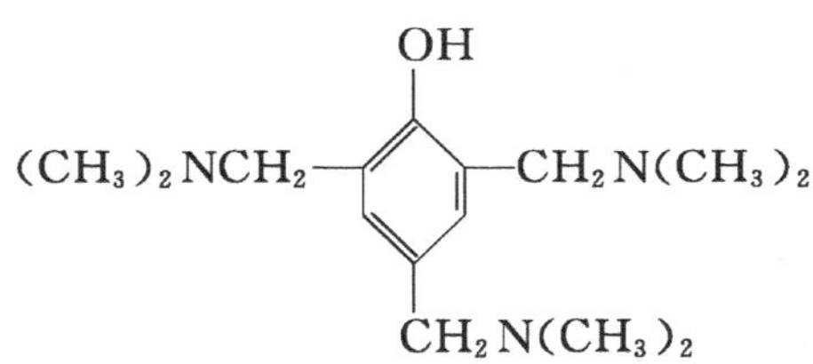

(Tripropylamine) [2,4,6-tri(dimethylamino methylene) phenol]

Alkyl Phenol Acetaldehyde Resin (Water-Based)

This kind of resin is generated by underground synthesizing method. For example, by injecting alkyl phenol, acetaldehyde, and catalyst (e.g., petroleum sulfonate) to the formation, a kind of branched resin is generated under around 100 °C.

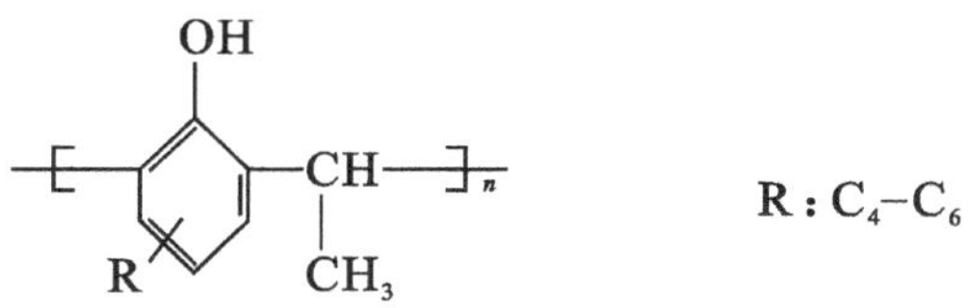

(Alkyl phenol acetaldehyde resin)

This kind of resin dissolves in oil but not in water, so it is a selective water shutoff agent.

Rosin Dimer Alcohol Solution (Alcohol-Based)
The rosin dimer is synthesized by polymerization of rosin under the function of sulfuric acid.

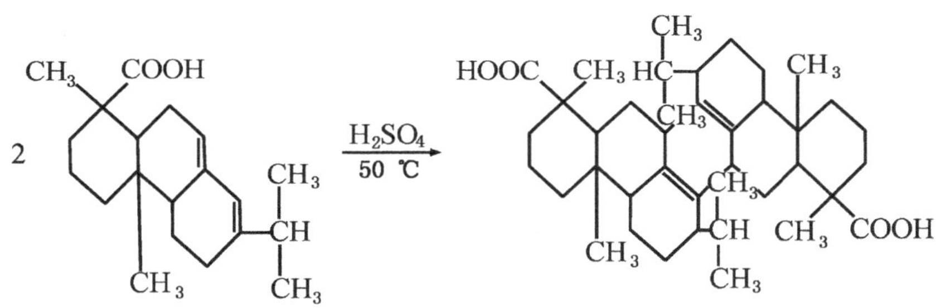

(Rosin) (Rosin dimer)

Rosin dimer dissolves in the low-molecular alcohols (such as methanol, ethanol, n-propanol, and isopropanol) but not in water. Therefore, when the rosin dimer alcohol solution meets with water, the water dissolves in the alcohol immediately, reducing the solubility of the rosin dimer and resulting in the saturated precipitation of the rosin dimer. Due to the high softening point (at least 100 °C) of rosin dimer, its saturated precipitation is in solid state, which has strong plugging ability for the water production layer.

The optimized mass concentration of the rosin dimer ranges from 40 to 60% in rosin dimer alcohol solution. The viscosity of the rosin dimer alcohol solution is too high at high mass concentration, while the water-plugging effect is not as good at low mass concentration.

Sodium Silicate (Water-Based) (Zhao et al. 2009)
The characteristics of the water glass are as follows.

- Thermal sensitivity
  A series of 40 wt% sodium silicate ($Na_2O \cdot 3.17SiO_2$) solutions are placed at different temperatures for different time before their precipitation amounts are measured by centrifugation method. The results are shown in Fig. 5.17.

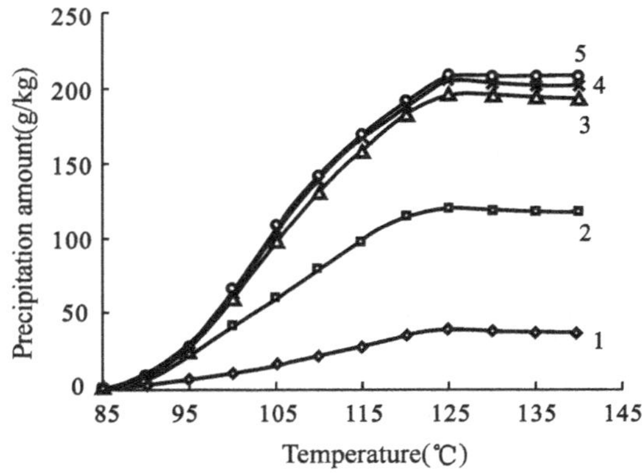

Fig. 5.17 Thermosensitivity of sodium silicate. 1—24 h; 2—48 h; 3—72 h; 4—96; 5—120 h

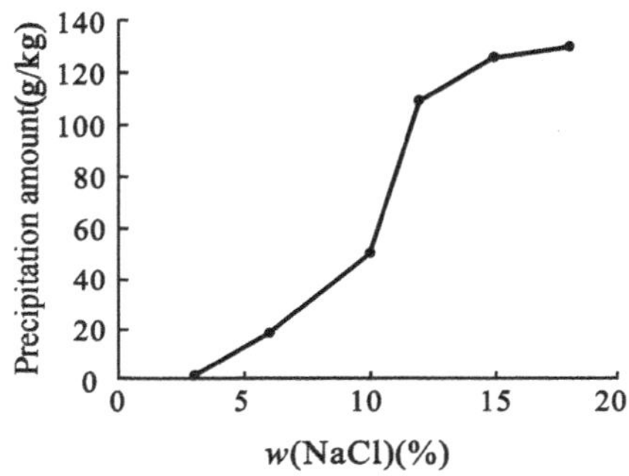

Fig. 5.18 Salinity sensitivity of sodium silicate

It can be seen from Fig. 5.17 that the sodium silicate has the characteristic of thermal sensitivity, because its degree of hydrolysis increases with temperature. The generated precipitation is the result of the condensation polymerization of the hydrolysate of sodium silicate.

- Salinity sensitivity

  If the sodium chloride solutions with different salinity are mixed with 40 wt% sodium silicate ($Na_2O{\cdot}3.17SiO_2$) solutions at mass ratio of 5:5, different centrifugation precipitation amounts can be obtained. The results are shown in Fig. 5.18.

  It can be seen from Fig. 5.18 that the sodium silicate has the characteristic of salinity sensitivity. This characteristic is generated because the solvation demand (hydration) of salt ions impels the condensation polymerization of the hydrolysate of sodium silicate (the water molecule is given out at the same time).

- Calcium and magnesium sensitivity

  When sodium silicate meets with calcium and magnesium ions in water, the precipitation of calcium silicate and magnesium silicate would be generated to perform the plugging function.

$$Na_2O \cdot mSiO_2 + Ca^{2+} \rightarrow CaO \cdot mSiO_2 \downarrow + 2\,Na^+$$

$$Na_2O \cdot mSiO_2 + Mg^{2+} \rightarrow MgO \cdot mSiO_2 \downarrow + 2\,Na^+$$

- Acid sensitivity
  When the sodium silicate meets with acidic water containing CO_2 and/or H_2S, it can generate silicic acid sol through the following reactions and then change into silicic acid gel, performing the plugging function.

$$Na_2O \cdot mSiO_2 + 2\,H^+ \rightarrow H_2O \cdot mSiO_2 + 2\,Na^+$$

Oil-Based Cement (Oil-Based) (Mckown and Rothenberger 1993)
Oil-based cement is the suspension of cement in oil. The cement surface is water-wet. When it enters into the water-producing layers, the water replaces the oil on the cement surface and solidifies the cement, thus blocking the layer. The cement used should be applicable to oil wells at corresponding depth. The available oil is gasoline, diesel, kerosene, and crude oil with low viscosity. In addition, surfactants such as carboxylate surfactants and sulfonate surfactants should also be added to adjust the mobility of the suspension. For example, by adding 300–800 kg oil well cement and 0.1–1.0 kg surfactant into 1 m^3 oil, an oil-based cement with the density of 1.05–1.65 g/cm^3 can be prepared for water shutoff.

Active Heavy Oil (Oil-Based)
This is a kind of heavy oil with emulsifier dissolved in it. The emulsifier should be w/o emulsifier (such as Span 80), which can help generate highly viscous emulsion when oil contacts with water.

As there are considerable amounts of w/o emulsifiers in heavy oil (such as naphthenic acid, resin, and asphaltene), heavy oil can be used directly as selective water shutoff agent. Oxidized asphalt can be dissolved in oil to prepare active heavy oil, where the asphalt is both w/o emulsifier and oil thickener.

Oil/Water Emulsion (Water-Based)
This kind of water shutoff agent is prepared by emulsifying heavy oil in water with o/w emulsifier. As water is external phase, the emulsion has low viscosity, and therefore, it is more likely to enter the water layers. In the water layers, with the adsorption of emulsifier onto the rocks, the emulsion is demulsified and the oil droplets coalesce into highly viscous heavy oil, resulting in high flow resistance and reduction of water cut. Cationic surfactant is the most appropriate emulsifier because it is easy to absorb onto the electronegative sandstone surface and break the emulsion.

Coupled Heavy Oil (Oil-Based)
This kind of water shutoff agent is prepared by dissolving the low polymerization degree phenol formaldehyde resin, phenol furfural resin, or their mixture as

coupling agent in the heavy oil. The generation of chemical adsorption by the reaction between these resins and the formation surface will strengthen the combination (coupling) between the formation surface and the heavy oil. This will make it difficult to be discharged, thusly extending the valid period.

Acid Sludge
The acid sludge generated in the process of sulfuric acid refining petroleum distillate can be used for selective water shutoff, because this kind of acid sludge can separate out insoluble matter when contacted with water. Meanwhile, sulfuric acid can react with Ca^{2+}, Mg^{2+} in the formation brine to generate precipitation, blocking the water-producing layers.

Among all the selective water shutoff agents, water-based agents can preferentially enter the water-producing layers (while oil-based agents cannot) and are cheaper than alcohol-based ones; therefore, water-based agents are more desirable than the other two kinds of water shutoff agents.

5.2.3.3 Important Non-selective Water Shutoff Agents

Resin-Typed Water Shutoff Agent
This is a kind of insoluble and infusible polymeric water shutoff agent which is generated by condensation polymerization of lower molecular-weight substances. Phenolic resin, urea-formaldehyde resin, and epoxy resin all belong to this category.

The most common resin-typed water shutoff agent is phenolic resin. When phenolic resin is used for water shutoff, the thermosetting phenolic resin and curing agent (a catalyst which accelerates the curing, e.g., oxalic acid) are mixed and injected into the water zone. Under the function of water zone temperature and curing agent, the thermosetting phenolic resin can cross-link to form insoluble and infusible phenolic resin within a certain time, which can plug the water zone.

Gel-Typed Water Shutoff Agent
This kind of water shutoff agent is prepared from the gelation of sols. The most commonly used gel-typed water shutoff agent is silicic acid gel. When silicic acid gel is used for water shutoff, sodium silicate and activator are mixed (adding the former one to the latter one will produce acidic silica sol while adding the latter one to the former one will generate alkaline silica sol) before injected into the formation. Or sodium silicate and activator can be used as slugs separated by spacer fluid, which are injected into water zone alternately so that they are mixed after a certain distance in the water zone. After the mixing of sodium silicate and activator, the silicic acid sol is generated first before transformed into silicic acid gel.

Precipitation-Typed Water Shutoff Agent
This kind of water shutoff agent consists of two substances which can generate precipitation. Some examples are shown as follows.

$$Na_2O \cdot mSiO_2{-}CaCl_2 \quad Na_2O \cdot mSiO_2{-}FeSO_4 \quad Na_2O \cdot mSiO_2{-}FeCl_3$$

When the solutions containing two corresponding substances are used as slugs separated by spacer fluid, and injected into water zone alternately, they can be mixed after a certain distance in the formation, producing precipitation and plugging the formation.

As the $CaO{\cdot}mSiO_2$ precipitation produced by the reaction between $Na_2O{\cdot}mSiO_2$ and $CaCl_2$ has strong plugging capacity, it is the most commonly used precipitation-typed water shutoff agent. When $Na_2O{\cdot}mSiO_2$–$CaCl_2$ is used as non-selective water shutoff agent, $w(Na_2O{\cdot}mSiO_2)$ ranges between 20% and 40%, and $w(CaCl_2)$ ranges between 15% and 42%.

Dispersion-Typed Water Shutoff Agent
The dispersion mainly refers to solid dispersion for blocking the layers with ultra-high permeability. For example, the above-mentioned clay/cement, calcium carbonate/cement, and fly ash/cement dispersion systems can be used to plug the extra-high permeable strata.

5.2.4 Selection of Water Shutoff Agent

Water shutoff agents should be selected based on the following principles.

5.2.4.1 Water-Based Water Shutoff Agents

Because water-based water shutoff agents can preferentially enter the layers with high water saturation.

5.2.4.2 Single-Fluid Water Shutoff Agents

Because the injection of single-fluid water shutoff agents is construction convenient.

Table 5.5 Water shutoff agent formula for well 22N169 in Shengtuo oilfield

Water shutoff agent	Formula
Agent for far wellbore zone	0.25% HPAM + 0.80% phenolic resin
Agent for transition zone	0.30% HPAM + 1.00% phenolic resin
Agent for near wellbore zone	0.35% HPAM + 0.10% $Na_2Cr_2O_7$ + 0.20% Na_2SO_3

Note The volume ratio of the three agents is 7:2:1

5.2.4.3 Gel-Typed Water Shutoff Agent

Because gel-typed water shutoff agent has good selectivity and is applicable for the low-temperature and low-salinity reservoirs.

5.2.4.4 Sodium Silicate

Because sodium silicate has ideal selectivity in high-temperature and high-salinity formations or formations with high acid gas content.

5.2.5 Field Tests of Water Shutoff

Example 1 Well 22N169 of Shengtuo oilfield (Cui et al. 2005)

The middle depth of this well is 1,994.2 m, the reservoir thickness is 6.6 m, the perforated thickness is 1.2 m, the reservoir temperature is 75 °C, and the salinity of the formation water is 1.6×10^4 mg/L.

Due to the invasion of the injection water into the producing well along the high-permeability layer at bottom of the reservoir, the water cut of the produced fluid reaches 96.9%.

The formula of the water shutoff agents is shown in Table 5.5. A total volume of 585 m^3 water shutoff agents were injected, which were displaced to the reservoir by 60 m^3 over-displacing fluid (0.40% HPAM), and the field result was shown in Fig. 5.19.

Example 2 C4 well of Wei 11-4 oilfield (Dai et al. 2005)

C4 well is a horizontal well with the length of the horizontal section of 305.6 m and the vertical depth of 964.8 m. The reservoir temperature is 77 °C, and the salinity of the formation water is 1.57×10^4 mg/L. The bottom water is cresting into the producing well, and the well is flooded by water.

In order to control cresting of the bottom water, the gel-typed water shutoff agents were injected into the horizontal well. The formula of the water shutoff agents is shown in Table 5.6. A total volume of 768 m^3 water shutoff agents were injected, which were displaced to the reservoir by 614 m^3 over-displacing fluid (0.45% HPAM), and the field result was shown in Table 5.7.

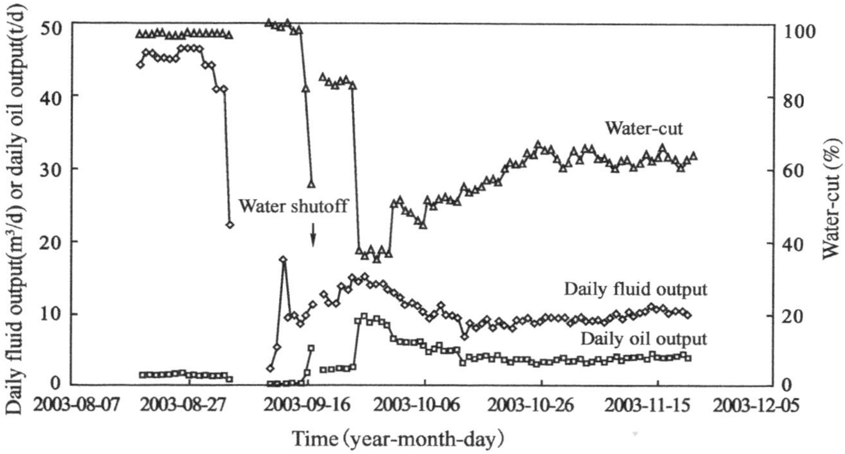

Fig. 5.19 Production curve of well 22N169 in Shengtuo oilfield

Table 5.6 Formula of water shutoff agents for C4 well in Wei 11-4 oilfield

Water shutoff agent	Formula
Agent for far wellbore zone	0.20% HPAM + 0.60% phenolic resin
Agent for transition zone	0.30% HPAM + 0.80% phenolic resin
Agent for near wellbore zone	0.40% HPAM + 0.15% $Na_2Cr_2O_7$ + 0.60% mixed reducing agent

Note The volume ratio of the three agents is 3:2:1
The mixed reducing agents were composed of sodium sulfite and thiourea

Table 5.7 Water shutoff effect of C4 well in Wei 11-4 oilfield

Time		Daily liquid production (m^3/d)	Daily oil production (m^3/d)	Water cut (%)
Before treatment	June 2001	300	0	100
After treatment	January 2003	191.3	16.3	91.5
	March 2003	294.8	24.8	90.9
	May 2003	293.0	25.0	91.5
	November 2003	301.2	19.3	93.6

References

Barbosa ICF, Ribeiro AM, Bonet EJ et al (1987) Process for the correction of oil well productivity and/or injectivity profiles. US Patent 4,643,254, 17 Feb 1987
Bai B, Li L, Liu Y et al (2004) Preformed particle gel for conformance control: factors affecting its properties and application. SPE 89389
Busolo MA, Mogollon JL, Rojas F et al (2001) Permeability modifications by in-situ cations hydrolysis. SPE 64990
Cui Z, Wang Y, Ma W et al (2005) A novel selective plugging agent and its field application. Oilfield Chem 22(1):35–37, 41
Dai C, Zhao F, Li Y et al (2005) Controlling technology of the bottom water coning in the horizontal well in offshore oilfield. Acta Pet Sin 26(4):69–72
Davies S, Hughes T, Lekker H et al (2002) An aqueous delayed-gelation solution and methods of use in hydrocarbon well. GB Patent 2392460A, 29 Aug 2002
Dolan DM, Thiele JL, Willhite GP (1998) Effects of pH and shear on the gelation of a xanthan—Cr(III) system. SPE Prod Facil 13(2):97–103
Dong B (1987) Swollen polyacrylamide for profile control in water injection well and water shutoff in oil well. Oilfield Chem 4(2):91–97
Eoff L, Dalrymple D, Everett D et al (2007) Worldwide field applications of a polymeric gel system for conformance applications. SPE Prod Oper 22(2):231–235
Frampton H (1997) Downhole fluid control processes. US Patent 5,701,955, 30 Oct 1997
Fielding RC Jr., Gibbons DH, Legrand FP (1994) In-depth drive fluid diversion using an evolution of colloidal dispersion gels and new bulk gels: an operational case history of North Rainbow Ranch Unit. SPE/DOE 27773
Grattoni CA, Jing XD, Zimmerman RW (2001) Disproportionate permeability reduction when a silicate gel is formed in-situ to control water production. SPE 69534
Hardy M, Botermans W, Hamouda A (1999) The first carbonate field application of a new organically crosslinked water shutoff polymer system. SPE 50738
Hirasaki GJ, Miller CA (2003) Reducing water permeability in subterranean formation. EP Patent 1312753A1, 21 May 2003
Hudson PS, Shioyama TK (1987) Permeability contrast correction employing propionate sequestered chromium(III) prepared by nitrite/dichromate redox. US Patent 4,636,572, 13 Sept 1987
Kang S, Lu X, Sun Y et al (2002) Profile control experiments of anionic and cationic polymer solutions. J Daqing Pet Inst 26(3):36–38
Karmakar GP, Grattoni CA, Zimmerman RW (2002) Relative permeability modification using an oil-soluble gellant to control water production. SPE 77414
Lakatos I, Lakatos-Szabo J, Kosztin B et al (2000) Application of iron-hydroxide-based well treatment techniques at Hungarian oil field. SPE 59321
Li Z, Sun M, Lin R et al (2007) Experimental studies on the foam plugging and selective flow-dividing. Acta Pet Sin 28(4):115–118
Liao J, Zhao F (1994) Experimental study on the ferrous sulfate single-liquid profile control agent. J Pet Univ (Nat Sci Ed) 18(3):48–52
Liang J, Sun H, Seright RS (1992) Reduction of oil and water permeabilities using gels. SPE/DOE 24195
Lu Q, Liu Z, Jia L et al (2004) Study on the delayed silica gel plugging agent. Oilfield Chem 21 (1):33–35
Mack JC, Smith JE (1994) In-depth colloidal dispersion gels improve oil recovery efficiency. SPE/DOE 27780
McCool CS, Green DW, Willhite GP (1991) Permeability reduction mechanisms involved in situ gelation of a polyacrylamide/chromium (VI)/thiourea system. SPE Reserv Eng 6(1):77–83
Mckown KW, Rothenberger D (1993) Methods of selectively reducing the water permeabilities of subterranean formations. US Patent 5,181,568, 26 Jan 1993

Mitchell TI, Donovan SC, Mcspadden HW (1984) Field application of a chemical heat and nitrogen generating system. SPE 12776

Nasr-El-Din HA, Taylor KC (2005) Evaluation of sodium silicate/urea gels used for water shut-off treatments. J Petrol Sci Eng 48(3–4):141–160

Natarajan D, Mccool CS, Green DW et al (1998) Control of in-situ gelation time for hydrolyzed polyacrylamide-chromium acetate systems. SPE 39696

Needham RB, Threlkeld CB, Gall JW (1974) Control of water mobility using polymers and multivalent cations. SPE 4747

Nilsson S, Stavland A, Jonsbraten HC (1998) Mechanistic study of disproportionate permeability reduction. SPE 39635

Seright RS (1997) Use of preformed gels for conformance control in fractured systems. SPE Prod Facil 12(1):59–65

Sun H, Wang T, Xiao J et al (2006) Deep profile control technology of novel polymeric microsphere. Pet Geol Recovery Effic 13(4):77–79

Willhite GP, Zhu H, Natarajan D et al (2000) Mechanisms causing disproportionate permeability in porous media treated with chromium acetate/HPAAM gels. SPE 59345

Wu A, Chen M, Gu S et al (1995) Reaction dynamics of $NaNO_2$ and NH_4Cl and their application in oilfield. Oil Drill Prod Technol 17(5):60–64

Wurster DE (1953) Method of applying coatings to edible tablets or the like. US Patent 2,648,609, 11 Aug 1953

Zaitoun A, Kohler N (1991) Thin polyacrylamide gels for water control in high-permeability production wells. SPE 22785

Zaitoun A, Kohler N, Guerrini Y (1991) Improved polyacrylamide treatments for water control in production well. JPT 43(7):862–867

Zhao F, Zhang G, Tao B et al (1987) Laboratory study on the precipitation-type double-liquid plugging agent. Oilfield Chem 42(2):81–90

Zhao F, Chen D, Zang J (1988) Alkaline silicate gel plugging agent. J Pet Univ (Nat Sci Ed) 12(4–5):54–62

Zhao F, Wang Y, Li A et al (1989) Improvement on the chromium gel plugging agent in Chengdong oilfield. J East China Pet Inst 12(2):28–38

Zhao F, Li L (1996) Single-liquid zirconium gel plugging agent. J Pet Univ (Nat Sci Ed) 20(1):43–47

Zhao J, Dai C, Wang L et al (2009) Study on the sodium silicate inorganic plugging agents. Oilfield Chem 26(3):269–272

Zhao F, Wang F, Yang N (1991) The research and application of the plugging agent used in the whole water shutoff in south region of west Chengdong oilfield. J Pet Univ (Nat Sci Ed) 12(1):37–48

Zhdanov SA, Amiyan AV, Surguchev LM et al (1996) Application of foam for gas and water shut off: review of field experience. SPE 36914

Zhu H, Yu C, Zhao F (1989) Zirconium gel water shutoff agent. Oilfield Chem 6(1):27–31

Chapter 6
Viscosity Reduction of Heavy Oil

Heavy oil resources are abundant in China. However, the high viscosity and poor fluidity of heavy oil bring great difficulties to its exploitation and transportation. To improve the exploitation and transportation of heavy oil, the properties and viscosity reduction methods of heavy oil must be studied.

6.1 Heavy Oil

Heavy oil refers to the oil with high viscosity and high relative density. The classifications of heavy oil are shown in Table 6.1.

The high content of asphaltene and resin fractions results in the high viscosity of heavy oil. Table 6.2 shows that the higher the content of asphaltene and resin fractions is, the larger the viscosity of heavy oil is.

Both the asphaltene and resin in crude oil are complex mixtures of non-hydrocarbon compounds with complicated structures (Liang 1995; Speight et al. 1996; Schabron et al. 1998). With relatively low molecular weight (5×10^2–1.5×10^3), resin can be soluble in oil. The asphaltene is formed by condensation of resin and has relatively high molecular weight (1.5×10^3–5×10^5), so it is insoluble in oil. The fused rings of asphaltene molecules are flakes which are connected by carbon chains or heteroatomic carbon chains (containing sulfur atom, nitrogen atom, etc.) to form complicated structures, which can be seen from Fig. 6.1 (Hirsch et al. 1970) and Fig. 6.2 (Murgich et al. 1999).

Hydroxyl, amino, carboxy, and other groups in asphaltene molecules can form hydrogen bonds with each other to stack fused rings, generating an asphaltene phase. Although the molecular weight of resin molecule is low, a complex structure including hydroxyl, amino, carboxy, and other groups still exists in the molecule.

C. Dai and F. Zhao, *Oilfield Chemistry*,
https://doi.org/10.1007/978-981-13-2950-0_6

Table 6.1 Classifications of heavy oil[a]

Types	Viscosity (mPa • s)	Relative density
Conventional heavy oil	100–1 × 10^4	>0.9200
Extra-heavy oil	1 × 10^4–5 × 10^5	>0.9500
Super-heavy oil	5 × 10^4–10 × 10^5	>0.9800
Ultra-heavy oil	>10 × 10^5	>1.0000

Note [a]The viscosity and relative density of heavy oil (degassed crude) are determined at 50 °C and 20 °C, respectively

Table 6.2 Properties of heavy oil

Well No.	Relative density	Viscosity (mPa • s)	Concentration		
			Resin	Asphaltene	Resin + asphaltene
1	0.9534	6,105	55.22	9.29	64.51
2	0.9521	4,875	39.27	10.24	49.51
3	0.9414	832	30.68	11.25	41.93
4	0.9331	457	31.71	4.96	36.67
5	0.8816	51	24.33	4.30	28.63

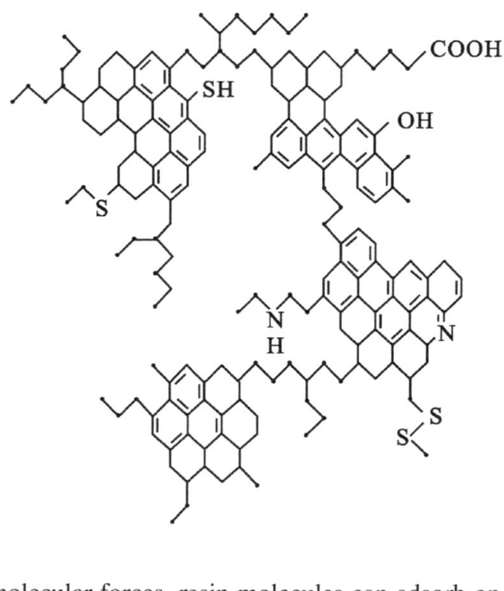

Fig. 6.1 Model of an asphaltene molecule with relative molecular weight of 2,606 (represent carbon atoms with corresponding number of hydrogen atoms)

Through hydrogen bonds and intermolecular forces, resin molecules can adsorb on the surface of asphaltene and protect the asphaltene phase. This adsorption behavior helps asphaltene disperse in oil phase and form a special colloid structure (Fig. 6.3) (Evedo et al. 1997).

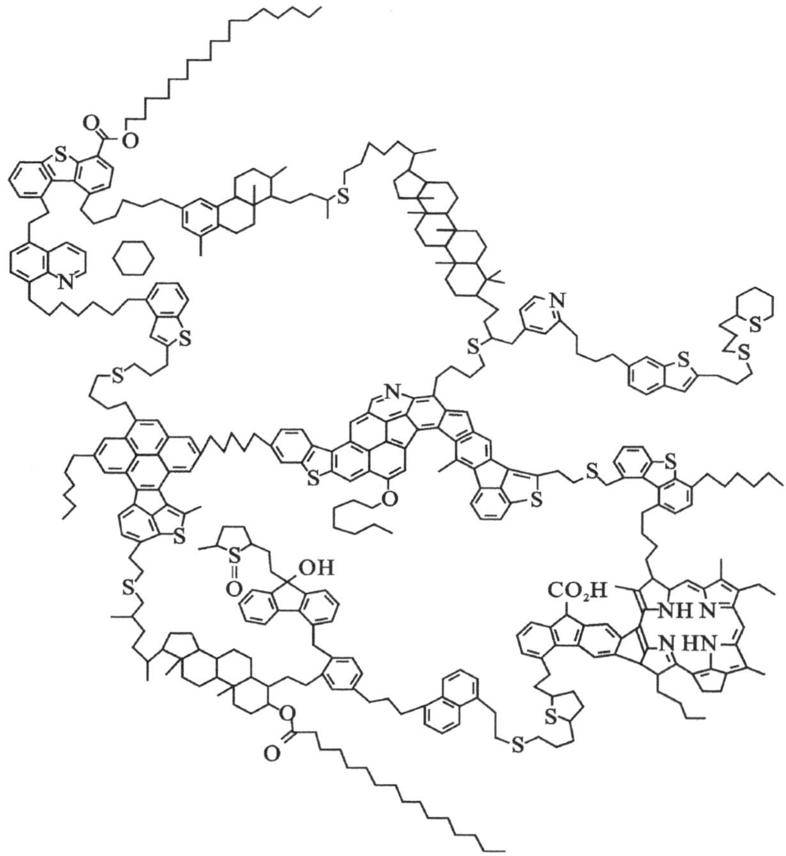

Fig. 6.2 Model of an asphaltene molecule with molecular weight of 6,134

When the heavy oil flows, the internal friction is caused by the relative motion between different layers of the liquid.

- Between oily molecules,
- Between resin molecules,
- Between asphaltene dispersed phases,
- Between oily molecules and resin molecules,
- Between oily molecules and asphaltene dispersed phases,
- Between resin molecules and asphaltene dispersed phases.

Here the oily molecules refer to the molecules besides resin in oil phase.

Due to the special structure of resin molecules and special dispersed phase structure of asphaltene, the internal frictional force generated by hydrogen bond and

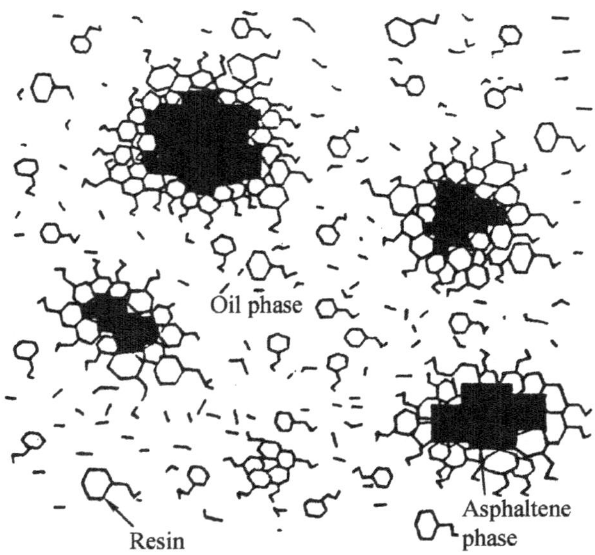

Fig. 6.3 Colloidal model of crude oil

intermolecular entanglement needs to be overcome when they have a relative motion. Therefore, the heavy oil containing high content of resin and asphaltene shall bear high viscosity.

Various methods are used to reduce viscosity of heavy oil, including thermal method, dilution, emulsification, oxidation, hydrothermal catalytic pyrolysis, etc.

6.2 Viscosity Reduction of Heavy Oil by Thermal Method

The temperature of heavy oil can be increased by injection of steam and electric heating. Figure 6.4 shows the viscosity change of heavy oil with temperature.

Figure 6.4 shows that within a certain temperature range, the viscosity decreases significantly with the temperature increase (the viscosity decreases by −50% when the temperature increases by 10 °C). However, the viscosity of heavy oil has little change beyond this temperature range.

The results of viscosity changes indicate that heavy oil has structures. Similar to the viscosity of polymer solution, the viscosity of heavy oil consists of structural viscosity and Newtonian viscosity. The structural viscosity is generated due to the existence of structures while the Newtonian viscosity is the inherent nature of heavy oil. When temperature increases, the forces that produce structures through hydrogen bonds and intermolecular entanglements between resin molecules, between asphaltene dispersed phases, and between resin molecule and asphaltene

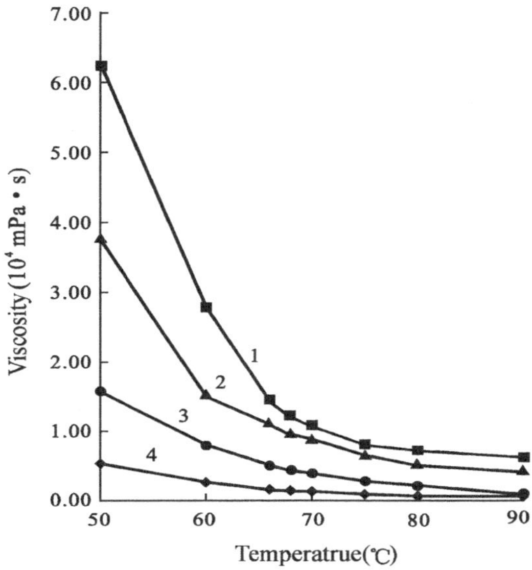

Fig. 6.4 Viscosity change of heavy oil with temperature. Initial heavy oil viscosity (50 °C, mPa · s): 1—6.24×10^4, 2—3.77×10^4, 3—1.58×10^4, 4—5.26×10^3

dispersed phase are reduced. The structures in heavy oil are destroyed, resulting in dramatic decrease in viscosity. While when the structure of heavy oil is completely destroyed, the viscosity changes little with temperature.

The variation of viscosity with temperature can be expressed by the Andrade equation, where μ is the viscosity of heavy oil, T is thermodynamic temperature, a and b are constants which can be calculated by experimental data.

$$\mu = ae^{b/T} \tag{6.1}$$

6.3 Viscosity Reduction of Heavy Oil by Dilution

The viscosity of heavy oil can be decreased by adding light oil such as kerosene, diesel, low-viscosity crude oil.

Table 6.3 shows the viscosity changes of heavy oil after adding kerosene. In these experiments, the initial viscosities of heavy oil and kerosene are 6.24×10^4 and 10.2 mPa · s, respectively. All the tests were conducted at 50 °C.

It can be seen from Table 6.3 that the viscosity of heavy oil decreases significantly after adding kerosene. This dilution effect increases the distance between

Table 6.3 Viscosity changes of heavy oil after adding kerosene

Mass ratio of heavy oil to kerosene	Viscosity (mPa • s)	Mass ratio of heavy oil to kerosene	Viscosity (mPa • s)
100:0	6.24×10^4	100:15	6.67×10^3
100:5	2.83×10^4	100:20	3.22×10^3
100:10	1.24×10^4	100:30	1.45×10^3

resin molecules and asphaltene dispersed phase and reduces their interaction force. Therefore, the structure of heavy oil is destroyed to certain degree and its viscosity is reduced.

The viscosity of heavy oil after dilution can be calculated by the empirical equation (Luo 1992), where μ_d is the viscosity of heavy oil after dilution, μ_l is the viscosity of light oil, μ_v is the viscosity of heavy oil, x is the mass ratio of light oil to heavy oil.

$$\lg(\lg \mu_d) = x \times \lg(\lg \mu_l) + (1 - x) \times \lg(\lg \mu_v) \tag{6.2}$$

This equation indicates that the lower the viscosity of light oil is, and the larger the mass ratio of light oil to heavy oil is, the lower the viscosity of heavy oil after dilution can be.

6.4 Viscosity Reduction of Heavy Oil by Emulsification

Under certain oil–water ratio, water-soluble surfactant can emulsify the heavy oil into heavy oil/water emulsion (Mccafferty and Mcclaflin 1992; Bertero et al. 1994; Browne et al. 1996). The viscosity of this heavy oil/water emulsion is much smaller than that of heavy oil. Moreover, its viscosity is independent of that of heavy oil.

The viscosity of heavy oil/water emulsion can be expressed by the classic Richardson equation, where μ is the viscosity of heavy oil/water emulsion, μ_0 is the viscosity of water, φ is the volume fraction of oil in emulsion, k is constant.

$$\mu = \mu_0 e^{k\varphi} \tag{6.3}$$

In Eq. 6.3, the constant k depends on φ. k is 7.0 when $\varphi \leq 74.0\%$; while k is 8.0 when $\varphi > 74.0\%$.

The water-soluble surfactants with HLB value of 7–18 can be used to reduce viscosity of heavy oil by emulsification (Ahmed et al. 1999).

$R{-}SO_3Na$

(Alkyl sulfonate) — $R: C_{12}–C_{18}$

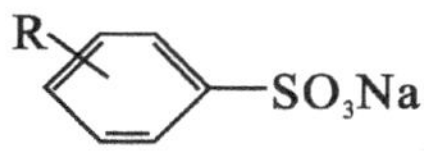

(Sodium alkylbenzene sulfonate) — $R: C_8–C_{14}$

$R{-}O{-}[CH_2CH_2O]_n{-}H$

(Polyoxyethylene alkyl alcohol ether) — $R: C_{12}–C_{18}; n: 5–30$

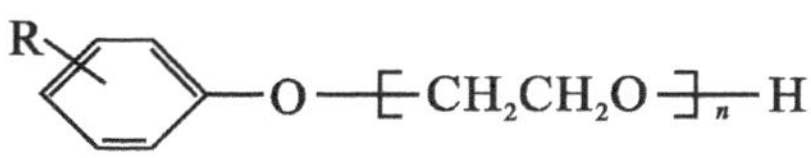

(Polyoxyethylene alkyl phenol ether) — $R: C_8–C_{14}; n: 5–30$

$CH_3{-}CH{-}O{-}(C_3H_6O)_m{-}(C_2H_4O)_n{-}H$

$\quad\;|$

$\quad CH_2{-}O{-}(C_3H_6O)_m{-}(C_2H_4O)_n{-}H$

(Polyoxyethylene polyoxypropylene propylene glycol ether) — $m: 17; n: 15–53$

$R{-}O{-}[CH_2CH_2O]_n{-}SO_3Na$

(Polyoxyethylene alkyl alcohol ether sulfate sodium salt) — $R: C_{12}–C_{18}; n: 1–10$

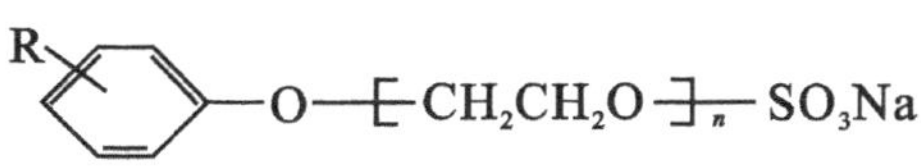

(Polyoxyethylene alkyl phenol ether sulfate sodium salt) — $R: C_8–C_{14}; n: 1–10$

$R{-}O{-}[CH_2CH_2O]_n{-}R'COONa$

(Polyoxyethylene alkyl alcohol ether sodium carboxylate) — $R: C_{12}–C_{18}; R': C_1–C_3; n: 1-10$

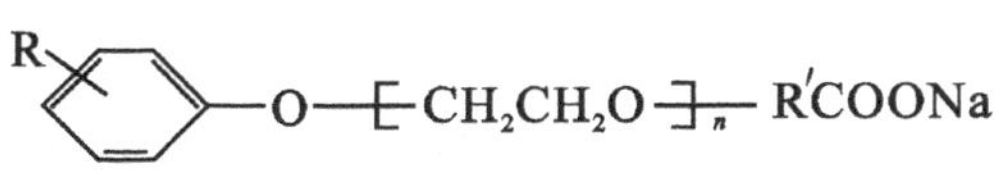

(Polyoxyethylene alkyl phenol ether Sodium carboxylate) — $R: C_8–C_{14}; R': C_1–C_3; n: 1-10$

Emulsifier is not always added into heavy oil. For example, the surfactant generated by the reaction between sodium hydroxide and petroleum acid in heavy oil can be used as oil/water emulsifier.

The mass fraction of surfactant in water is usually between 0.02 and 0.5%.

The volume ratio of heavy oil to water is usually between 70:30 and 80:20.

6.5 Viscosity Reduction of Heavy Oil by Oxidation

By adding oxidizer into the heavy oil, the carbon chains or other heteroatomic chains in fused rings of asphaltenes are broken through the following oxidation reactions (Hayashi et al. 1999; Hayashi and Chiba 1999), which reduces asphaltenes' capacity of forming structure, resulting in viscosity reduction.

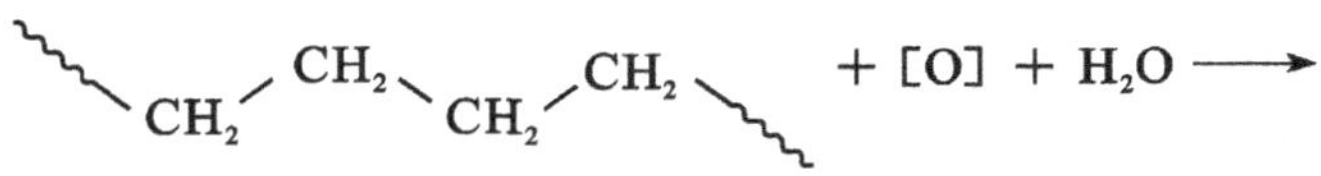

(C-C bond in fused ring of asphaltene)

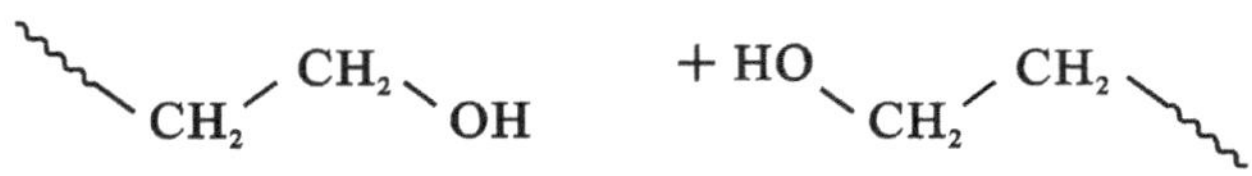

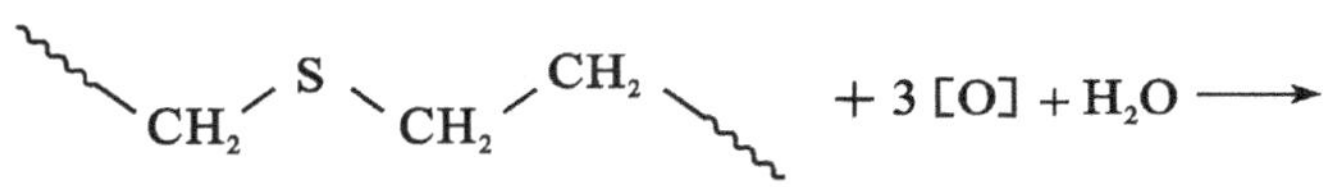

(C-S bond in fused ring of asphaltene)

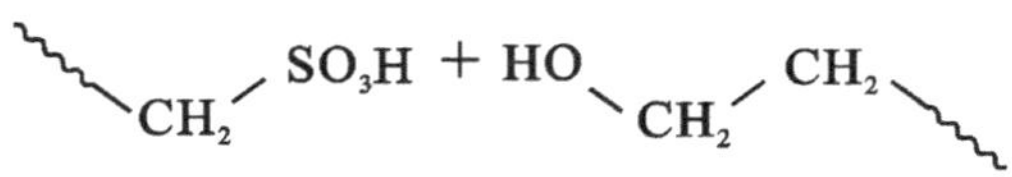

The alcohol produced from the above oxidation reaction can be further oxidized to aldehyde and then oxidized to acid. Therefore, the acid value of heavy oil increases after the oxidation.

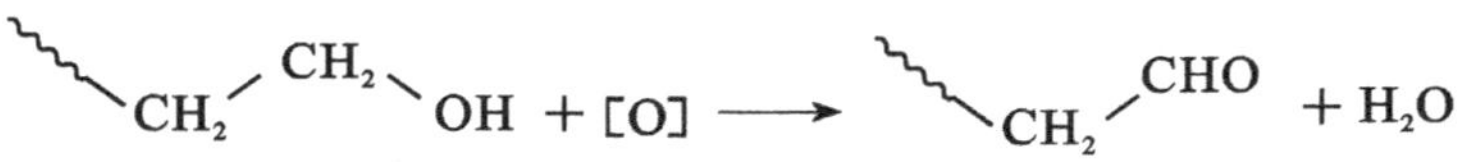

(Alcohol) (Aldehyde)

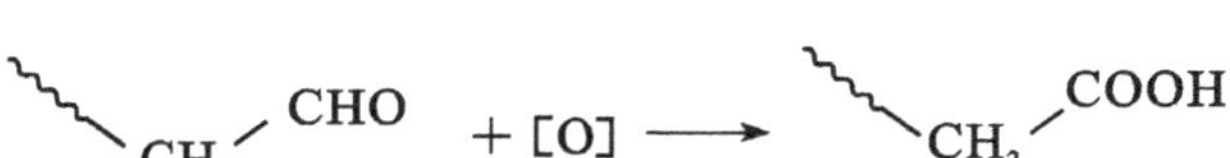

(Aldehyde) (Acid)

To make the oxidation reaction proceed easily, the agents which can provide H^+ are added for cationization of the –N–, –S–, and –SH of asphaltene.

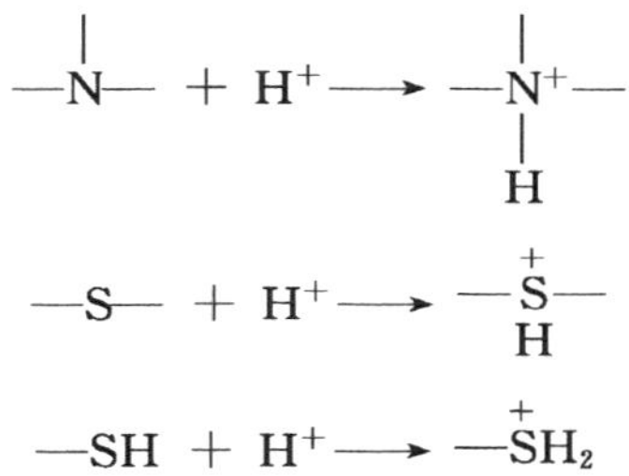

The cationized asphaltene sheets will loosen, which is beneficial to the oxidation reaction.

The available oxidants (main agent) are $NaIO_4$ and 30% H_2O_2 (Liao et al. 2003).

The agents (auxiliary agent) which can provide H^+ are NaH_2PO_4 and CH_3COOH.

The main agent and auxiliary agent can form two kinds of oxidation systems. One is oxidation system A consisting of $NaIO_4$ and NaH_2PO_4 (mixed with mass ratio of 1:1). The other is oxidation system B consisting of 30% H_2O_2 and CH_3COOH (mixed with volume ratio of 1:10). Several kinds of heavy oil are oxidized using these two oxidation systems for 12 hours at 30 °C. The results are shown in Table 6.4.

Table 6.4 shows that both molecular weight and viscosity of heavy oil decrease after oxidation reaction. Meanwhile, the asphaltene content of heavy oil decreases while resin content increases after the oxidation, indicating macromolecule asphaltenes are converted to small molecule resins after oxidation reaction.

Table 6.4 Viscosity changes of heavy oil before and after oxidation

Oil sample	Oxidation system	Saturate (%)	Aromatic (%)	Resin (%)	Asphaltene (%)	Molecular weight of asphaltene	Viscosity (mPa • s)
1#	—	40.13	8.56	42.58	8.64	3,866	6.24×10^4
	A	41.01	9.53	46.72	3.21	1,782	3.96×10^4
	B	40.95	9.03	44.40	5.62	2,138	4.13×10^4
2#	—	41.62	15.17	35.41	7.80	3,271	3.77×10^4
	A	41.80	18.62	37.43	2.16	1,429	2.32×10^4
	B	41.73	18.45	36.75	3.77	1,856	2.75×10^4
3#	—	49.51	19.06	26.35	5.08	2,894	1.58×10^4
	A	50.45	19.34	27.57	2.66	1,228	9.93×10^3
	B	49.65	19.22	27.51	3.22	1,552	1.12×10^4
4#	—	61.27	21.31	14.29	3.13	2,266	5.26×10^3
	A	61.46	21.38	15.41	1.75	982	3.54×10^3
	B	61.31	21.35	14.65	2.78	1,435	3.98×10^3

Note—without oxidants

6.6 Viscosity Reduction of Heavy Oil by Hydrothermal Catalytic Pyrolysis

Hydrothermal catalytic pyrolysis refers to a serious of reactions between the active ingredients (resin and asphaltene with sulfur bond in the carbon chain of the fused ring part) in heavy oil and water under the condition of high temperature and catalyst (Clark et al. 1990; Rivas et al. 1988; Campos and Hernandez 1994; Nares et al. 2007), which will reduce the viscosity of heavy oil.

With an oil/water mass ratio of 10:1, catalyst solutions (such as $FeSO_4$, $NiSO_4$, and $VOSO_4$) are added into heavy oil whose initial properties are shown in Table 6.5. The mixtures undergo hydrothermal catalytic pyrolysis at 250 °C for 24 hours. The properties of the resulting oil are illustrated in Table 6.6.

Two conclusions can be obtained from Table 6.6.

- The viscosity reduction rate of heavy oil with high sulfur content (sample 2#) is larger than that with low sulfur content (sample 1#), indicating that the catalytic thermal cracking mainly occurs in the carbon–sulfur bond. It can be attributed to the different bond energy of C–S bond (272 kJ/mol), C–O bond (360 kJ/mol), C–H bond (293 kJ/mol), and C–C bond (346 kJ/mol). The smaller the bond energy is, the easier the cracking reaction is.
- The mass fractions of resin and asphaltene of heavy oil decease after catalytic thermal cracking reaction. The viscosity of heavy oil also decreases.

  The whole process of catalytic thermal cracking reaction can be elaborated through the following reactions (Hyne 1986).

Table 6.5 Properties of heavy oil before hydrothermal catalytic pyrolysis

Oil sample	Density (g/cm^3)	Viscosity (mPa • s)	Resin (%)	Asphaltene (%)	Sulfur (%)
1#	0.9812	50,400	35.2	9.62	0.03
2#	1.0031	101,200	45.5	16.1	0.92

Table 6.6 Properties of heavy oil after hydrothermal catalytic pyrolysis

Oil sample	Catalyst[a]	Resin (%)	Asphaltene (%)	Sulfur (%)	Viscosity reduction rate[b] (%)
1#	$FeSO_4$	34.9	9.52	0.02	18.5
	$VOSO_4$	33.7	9.03	0.02	22.4
	$NiSO_4$	34.5	9.12	0.02	21.2
2#	$FeSO_4$	38.7	10.51	0.65	31.6
	$VOSO_4$	34.8	9.32	0.51	52.7
	$NiSO_4$	35.1	9.34	0.52	51.7

Note [a]Mass ratio of heavy oil to catalyst is 100:0.15

[b]Viscosity reduction rate = ((Initial viscosity-Viscosity after hydrothermal catalytic pyrolysis)/Initial viscosity) × 100%

- The active ingredients of heavy oil can produce enolate and mercaptan through dehydrogenation and hydrolysis under the conditions of catalysts (M^{n+}, including Fe^{2+}, Ni^{2+}, V^{4+}) and high temperature (150–300 °C).

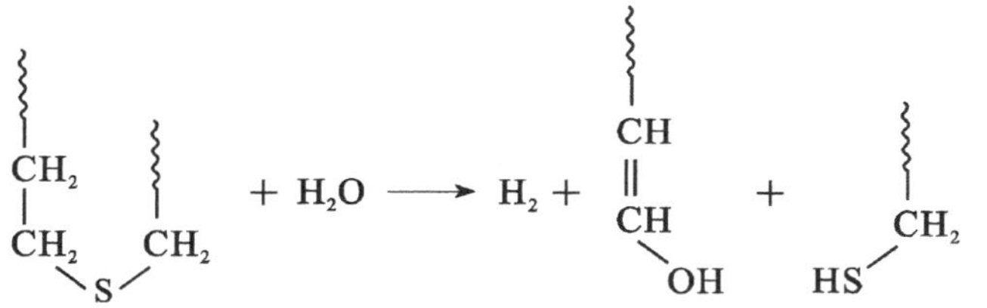

(1)

(Active ingredients) (Enolate) (Mercaptan)

- Enolate is oxidized to aldehyde and then produces carbon monoxide by decomposition reaction.

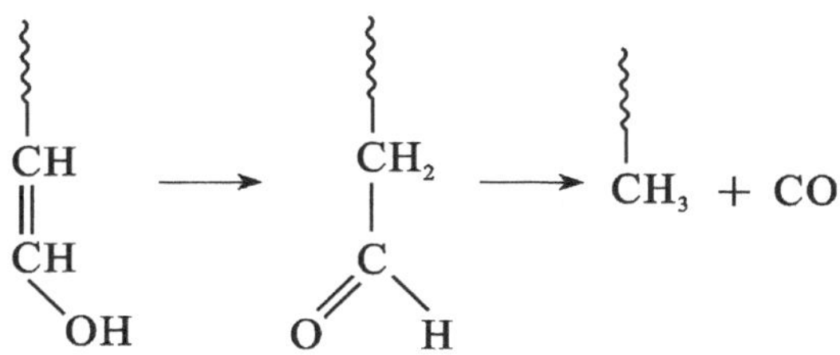

(2)

- CO can react with H_2O under catalysts to generate H_2.

$$CO + H_2O \rightarrow CO_2 + H_2 \tag{3}$$

- The active ingredients of heavy oil react with H_2 under catalysts, to break the bonds of active ingredients and release H_2S.

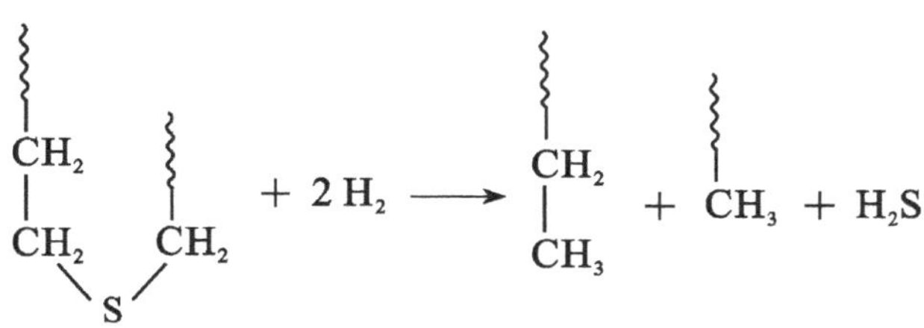

(4)

It can be seen from the whole reaction process that the viscosity reduction can be attributed to the reaction steps (1) and (4) in the hydrothermal catalytic pyrolysis.

The gas (CO_2, H_2, H_2S, Hydrocarbon gas, etc.) generated in the hydrothermal catalytic pyrolysis has been detected in lab experiments (Fan et al. 2001).

Hydrothermal catalytic pyrolysis viscosity reduction method has been successfully applied in oilfield (Liu et al. 2001).

Currently, the viscosity reduction of heavy oil is mainly achieved by thermal method, dilution, and emulsification.

To exploit extra-heavy oil, super-heavy oil, and ultra-heavy oil, combinations of different viscosity reduction methods are usually employed, such as the combination of thermal method and dilution method, thermal method and emulsion method, dilution method and emulsification method, and the combination of thermal, dilution, and emulsification methods. However, the combination of different viscosity reduction methods will increase cost as well as be more demanding to the follow-up processing. Therefore, the combined viscosity reduction methods must be optimized.

References

Ahmed NS, Nassar AM, Zaki NN et al (1999) Stability and rheology of heavy crude oil-in-water emulsion stabilized by an anionic-nonionic surfactant mixture. Pet Sci Technol 17(5–6): 553–576

Bertero L, Dillulo A, Lentini A (1994) An innovative way to produce and transport heavy oil through dispersion in water: laboratory study and field test results. SPE 28543

Browne GE, Hass GR, Sell RD (1996) Downhole emulsification: viscosity reduction increases production. J Can Pet Technol 35(4):25–31

Campos RE, Hernandez JA (1994) In situ reduction of oil viscosity during steam injection process in EOR. US Patent 5,314,615, 24 May 1994

Clark PD, Clark RA, Hyne JB et al (1990) Studies on the effect of metal species on oil sands undergoing steam treatments. AOSTRA J Res 53(6):53–64

Evedo S, Escobar G, Ranaudo MN (1997) Observations about the structure and dispersion of petroleum asphaltenes aggregates obtained from dialysis fractionation and characterization. Energy & Fuels 11(40):774–778

Fan H, Liu Y, Zhao X et al (2001) Study on effect of metal ions on aquathermolysis reaction of Liaohe heavy oils under steam treatment. J Fuel Chem Technol 25(3):56–59

Hayashi J, Aizawa S, Kumagai H et al (1999) Evaluation of macromolecular structure of a brown coal by means of oxidative degradation in aqueous phase. Energy & Fuels 13(1):69–76

Hayashi J, Chiba T (1999) Quantitative description of oxidative degradation of brown coal in aqueous phase on the basis of Bethe lattice statistics. Energy & Fuels 13(6):1230–1238

Hirsch E, Altgelt KH (1970) A method for the determination of average structural parameters of petroleum heavy ends. Anal Chem 42(12):1330–1339

Hyne JB (1986) Aquathermolysis—a synopsis work on the chemical reaction between water (steam) and heavy oil sand during simulated stimulation. Synopsis Report No. 50, AOSTRA

Liang W (1995) Petroleum chemistry. University of Petroleum Press, pp 58–67

Liao Z, Geng A (2003) The mild oxidative degradation of the heavy constituents in viscous crude oils/oil sands and its prospect in oil recovery. Pet Sci Technol 21(5–6):777–787

Liu Y, Fan H, Zhong L et al (2001) Fundamental research on aquathermolysis for heavy oil recovery technology. J Daqing Pet Inst 25(3):56–59

Luo J (1992) The gathering and transportation technology of heavy oil in Liaohe oilfield. Pet Plan & Eng 3(1):48–51

Mccafferty JF, Mcclaflin GG (1992) The field application of a surfactant for the production of heavy, viscous crude oils. SPE 24850

Murgich J, Abanero A, Strausz OP (1999) Molecular recognition in aggregates formed by asphaltene and resin molecules from the Athabasca oil sand. Energy & Fuels 13(2):278–286

Nares HR, Schacht-Hernandez P, Ramirez-Garnica MA et al (2007) Heavy crude oil upgrading with transition metals. SPE 107837

Rivas OR, Campos RE, Borges LG (1988) Experimental evaluation of transition metals salt solutions as additives in steam recovery processes. SPE 18076

Schabron JF, Speight GG (1998) The solubility and three dimensional structure of asphaltenes. Pet Sci Technol 16(3):361–375

Speight JG, Long RB (1996) The concept of asphaltenes revisited. Fuel Sci Technol INT'L 14(1–2):1–12

Chapter 7
Acids and Additives Used in Acidizing

Acidizing (King 1986) is a measure of increasing production and injection by injecting acid into formation through oil–water wells to remove formation blockages (such as ferric oxide, ferrous sulfide, and clay) and/or dissolve the rocks to recover and/or enhance formation permeability.

7.1 Acids Used in Acidizing

The following acids can be used for the acidizing of the oil–water well.

7.1.1 Hydrochloric Acid

Hydrochloric acid can dissolve the corrosion products that block the water wells and recover formation permeability.

$$\underset{\text{(Ferric oxide)}}{Fe_2O_3} + 6HCl \rightarrow 2FeCl_3 + 3H_2O$$

$$\underset{\text{(Ferrous sulfide)}}{FeS} + 2\,HCl \rightarrow FeCl_2 + H_2S \uparrow$$

It can also dissolve limes (limestone and dolomite) and improve formation permeability.

C. Dai and F. Zhao, *Oilfield Chemistry*,
https://doi.org/10.1007/978-981-13-2950-0_7

$$CaCO_3 + 2\,HCl \rightarrow CaCl_2 + CO_2 \uparrow + H_2O$$
(Limestone)
$$CaCO_3 \cdot MgCO_3 + 4\,HCl \rightarrow CaCl_2 + MgCl_2 + 2CO_2 \uparrow + 2H_2O$$
(Dolomite)

The hydrochloric acid used for acidizing includes concentrated acid and diluted acid. Diluted acid is the hydrochloric acid with mass fraction of 3–5%. Concentrated acid is the hydrochloric acid with mass fraction of 15–37%. Generally, diluted acid is used; concentrated acid is used for reducing the dilution effect of formation water to acid, so that acid can acidize deep formation. Meanwhile, the acidizing of concentrated acid can produce a large amount of carbon dioxide and increase the viscosity of spent acids (the acid after acidizing), so that the concentrated acid and the rock fragments that suspend in it can be easily discharged from formation.

In order to make hydrochloric acid acidize deep formation, potential hydrochloric acid can be used.

There are some available potential hydrochloric acids.

Carbon Tetrachloride (Watanabe 1981)
Carbon tetrachloride can hydrolyze and produce hydrochloric acid between 120 and 370 °C.

$$CCl_4 + 2H_2O \rightarrow 4HCl + CO_2 \uparrow$$

Cellon (Vivian 1982)
Cellon can hydrolyze and produce hydrochloric acid between 120 and 260 °C.

$$CHCl_2 - CHCl_2 + 2\,H_2O \rightarrow 4\,HCl + OHC - CHO$$

Ammonia Chloride + Formaldehyde (Roddy et al. 2005)
Ammonia chloride + formaldehyde can react and produce hydrochloric acid between 80 and 120 °C.

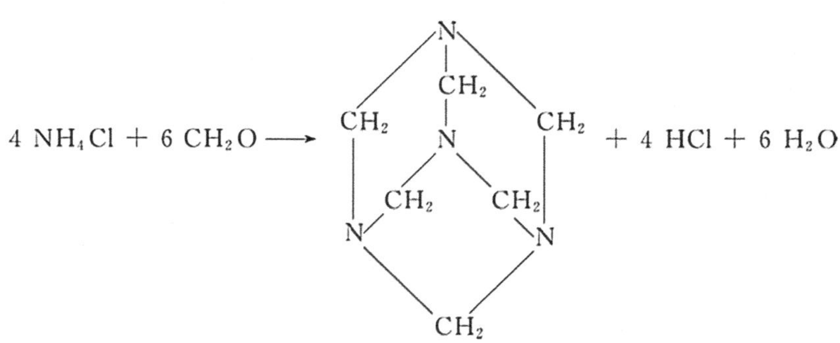

(Hexamine)

7.1.2 Hydrofluoric Acid

Hydrofluoric acid can remove the clay blockage in the formation infiltration surface to recover the formation permeability.

$$Al_4[Si_4O_{10}](OH)_8 + 48\,HF \rightarrow 4\,H_2SiF_6 + 4\,H_3AlF_6 + 18\,H_2O$$
(Kaolinite)
$$Al_4[Si_8O_{20}](OH)_4 + 72\,HF \rightarrow 8\,H_2SiF_6 + 4\,H_3AlF_6 + 24\,H_2O$$
(Montmorillonite)

Hydrofluoric acid can also dissolve sandstone to improve the formation permeability.

$$SiO_2 + 6\,HF \rightarrow H_2SiF_6 + 2\,H_2O$$
(Quartz)
$$Na_2O \cdot Al_2O_3 \cdot 6\,SiO_2 + 50\,HF \rightarrow 2\,NaF + 6\,H_2SiF_6 + 2\,H_3AlF_6 + 16\,H_2O$$
(Albite)

Hydrofluoric acid cannot be used in every situation. For example, hydrofluoric acid cannot be used for handling limestone and dolomite because hydrofluoric acid can react with them to produce deposits that can block formation.

$$CaCO_3 + 2\,HF \rightarrow CaF_2 \downarrow + CO_2 \uparrow + H_2O$$
$$CaCO_3 \cdot MgCO_3 + 4\,HF \rightarrow CaF_2 \downarrow + MgF_2 + 2\,CO_2 \uparrow + 2\,H_2O$$

Even sandstone formation contains certain amount of carbonate, so formation must be pre-treated with hydrochloric acid before acidized by hydrofluoric acid to remove carbonate and reduce unfavorable impact that caused by precipitation reaction mentioned above.

In order to make hydrofluoric acid acidize deep formation, potential hydrofluoric acid can be used.

The following are some potential hydrofluoric acids.

Boratofluoric Acid

Boratofluoric acid (Kunze et al. 1980) can hydrolyze to produce hydrofluoric acid.

$$HBF_4 + 3\,H_2O \rightarrow 4\,HF + H_3BO_3$$
(Boratofluoric acid)

Tetrafluoroethane

Tetrafluoroethane can hydrolyze to produce hydrofluoric acid.

$$CHF_2 - CHF_2 + 2\ H_2O \rightarrow 4\ HF + OHC - CHO$$

Ammonium Fluoride + Formaldehyde

Ammonium fluoride + formaldehyde can react and produce hydrofluoric acid.

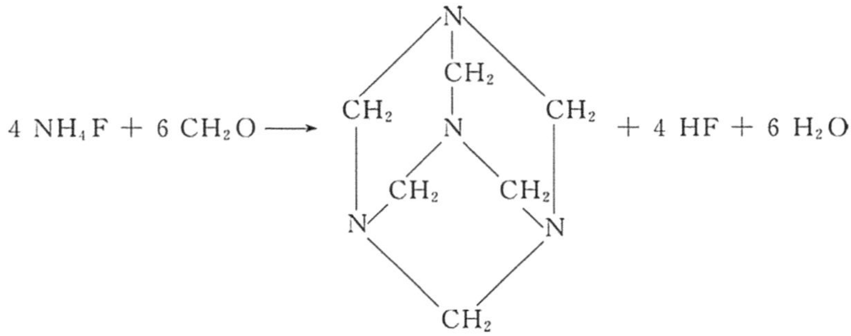

Ammonium Fluoride + Phosphoric Acid (Lullo et al. 1996; Lullo 1996)

Ammonium fluoride + phosphoric acid can react and produce hydrofluoric acid.

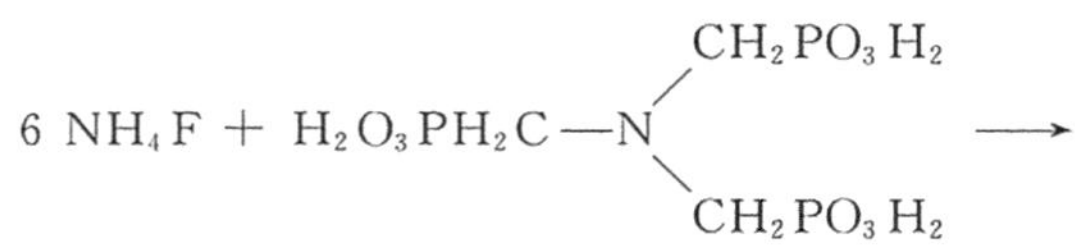

(Nitrilotrimethylene phosphonic acid)

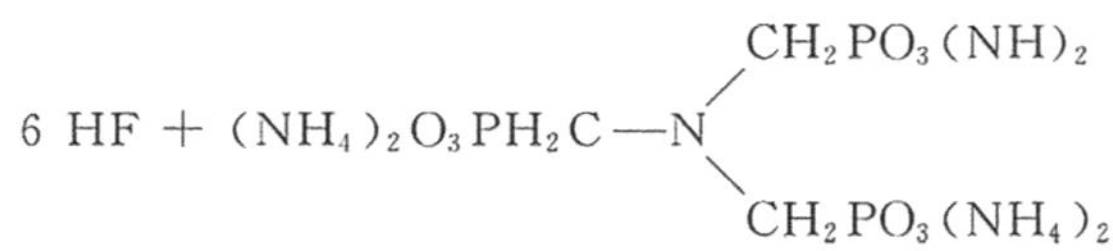

(Nitrilotrimethylene phosphonic amine)

Hydrofluoric acid is usually used in combination with hydrochloric acid. The mixed acid composed of hydrochloric acid and hydrofluoric acid is called mud acid. In mud acid, the mass fraction of hydrochloric acid is 6–15%, and the mass fraction of hydrofluoric acid is 3–15%.

In order to make mud acid acidize deep formation, we can use potential mud acid. The typical potential mud acid is 1,2-dichloro-1,2-difluoroethane which can generate mud acid.

$$CHClF - CHClF_2 + 2\,H_2O \rightarrow \underbrace{2\,HCl + 2\,HF}_{\text{(Mud acid)}} + OHC - CHO$$

The combination of ammonia chloride + ammonium fluoride + formaldehyde can also be used to generate mud acid.

$$2\,NH_4Cl + 2\,NH_4F + 6\,CH_2O \longrightarrow$$

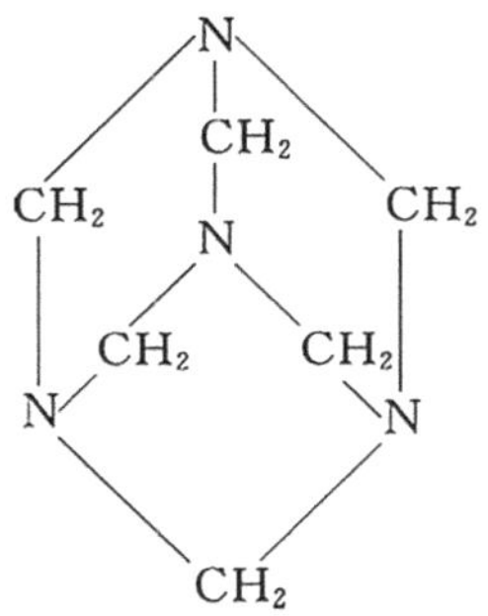

$$+\ \underbrace{2\,HCl + 2\,HF}_{\text{(Mud acid)}} + 6\,H_2O$$

The formation must be pre-treated by hydrochloric acid before acidized by mud acid.

7.1.3 Phosphoric Acid

Phosphoric acid can remove blockages from corrosion products (Clark et al. 1982).

$$FeS + 2\,H_3PO_4 \rightarrow Fe(H_2PO_4)_2 + H_2S \uparrow$$
$$Fe_2O_3 + 6\,H_3PO_4 \rightarrow 2\,Fe(H_2PO_4)_3 + 3\,H_2O$$

Phosphoric acid can also dissolve limestone.

$$CaCO_3 + 2\,H_3PO_4 \rightarrow Ca(H_2PO_4)_2 + CO_2 \uparrow\ + H_2O$$
$$CaCO_3 \cdot MgCO_3 + 4\,H_3PO_4 \rightarrow Ca(H_2PO_4)_2 + Mg(H_2PO_4)_2 + CO_2 \uparrow\ + 2\,H_2O$$

Usually, phosphoric acid with a mass fraction of 15% is used to acidize formation. The mass fraction of phosphoric acid decreases during acidizing but the corresponding pH value does not change much (Table 7.1). This is because H_3PO_4 and H_2PO_4 form a buffer system through the following reactions.

$$H_3PO_4 \rightleftharpoons H^+ + H_2PO_4^-$$

Table 7.1 Solution pH during the acidizing of phosphoric acid

w (H_3PO_4) (%)	pH
15	1.00
10	1.06
5	1.20
1	1.57
0.5	1.72

Compared with that of hydrochloric acid, the reaction rate of phosphoric acid with formation is much slower. Therefore, phosphoric acid can be used for deep formation acidizing.

7.1.4 Sulfuric Acid

Sulfuric acid is a special acid used for acidizing water injection wells. It can recover and/or improve formation permeability by dissolving the blockages or carbonates in infiltration surfaces and areas near wellbore.

$$FeS + H_2SO_4 \rightarrow FeSO_4 + H_2S \uparrow$$
$$Fe_2O_3 + 3\,H_2SO_4 \rightarrow Fe_2(SO_4)_3 + 3\,H_2O$$
$$CaCO_3 + H_2SO_4 \rightarrow CaSO_4 + CO_2 \uparrow + H_2O$$
$$CaCO_3 \cdot MgCO_3 + 2H_2SO_4 \rightarrow CaSO_4 + MgSO_4 + 2\,CO_2 \uparrow + 2\,H_2O$$

The above-mentioned reaction products mainly accumulate in high-permeability layer after getting into formations. They generate blockage through the following hydrolysis reactions or saturation precipitation (such as calcium sulfate and magnesium sulfate) to perform profile control effects.

$$FeSO_4 + 2\,H_2O \rightarrow Fe(OH)_2\downarrow + H_2SO_4$$
$$Fe_2(SO_4)_3 + 6\,H_2O \rightarrow 2\,Fe(OH)_3\downarrow + 3\,H_2SO_4$$

Besides, the heat of dilution generated by sulfuric acid near wellbore can increase the formation temperature, which contributes to pushing the oil that blocks the near wellbore area to the far wellbore area. Therefore, the near wellbore permeability and the effect of in-depth profile control can be improved.

7.1.5 Carbonic Acid

Carbonic acid is generated by dissolving carbon dioxide into water.

$$CO_2 + H_2O \rightarrow H_2CO_3$$

Carbonic acids can react with the carbonate rocks in the formation to generate water-soluble bicarbonates. So, carbonic acid can be used for acidizing carbonate rocks.

$$CaCO_3 + H_2CO_3 \rightarrow Ca(HCO_3)_2$$
$$CaCO_3 \cdot MgCO_3 + 2H_2CO_3 \rightarrow Ca(HCO_3)_2 + Mg(HCO_3)_2$$

Furthermore, the carbonic acids which get into the formations can separate out carbon dioxide through chemical equilibrium. Carbon dioxide can dissolve in oil to decrease the oil viscosity, making the oil easily be discharged to the ground or pushed to the in-depth formation to improve the acidizing performance.

If fluid unloading is needed after the acidization of the formation with carbonic acids, anti-scaling would be necessary because after the pressure reduction, bicarbonate will separate out carbonate again and deposit on the surface of line pipes and equipment.

$$Ca(HCO_3)_2 \rightarrow CaCO_3 \downarrow + CO_2 \uparrow H_2O$$
$$Mg(HCO_3)_2 \rightarrow MgCO_3 \downarrow + CO_2 \uparrow + H_2O$$

7.1.6 Aminosulfonic Acid

Aminosulfonic acid is a kind of solid acid produced in the form of powder. Since it can dissolve blockages and limestones, it can be used for formation acidizing (Deng 1985).

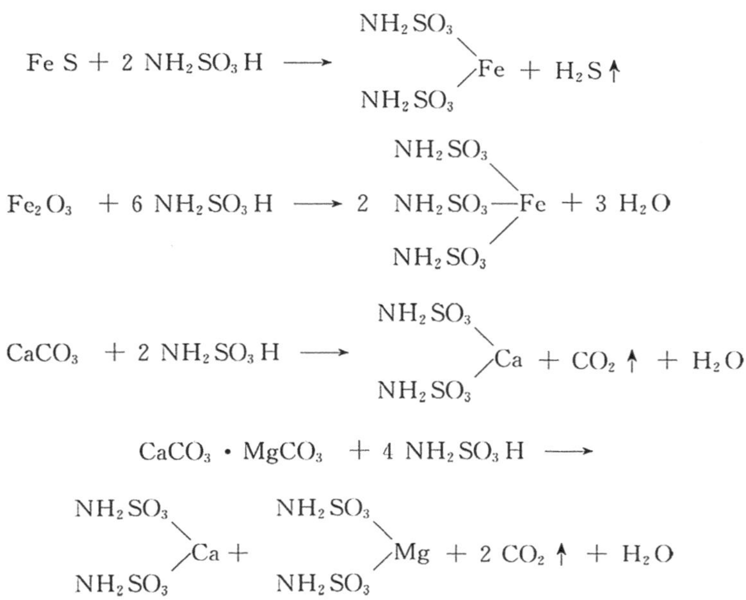

Aminosulfonic acid cannot be used in formations, where temperatures exceed 90 °C; otherwise, the following reactions will happen to deprive its acidizing ability.

$$NH_2SO_3H + H_2O \rightarrow NH_4HSO_4$$

7.1.7 Low Molecular Carboxylic Acid

The general formula of low molecular carboxylic acid is R–COOH. There are some examples such as formic acid (R is H), acetic acid (R is CH_3), propionic acid (R is CH_3–CH_2), or mixture of them.

During acidizing, low molecular carboxylic acid can be used to dissolve limestone.

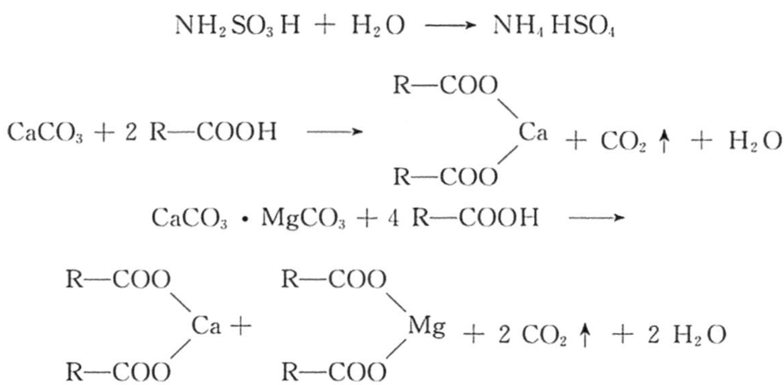

Due to the low solubility of calcium carboxylate and magnesium carboxylate as reaction productions, the mass fraction of low molecular carboxylic acid should not be too high. For example, the mass fraction of formic acid should not exceed 11%, the mass fraction of acetic acid should not exceed 18%, and the mass fraction of propionic acid should not exceed 18%.

During acidizing, the pH value of low molecular carboxylic acid has little change (Table 7.2). This is because low molecular carboxylic acid and carboxylate can form buffer system as shown below.

$$R - COOH \rightleftharpoons H^+ + R - COO^-$$

Therefore, the reaction rate of low molecular carboxylic acid with limestone formation is far lower than that of hydrochloric acid. Low molecular carboxylic acid can be used to acidize in-depth formation.

Table 7.2 Solution pH during acidification using low molecular carboxylic acid

Carboxylic acid	w(Carboxylic acid) (%)	pH
Formic acid	10	1.71
	8	1.76
	6	1.82
	4	1.91
	2	2.06
Acetic acid	18	1.61
	14	1.69
	10	1.77
	6	1.88
	2	2.12
Propionic acid	28	1.15
	20	2.22
	15	2.28
	10	2.37
	5	2.52

Potential acid of low molecular carboxylic acid can be used to acidize deeper formation. The potential acid listed below can generate low molecular carboxylic acid.

Ester

Ester such as ethyl formate, acetic ether, and ethyl propionate can produce corresponding low molecular carboxylic acids by hydrolyzing.

$$HCOOC_2H_5 + H_2O \rightarrow HCOOH + C_2H_5OH$$
$$CH_3COOC_2H_5 + H_2O \rightarrow CH_3COOH + C_2H_5OH$$
$$CH_3CH_2COOC_2H_5 + H_2O \rightarrow CH_3CH_2COOH + C_2H_5OH$$

Acid Anhydride

For example, acetic anhydride and propionic anhydride can produce corresponding low molecular carboxylic acids by hydrolyzing.

$$(CH_3CO)_2 + H_2O \rightarrow 2\,CH_3COOH$$
$$(CH_3CH_2O)_2O + H_2O \rightarrow 2\ CH_3CH_2COOH$$

7.2 Additives Used in Acidizing

Many additives have been used to improve the performance of acidizing.

7.2.1 Retardant

The chemical added in acids to slow the reaction rate of acids and formations is called retardant (Nasr-El-Din et al. 2006; Zhao 1986).

Retardants can be divided into two types:

Surfactant

This kind of retardant works based on adsorption mechanism. After adsorbed on formation surface, surfactant can decrease the reaction rate between acid and formation, and achieve the purpose of retardance. Surfactant is a kind of ideal retardant because it has a high concentration, large adsorption capacity, and strong capability of decreasing reaction rate when it just contacts with formation. While in the interior of formation, the low concentration and low adsorption capacity of surfactant result in the poor capability in decreasing the reaction rate, which is suitable for the function of low concentration acid. There are two kinds of suitable surfactants, cationic surfactant (such as fatty amine hydrochloride, quaternary ammonium salt, and pyridinium salt) and amphoteric surfactant (such as sulfonate, carboxylate, phosphate, and sulfate polyoxyethylene alkyl phenol ether). The mass fraction of surfactant in use is between 0.1% and 4%.

Surfactant can also be used as emulsifier. It can emulsify acid into oil, producing acid-in-oil emulsion and generating the effect of retardance (Al-Anazi et al. 1998; Buijse et al. 1998). The oil used to prepare acid-in-oil emulsion can be toluene, xylene, kerosene, diesel oil, light crude oil, or mixture of xylene and crude. The available emulsifiers include dodecyl sulfonic acid and its alkyl amine salt. The amount of emulsifier accounts 1–10% of oil mass. The volume ratio of oil to acid can be chosen between 7:93 and 45:55.

Polymer

This kind of retardant takes effect by thickening mechanism. Polymer dissolves into acid. When its concentration surpasses certain value, it can form structure in acid to thicken the acid, reducing the diffusion rate of hydrogen ion to the formation surface, thus controlling the reaction rate of acid with formation surface. The polymers that can be used for thickening acids are shown as below.

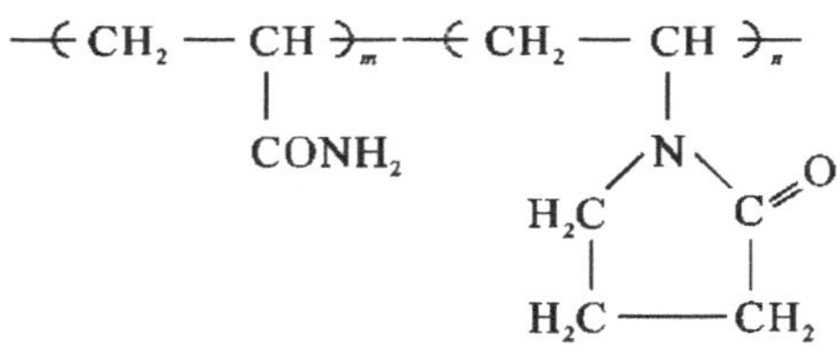

(Copolymer of acrylamide and *N*-vinyl pyrrolidone)

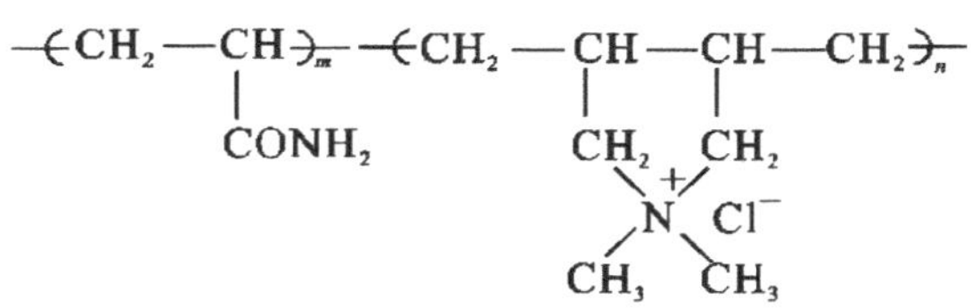

(Copolymer of acrylamide and diallyl dimethyl ammonium chloride)

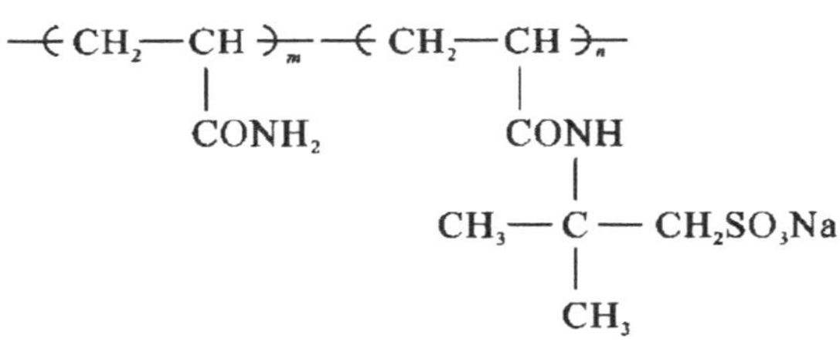

[Copolymer of acrylamide and (2-acrylamido-2-methyl) propyl sodium sulfonate]

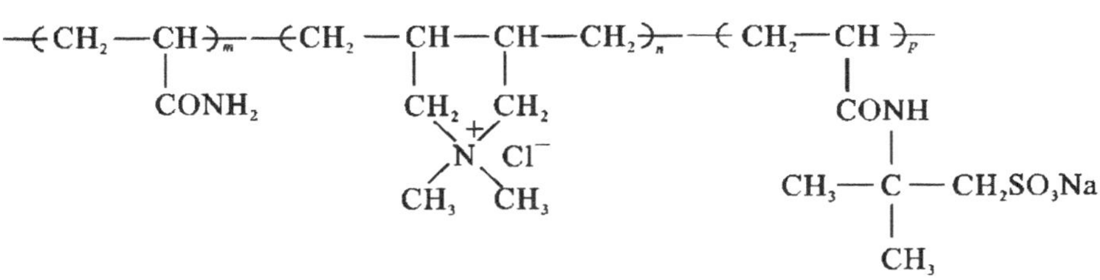

[Copolymer of acrylamide, diallyl dimethyl ammonium chloride, and (2-acrylamido-2-methyl) propyl sodium sulfonate] (Wang et al. 2007)

7.2.2 Corrosion Inhibitor

The chemical that can greatly reduce metal corrosion with a small amount is called corrosion inhibitor. Acids used to acidize formation need acidic medium inhibitors.

Acidic medium inhibitors can be divided into two types according to their function mechanism.

Adsorption Film-Type Corrosion Inhibitor

These corrosion inhibitors contain nitrogen, oxygen, and/or sulfur. All of these elements have nonbonding electron pairs in the outermost layer, which can get into empty tracks of metal structure to form ligands and form adsorption layers of inhibitor molecules on the metal surface, prohibiting the metal corrosion.

There are some important adsorption film-type corrosion inhibitors.

$R—NH_2$

(Alkylamine)

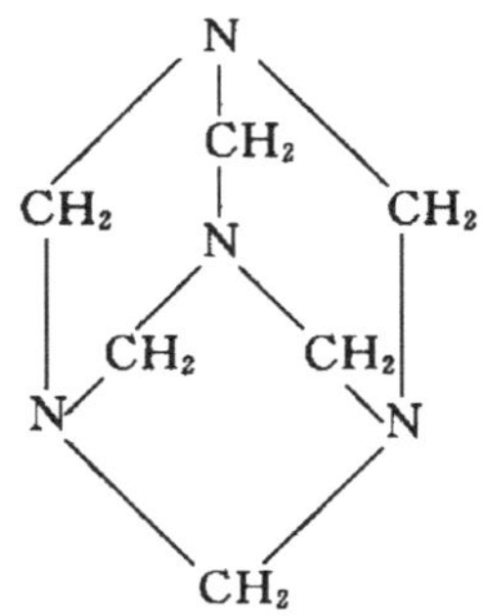

(Hexamine)

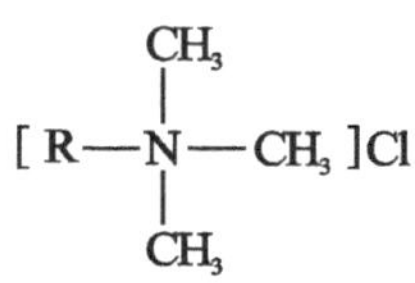

(Alkyl trimethylammonium chloride)

$OHC—(CH_2)_3—CHO$

(Glutaraldehyde)

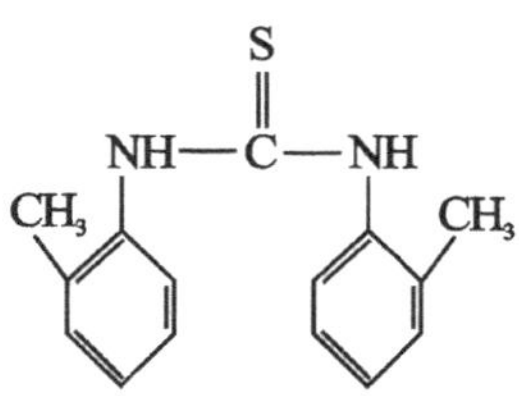

(Dio-toluenylthiourea)

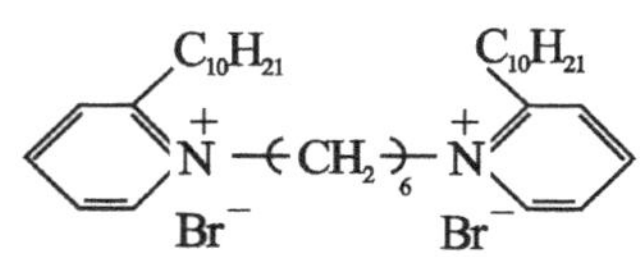

[Hexamethylenebic (bromination α-decylpyridinium) bromide] (Li et al. 2005)

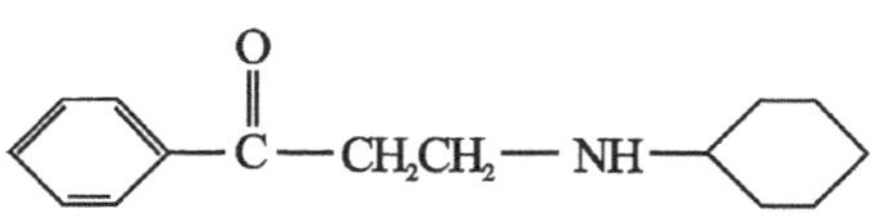

(Reaction product of benzophenone, formaldehyde and cyclohexylamine) (Zhao et al. 2007; Coffey et al. 1985)

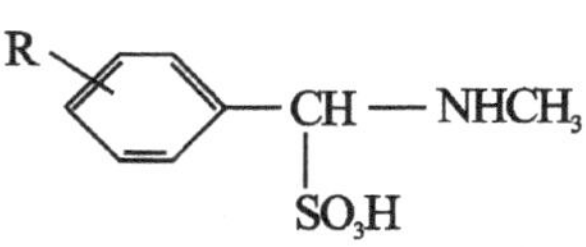

(α-methylaminoalkylbenzylidenesulfoacid)

Intermediate-Type Corrosion Inhibitor

This type of corrosion inhibitor takes effect by forming intermediate phase. For example, octyne alcohol can produce intermediate phase (Frenier et al. 1988) through the following reactions.

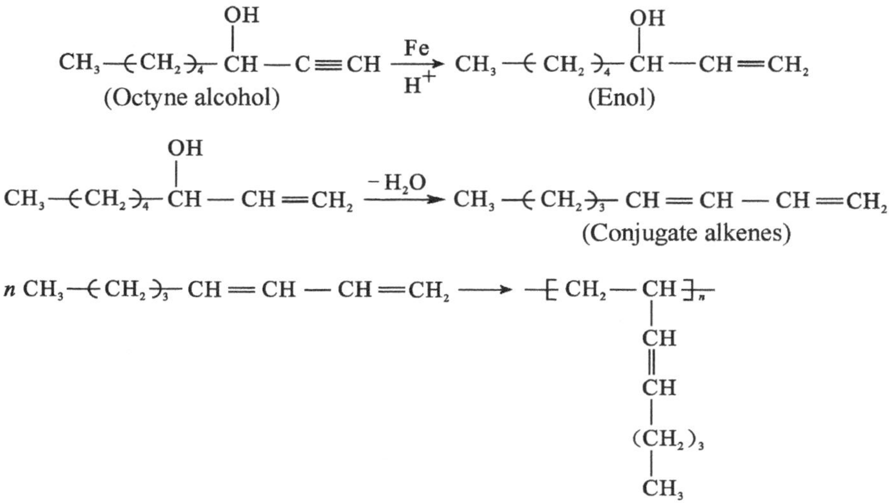

(Intermediate phase that takes corrosion inhibition effect)

Besides octynol, the following alkynols can be used as intermediate-type corrosion inhibitor.

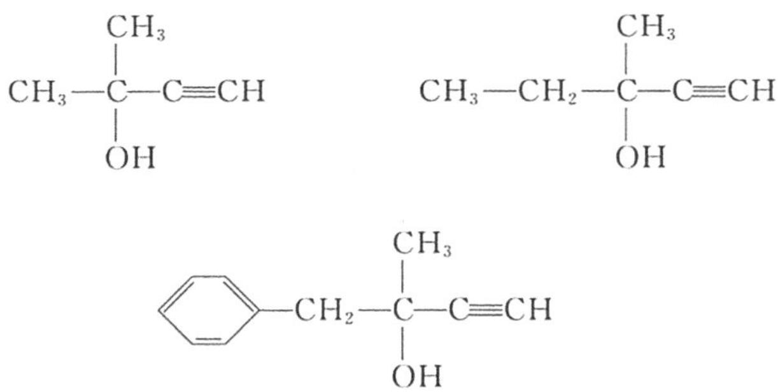

Acid corrosion inhibitors are often used by combination (Vorderbruggen et al. 2000).

7.2.3 *Iron Stabilizer*

Corrosion products of iron and steel (such as ferric oxide, ferrous sulfide) and iron minerals (such as siderite, hematite) can generate Fe^{2+} and Fe^{3+} in the spent acid after dissolving in acid.

The pH value of spent acids is generally in the range of 4–6. With a general mass fraction of 0.60%, Fe^{2+} and Fe^{3+} hydrolyze, respectively, when $pH > 7.7$ and $pH > 2.2$, to form precipitates again (also called secondary precipitation) to block the formation. Therefore, only Fe^{3+} has stabilization problems in spent acids.

$$Fe^{2+} + 2\,H_2O \rightarrow Fe(OH)_2\downarrow + 2\,H^+$$
$$Fe^{3+} + 3\,H_2O \rightarrow Fe(OH)_3\downarrow + 3\,H^+$$

The chemical that can stabilize Fe^{3+} in spent acids is called iron stabilizer (Crowe 1985).

Iron stabilizers can be divided into two types.

Complexant or Chelator

This kind of iron stabilizer can form complex or chelation with Fe^{3+}, so that it does not hydrolyze in spent acids. For example, Fe^{3+} can react with acetic acid (complexant) and EDTA (chelator), respectively, to form the following structures to stabilize Fe^{3+}.

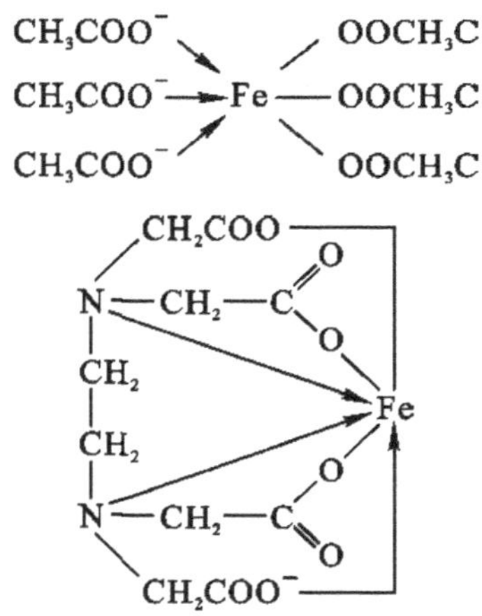

Furthermore, the following chemicals can also be used as iron stabilizers.

(Oxalic acid)

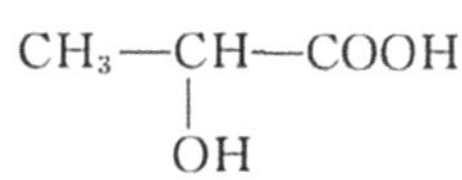
(Lactic acid)
(Taylor et al. 1998)

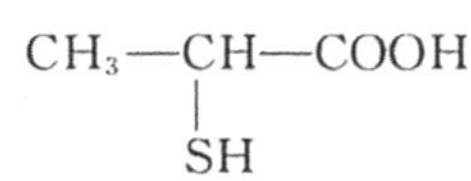
(Sulfydryl lactic acid)
(Brezinski 2002)

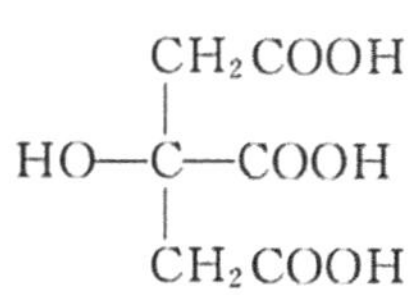
(Citric acid)

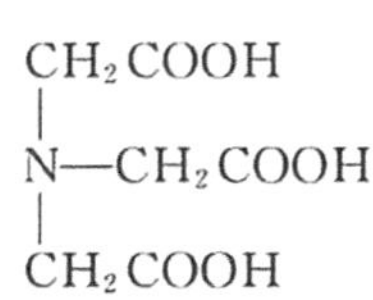
(Nitrilo- trimethylamine, NTA)

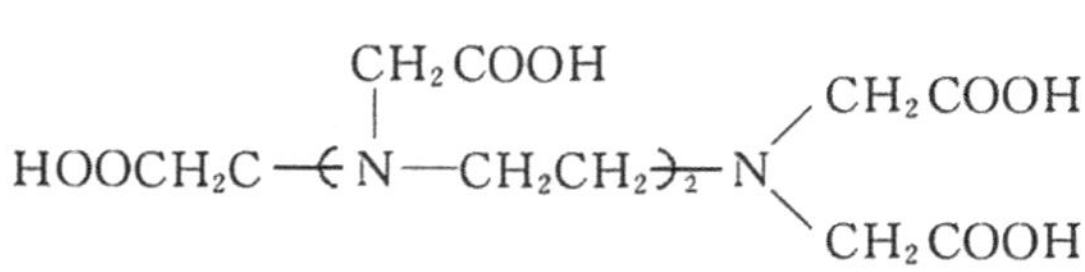
(Divinyl penteticacid, DTPA)

Reducer

Chemicals below can be used as reducers to reduce Fe^{3+} to Fe^{2+} in order to stabilize Fe in the pH range of spent acids.

CH_2O

(Formaldehyde)

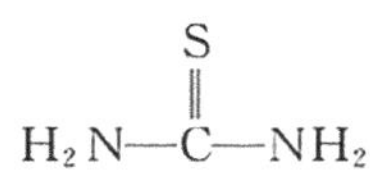

(Thiourea)

$NH_2—NH_2$

(Hydrazine)

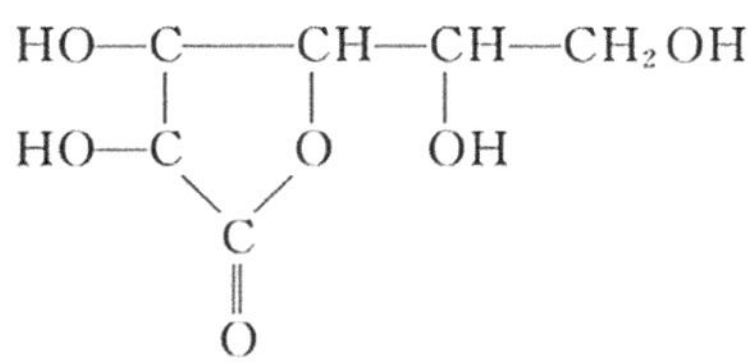

(Isoascorbic acid)

Among these reducers, isoascorbic acid is the most effective one, which can reduce Fe^{3+} to Fe^{2+} through the following reactions.

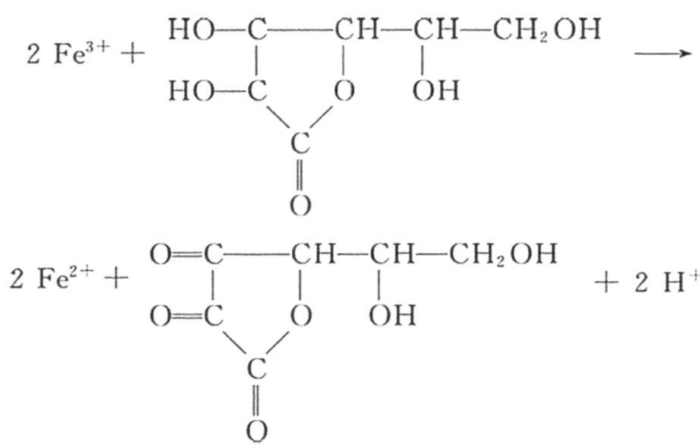

7.2.4 *Emulsion Inhibitor*

The natural surfactant in crude oil, surfactant in acid additives and rock particles generated during acidizing (particle size < 1 μm) all have certain emulsification effect. They can make crude oil and acid form emulsion and affect the flowback of spent acid.

Chemicals that can prevent crude oil and acid from forming emulsion are called emulsion inhibitors.

There are two types of emulsion inhibitors.

Surfactant with Branch Structure

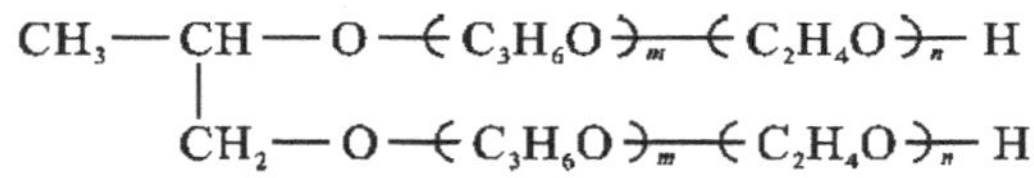

(Polyoxyethylene polyoxypropylene propylene glycol ether)

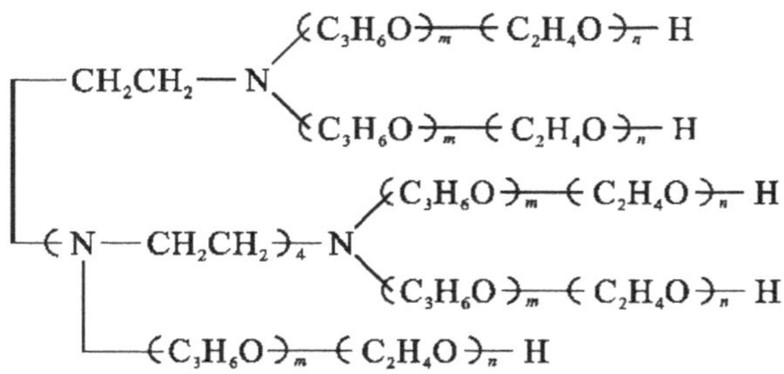

(Polyoxyethylene polyoxypropylene pentaethylenehexamine)

These surfactants can adsorb on the interface of crude oil and acid, but their branch structures cannot stabilize any kind of emulsion. Therefore, the liquid drops formed during acidizing are easy to coalescence, preventing the generation of emulsion.

Mutual Solvent

$$C_4H_9-O-CH_2CH_2OH$$

(Ethylene glycol butyl ether)

$$C_2H_5-O-[CH_2CH_2O]_2-H$$

(Diethylene glycol ethyl ether)

$$C_4H_9-O-[CH_2CH_2O]_2-H$$

(Diethylene glycol butyl ether)

$$C_4H_9-O-[CH_2CH_2O]_3-H$$

(Triethylene glycol butyl ether)

Adding mutual solvent into acid can reduce the adsorption of surfactant on the interface of crude oil and acid, facilitating the coalescence of the liquid drops formed during acidizing, thus has the effect of anti-emulsification.

7.2.5 Clay Stabilizer

The chemicals that can inhibit the swelling and particle movement of clay are called clay stabilizers (Cui 1989).

Acid is a kind of clay stabilizer which can stabilize clay by transforming the sodium soil with strong expansibility into H-clay with poor expansibility. However when the formation is subject to water injection or oil production, the sodium ions in formation water can progressively transform the H-clay into sodium soil by ion exchange to recover its expansibility. Therefore, acid belongs to impermanent clay stabilizer.

Organic cationic polymers can be added into acid to effectively stabilize the clays. Since organic cationic polymers stabilize clay by chemical adsorption, they have great resistance to temperature, acid, salinity, and fluid flushing; therefore, they belong to permanent clay stabilizer.

The following organic cationic polymers can be used as clay stabilizers.

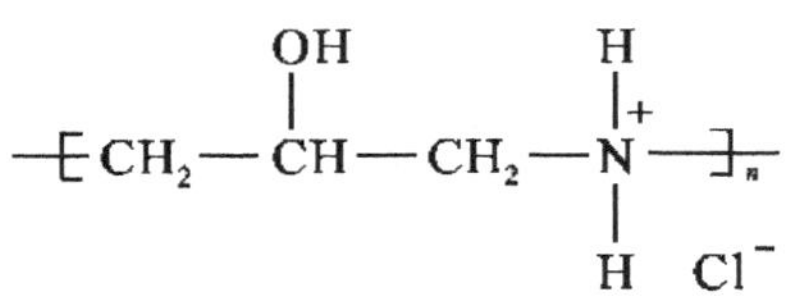

(Poly2- hydroxy-1,3- propylidene ammonium chloride)

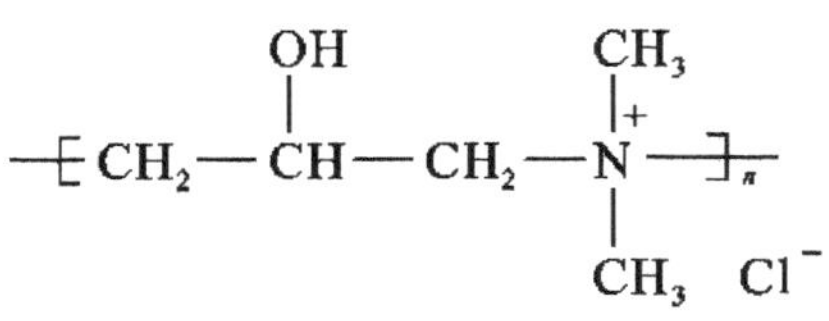

(Poly2- hydroxy-1,3- propylidene dimethyl ammonium chloride)
(Smith et al. 1983)

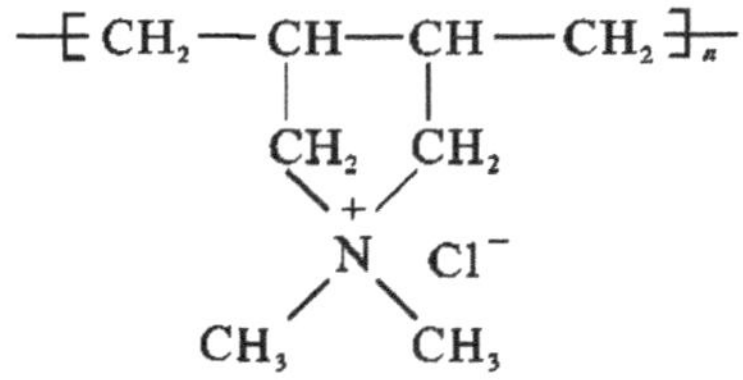

(Polydiallyl dimethyl ammonium chloride)

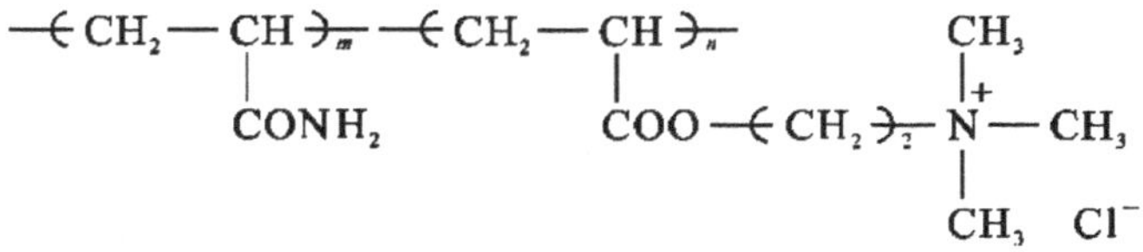

(Copolymer of polyacrylamide and crylic acid-1,2- ethylidene ester trimethyl ammonium chloride)

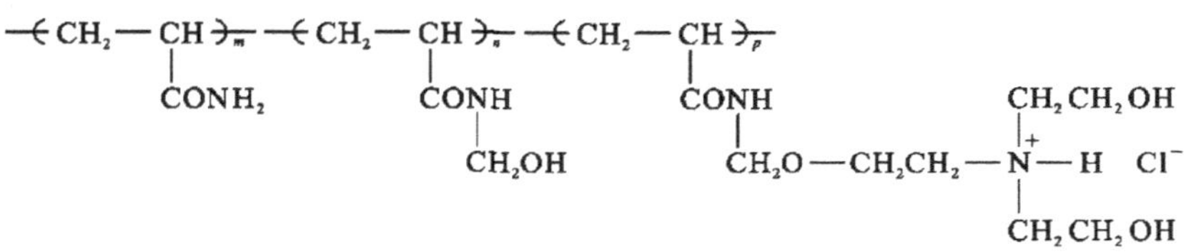

(Reaction product of hydroxymethyl acrylamide and triethanolamine hydrochloride) (Cui 1990)

7.2.6 Cleanup Additive

The chemicals that can reduce formation damage caused by secondary sedimentation and can facilitate the discharge of spent acids from formation are called cleanup additives.

There are two types of cleanup additives.

Surfactant

This kind of surfactants is acid and salt-resistant. Even in conditions of concentrated acid and high salinity, it can effectively reduce the interfacial tension and Jamin effect so that the spent acids can be discharged from the formation easily.

The available surfactants include ammonium surfactant, quaternary ammonium surfactant, pyridine salt surfactant, nonionic–anionic surfactant (Mokadam et al. 1996), and fluorine-containing surfactant.

Since fluorine-containing surfactant has great capability of reducing interfacial tension under the acidizing conditions, it is an ideal kind of cleanup additive. The following fluorine-containing surfactants can be used as cleanup additives.

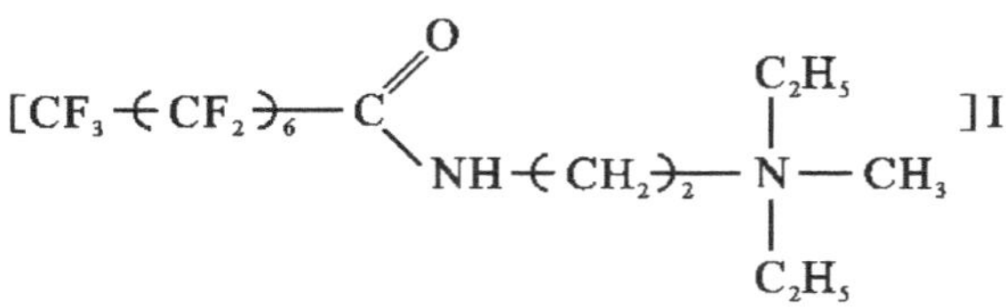

(Perfluorooctylamide-1,2- ethylidene methyl diethyl amine iodide)

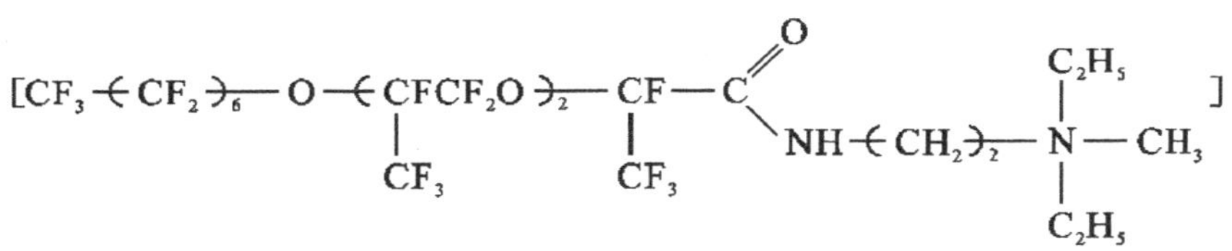

(Perfluoropolyoxypropylene heptanol ether perfluoropropanamide-1,2- ethylidene methyl diethyl amine iodide)

Energizer

Injecting a slug of energizer into formation before injecting acid can increase the near borehole pressure and facilitate the discharge of spent acids from the formation. High-pressure nitrogen is the most commonly used energizer. This kind of cleanup additive is mainly used to acidize formation with low pressure.

7.2.7 Sludge Inhibitor

Acidification sludge is generated by the reactions between the H^+ or Fe^{2+} and Fe^{3+} in acids, and the sulfurated and nitrogenous parts in colloid and asphaltene of crude oil, through the following reactions.

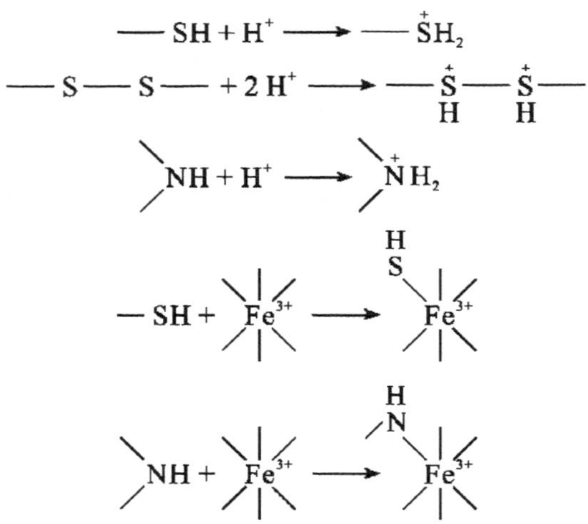

These reactions or complexations increase their polarity, decrease the solubility of colloid in oil and its stability to asphaltene solid particles, therefore colloid and asphaltene can be separated out from oil in the form of sludges. The higher the temperature and the higher the acid concentration and iron content (especially Fe^{3+}) is, the easier the sludges can be formed. Compared with formic acid and acetic acid, hydrochloric acid is more prone to form sludges.

The chemicals that can prevent acid from forming sludges when contact with crude oil are called sludge inhibitors (Dill et al. 1991; Dill 1992; Vinson 1996; Wong et al. 1997; Ge et al. 2000; Zheng et al. 2005).

There are three types of sludge inhibitors. One is oil-soluble surfactant such as fatty acid and alkyl benzene sulfonate. It can adsorb on the acid–oil interface to reduce the contact between acid and oil. While the surfactant getting into oil can combine with the sulfur parts and nitrogen parts in colloid and asphaltene according to the principle of similar polarities, so that it can reduce the reaction of colloid and asphaltene with acid and also reduce the complexation with iron ion, thus preventing the generation of sludges. The other one is iron stabilizer. It can reduce the formation of sludges through chelating the Fe^{2+} and Fe^{3+} in acids or by reducing Fe^{3+} into Fe^{2+}. The third type is aromatic hydrocarbon solvent such as benzene, toluene, xylene, or mixture of them. It can be used as buffer plug between acid and crude oil to reduce their contact and therefore reduce the sludge generation.

7.2.8 Wettability Reversal Agent

The chemicals that can change the wettability of solid surface are called wettability reversal agents.

The corrosion inhibitor in acid is a kind of wettability reversal agent. It is adsorbed in the near borehole regions and turns the hydrophilic surface into hydrophobic surface. In this way, it can reduce the formation permeability to oil and influence the effect of acidizing.

Wettability reversal agents can be applied to turn the hydrophobic surfaces in the near borehole regions back to hydrophilic surfaces.

There are two types of wettability reversal agents used for acidizing. One is surfactant such as polyoxyethylene polyoxypropylene alkanol ether, phosphate salt of polyoxyethylene polyoxypropylene alkanol ether, and their mixture (Crema 1987). Surfactant takes effect of wettability reversal in the formation surface by adsorbing on the second adsorption layer according to the principle of similar polarities. The other type is mutual solvent such as ethylene glycol butyl ether, diethylene glycol ethyl ether, and their mixture (Crema 1987). Mutual solvent takes effect by desorbing the corrosion inhibitor adsorbed on formation surface to recover the hydrophilicity of formation surface. These two types of wettability reversal agents are added into post-treatment fluid to treat the formation.

7.2.9 Diverting Agent

Diverting agent refers to the chemical that can plug high-permeability layers temporarily and make acids flow into low permeability layers to improve the effect of acidizing.

There are four types of diverting agents.

Grain Diverting Agent

This kind of diverting agent takes effect by forming filter cake at the entrance of high-permeability layer. The plugging is then removed by water-dissolving or oil-dissolving method. Water-soluble grain diverting agents include benzoic acid, boric acid particles, etc. Oil-soluble grain diverting agents include naphthalene, styrene, and vinyl acetate copolymer particles, etc.

Gel diverting Agent

Gel diverting agent is a kind of gel (e.g., chromium gel and zirconium gel) added with gel breaker (e.g., ammonium persulfate). After diverting, the gel breaker will degrade the gel and remove the plugging (Wiley 1981; Terracina et al. 1992).

Foam Diverting Agent

This kind of diverting agent plugs high-permeability layers by Jamin effect generated by the bubbles in high-permeability layers. After diverting, foam diverting agent is destroyed by oil to remove plugging (Kennedy et al. 1992; Zerhboub et al. 1994; Morphy et al. 1998; Logan et al. 1997; Kam et al. 2007).

Viscoelastic Surfactant Diverting Agent

Surfactants that are suitable to prepare this kind of diverting agent are cationic–anionic surfactants with long chains (Chang et al. 2001; Safwat et al. 2002; Al-Mutawa et al. 2003; Al-Anzi et al. Winter 2003–2004; Nasr-El-Din et al. 2007; Lungwitz et al. 2007; Zeiler et al. 2006).

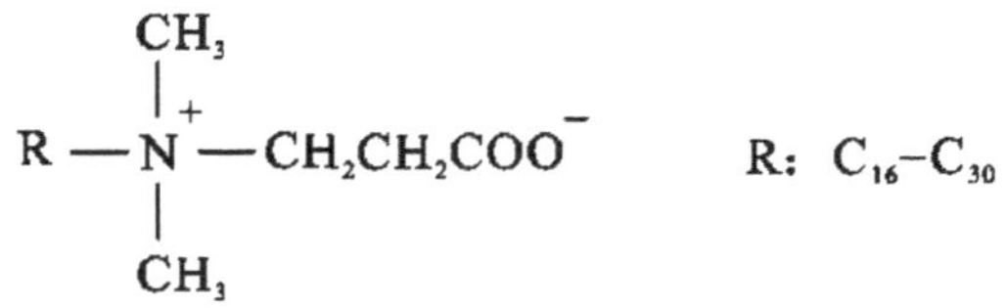

(Alkyl dimethylamino propionate)

In addition to surfactant, cosurfactant is also required.

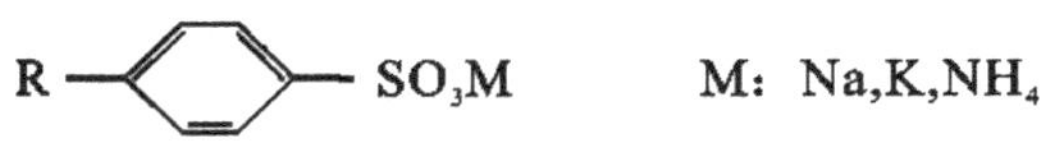

(Alkane sulfonate)

Surfactant and cosurfactant are both used by dissolving in acids. They can undergo the following chemical reactions in acids.

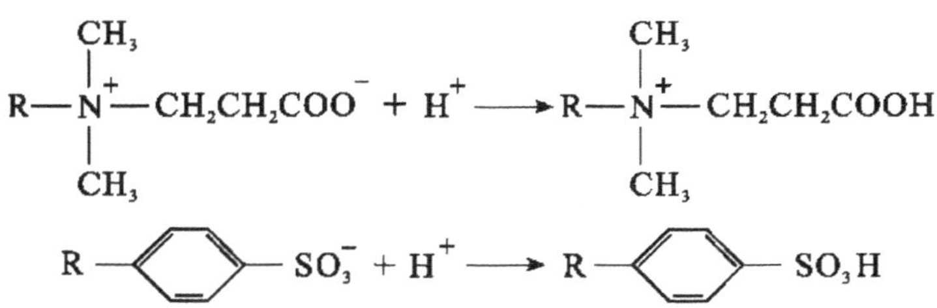

During acidizing, acids can react with the formation to produce high-valent metal ions (e.g. Ca^{2+}). Meanwhile, as the concentration of H^+ decreases gradually the balances of the two reactions move to the left. Therefore, the surfactant, cosurfactant and high valent metal ions will form structures shown below to thicken the spent acids and plug the formation, which will divert the subsequent acid to formations that are not acidized. The generated structures can be destroyed by the dissolution of formation oil or dilution of formation water after the acidizing.

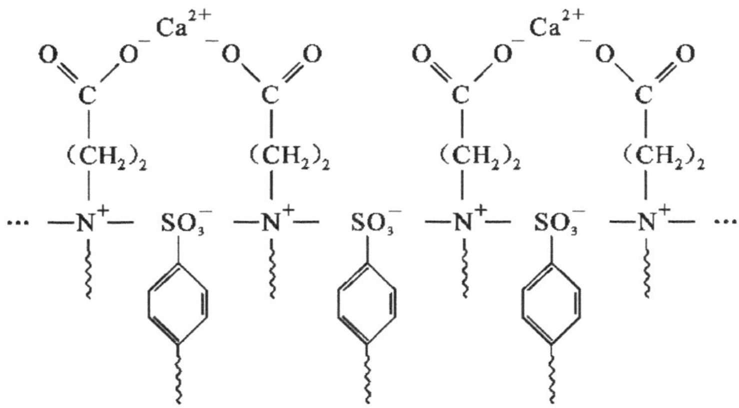

References

Al-Anazi HA, Nasr-El-Din HA, Mohamed SK (1998) Stimulation of tight carbonate reservoirs using acid in diesel emulsion: field application. SPE 39418
Al-Anzi E, Al-Mutawa M, Al-Habib N et al (Winter 2003–2004) Positive reactions in carbonate reservoir stimulation. Oilfield Rev 15(4):28–45
Al-Mutawa M, Al-Anzi E, Ravula C et al (2003) Field cases of a zero damaging stimulation and diversion fluid from the carbonate formations in Norh Kuwait. SPE 80225
Brezinski MM (2002) Well acidizing compositions. EP Patent 1,318,271,A1 12 Mar 2002
Buijse MA, Van Domelen MS (1998) Novel application of emulsified acids to matrix stimulation of heterogeneous formations. SPE 39583
Chang FF, Miller MJ (2001) A novel fluid system having controllable reversible viscosity. WO Patent 29,369,A1 26 Apr 2001
Clark GJ, Wong TCT, Mungan N (1982) New acid systems for sandstone stimulation. SPE 1066
Coffey MD, Kelly MY, Kennedy WC Jr (1985) Method and composition for corrosions. US Patent 4,493,775, 15 Jan 1985
Crema SC (1987) Acidizing concentrates for oil well acidizing system. US Patent 4,676,916, 30 Jun 1987
Crowe CW (1985) Evaluation of agents for preventing precipitation of ferric hydroxide from spent treating acid. JPT 37(4):691–695
Cui G (1989) Anti-clay swelling agent. J Pet Univ (Nat Sci Ed) 13(1):102–109
Cui G (1990) Study on the polyhydroxy cationic polymer anti-clay swelling agent. J Pet Univ (Nat Sci Ed) 14(1):26–30
Deng Y (1985) Sulfamic acid treatment of oil-water well. Oil Drill Prod Technol 7(3):83–86
Dill WR, Lane JL Jr (1991) Method for preventing the formation of sludge in crude oil. US Patent 5,058,678, 23 Oct 1991
Dill WR (1992) Method and composition for preventing the formation of sludge in crude oil. US Patent 5,084,192, 28 Jan 1992

Frenier WW, Growcock FB, Lop VR (1988) Mechanisms of corrosion inhibitors used in acidizing wells. SPE 14092

Ge J, Zhao F (2000) The generation, inhibition and cleanup of acidizing sludge. Oilfield Chem 17 (4):378–382

Kam SI, Frenier WW, Davies SN et al (2007) Experimental study of high-temperature foam for acid diversion. J Petrol Sci Eng 58(1–2):138–160

Kennedy DK, Kitziger FW, Hall BE (1992) Case study on the effectiveness of nitrogen foam and water zone diverting agents in multistage matrix acid treatments. SPE Pet Eng 7(2):203–211

King GE (1986) Acidizing concepts—matrix vs. fracture acidizing. JPT 38(5):507–508

Kunze KR, Shaughnessy CM (1980) Acidizing sandstone formation with fluoboric acid. SPE 9387

Li H, Jiang X, Zhong F et al (2005) Corrosion inhibition performance of two kinds of bipyridinium salts to A3 steel in hydrochloric acid. Oilfield Chem 22(1):42–43, 84

Logan ED, Bjomen KH, Sarver DR (1997) Foamed diversion in the chase series of Hugoton field in the mid-continent. SPE 37432

Lullo GD, Ahmad AJ (1996) Acid treatment method for siliceous formations. US Patent 5,529,125, 25 Jun 1996

Lullo GD (1996) A new acid for true stimulation of sandstone reservoirs. SPE 37015

Lungwitz B, Fred DC and Brady M et al (2007) Diversion and cleanup studies of viscoelastic surfactant-based self-diverting acid. SPE Prod & Oper 22(1):121–127

Mokadam AR, Strubelt CE, Williams DA et al (1996) Surfactant additive for oilfield acidizing. CA Patent 2,175,613, 6 Aug 1996

Morphy PH, Greenwald KG, Herries PE (1998) Operational experience with foam diverted acid jobs in the Gulf of Mexico. SPE 39423

Nasr-El-Din HA, Al-Habib NS, Al-Mumen AA et al (2006) A new effective stimulation treatment for long horizontal wells drilled in carbonate reservoirs. SPE Prod & Oper 21(3):330–338

Nasr-El-Din HA, Samuel M (2007) Lessons learned from using viscoelastic surfactants in well stimulation. SPE Prod & Oper 22(1):112–120

Roddy CW, Koch RR, Todd BL (2005) Composition and methods for the delivery of chemical components in subterranean well bores. US Patent 0,167,104,A1, 4 Aug 2005

Safwat M, Nasr-El-Din HA, Dossary K et al (2002) Enhancement of stimulation treatment of water injection wells using a new polymer-free diversion system. SPE 78588

Smith CW, Borchardt JK (1983) Clay stabilization during oil and gas well cementing operations. US Patent 4,393,939, 19 Jul 1983

Taylor KC, Nasr-El-Din HA, Al-Alawi MJ (1998) A systematic study of iron control chemicals used during well stimulation. SPE 39419

Terracina JM, Mccabe MA, Gramer JW et al (1992) Temporary chemical block brings efficiency and economy to eastern U.S. workover operations. SPE Prod Eng 7(1):70–74

Vinson EF (1996) A novel reducing agent for combating iron induced crude oil sludging: development and case histories. SPE 31127

Vivian TA (1982) Acidification of subterranean formations employing halogenated hydrocarbons. US Patent 4,320,014, 16 Mar 1982

Vorderbruggen MA, Williams DA (2000) Acid corrosion inhibitor. GB Patent 2,351,077,A, 20 Dec 2000

Wang J, Jia J, Zhao X et al (2007) The study on the high temperature acidification of gelling agent YJN-1. Oilfield Chem 24(2):113–116

Watanabe DJ (1981) Method for acidizing high temperature subterranean formations. US Patent 4,267,887, 19 Apr 1981

Wiley CB (1981) Success of a high friction diverting gel in acid stimulation of a carbonate reservoir—Cornell Unit, San Andres field. JPT 31(11):2196–2200

Wong TC, Hwang RJ, Beaty DW (1997) Acid sludge characterization and remediation improve well productivity and save costs in the Permain Basin. SPE Prod & Facil 12(1):51–58

Zeiler C, Alleman D, Qu Q (2006) Use of viscoelastic-surfactant-based diverting agents for acid stimulation: case histories in GOM. SPE Prod & Oper 21(4):448–454

Zerhboub M, Ben-Naceur K, Touboul E et al (1994) Matrix acidizing: a novel approach to foam diversion. SPE Prod Facil 9(2):121–126

Zhao F (1986) Retarded acid. Oilfield Chem 3(3):179–190

Zhao X, Yang Y, Qiu G (2007) Study on the aromatic ketone mannich base acidizing corrosion inhibitor. Mater Prot 40(2):60–62

Zheng Y, She Y, Cheng H (2005) Anti-sludge agents which can enhance the acidizing effect. J Oil Gas 27(5):658–660

Chapter 8
Fracturing Fluids and Fracturing Fluid Additives

Fracturing is one of the measures to increase production and injection by cracking the formation with high pressure. The cracks are then supported by proppants to reduce the flow resistance.

8.1 Fracturing Fluids

The liquids used in the fracturing process are fracturing fluids (Jennings 1996). A good fracturing fluid should meet the requirements of high viscosity, low friction, low filtrate volume, minimum formation damage, easy preparation, vast material source, low cost, etc. There are mainly two types of currently used fracturing fluids, which are water-base fracturing fluids and oil-base fracturing fluids.

8.1.1 Water-Base Fracturing Fluids

The fracturing fluids using water as solvent or dispersion medium are water-base fracturing fluids (Westland et al. 1993; Funkhouser et al. 2003). The following fracturing fluids are water-base fracturing fluids.

8.1.1.1 Thickened Water Fracturing Fluids

Thickened water fracturing fluids are prepared by dissolving thickening agent in water. There are lots of thickening agents available. Some important thickening agents are given in Table 8.1. The dosage of thickening agent is decided by the required viscosity of fracturing fluid, whose mass fraction usually lies in the range of 0.5–5%.

C. Dai and F. Zhao, *Oilfield Chemistry*,
https://doi.org/10.1007/978-981-13-2950-0_8

Table 8.1 Some important thickening agents

Synthetic polymer	Natural polymer and its modified product		Biopolymer
	From cellulose (C)	From galactomannan (GM) such as guar gum (GG)	
Polyoxyethylene (PEO)	Methylcellulose (MC)	Methyl galactomannan (MGM)	Xanthan gum (XC)
Polyvinyl alcohol (PVA)	Carboxymethyl cellulose (CMC)	Carboxymethyl galactomannan (CMGM)	Scleroglucan (SG)
Polyacrylamide (PAM)	Hydroxyethyl cellulose (HEC)	Hydroxyethyl galactomannan (HEGM)	Reticular bacterial cellulose (RBC)
Partially hydrolyzed polyacrylamide (HPAM)	Carboxymethyl hydroxyethyl cellulose (CMHEC)	Carboxymethyl hydroxyethyl galactomannan (CMHEGM)	
Copolymer of acrylamide and acrylate (AM–AA)	Hydroxyethyl carboxymethyl cellulose (HECMC)	Hydroxyethyl carboxymethyl galactomannan (HECMGM)	
Copolymer of acrylamide and acrylic ester (AM–AAE)		Hydroxypropyl galactomannan (HPGM)	
Copolymer of acrylamide and (2-acrylamide-2-methyl) propyl sulfonate (AM–AMPS)		Carboxymethyl hydroxypropyl galactomannan (CMHPGM)	
Copolymer of acrylamide, acrylate and (2-acrylamide-2-methyl) propyl sulfonate (AM–AA–AMPS)		Hydroxypropyl carboxymethyl galactomannan (HPCMGM)	

During the preparation of thickened water fracturing fluid, the dosage of thickening agent can be reduced through the synergistic effect of mixed thickening agents, because the thickening capacity of mixed thickening agents is higher than that of any single thickening agent used alone under the same conditions (Zhao et al. 1987). This effect can be observed in the mixed using of guar gum (GG) and partially hydrolyzed polyacrylamide (HPAM) (Table 8.2).

8.1.1.2 Water-Based Gel Fracturing Fluids

Water-based gel fracturing fluid is prepared mainly by water, gelling agent, cross-linker, and gel breaker. Gelling agent can be any one of the thickening agents

Table 8.2 The synergism between GG and HPAM

Introduction		Viscosity[a] (mPa • s)
Solution	w(GG) = 1%	1,720
	w(HPAM) = 1%	2,240
The volume ratio of GG solution to HPAM solution	1:1	3,750
	2:1	3,670
	5:1	3,400

Note [a]Measure by Brookfield LVF viscometer (25 °C, 20 r/min)

listed in Table 8.1. The choice of cross-linker is dependent on the cross-linkable groups and the cross-linking conditions of thickening agents (Table 8.3). Gel breakers are mainly peroxides which can oxidize and degrade gel. Water-based gel fracturing fluids have many advantages such as high viscosity, low friction, low filtration, and minimum formation damage (Zhao 1989; Nimerik 1997; Gawiezel et al. 1992; Moorhouse et al. 1997; Mitchell et al. 2001; Dobson et al. 2003; Willberg et al. 2004; Dawson et al. 2005).

Three representative cross-linking reactions are shown in Table 8.3. The first one is the cross-linking reaction between formaldehyde and PAM.

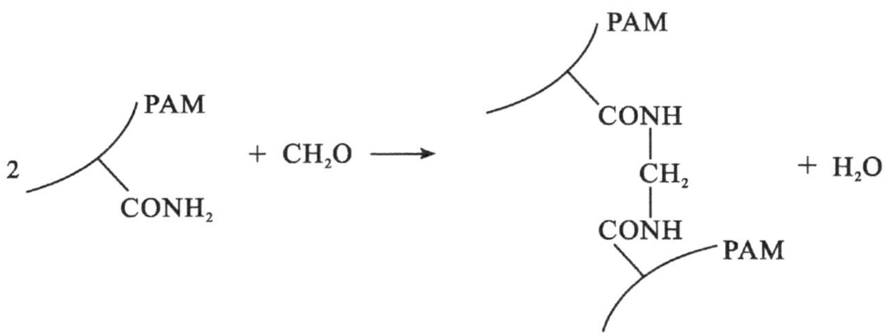

The second one is the cross-linking reaction between boric acid and GM.

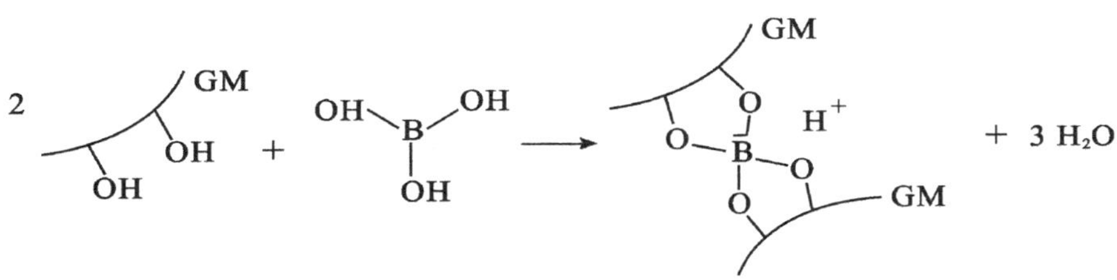

The third one is the cross-linking reaction between multinuclear olation complex ions of chromium and HPAM.

Table 8.3 The cross-linking groups, cross-linkers, and cross-linking conditions of thickening agents

The cross-linking groups of thickening agents	Typical polymer	Cross-linkers	Cross-linking conditions
$-CONH_2$	HPAM PAM	Aldehyde, dialdehyde and hexamine	Cross-linked in acidic condition
$-COO^-$	HPAM CMC CMGM XC	$AlCl_3$, $CrCl_3$, $K_2Cr_2O_7 + Na_2SO_3$, $ZrOCl_2$, $TiCl_4$	Cross-linked in acidic or neutral condition
Ortho-*cis*-hydroxy	GM CMGM HPGM CMHPGM PVA	Boric acid, sodium tetraborate, sodium pentaborate, organoboron, organic zirconium and organic titanic	Cross-linked in alkaline condition

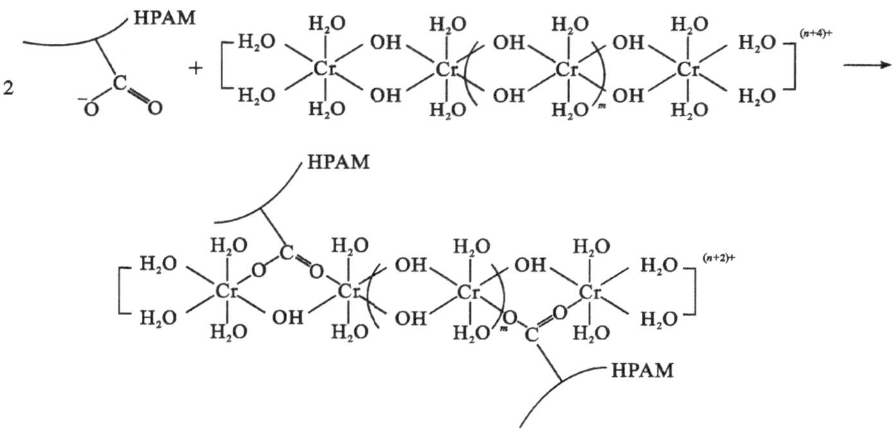

8.1.1.3 Viscoelastic Surfactant Fracturing Fluids

Viscoelastic surfactant fracturing fluids are prepared mainly by water, surfactant with long carbon chain, water-soluble salt, and/or alcohol.

The surfactant with long carbon chain can be anionic, cationic, non-ionic, and amphiprotic. Some examples are given below (Mathew et al. 1997; Liu et al. 2001; Hughes et al. 2001; Yang J 2002; Qu et al. 2002; Colaco et al. 2006; Dahayanake et al. 2007).

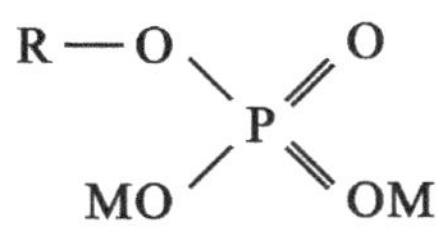

R: C_{16}–C_{30};
M: Na,K,NH_4

(Alkyl phosphate)

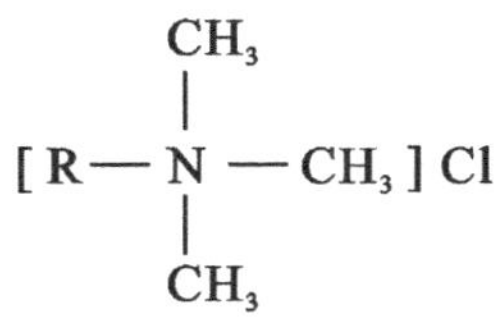

R: C_{16}–C_{30}

(Alkyltrimethylammonium chloride)

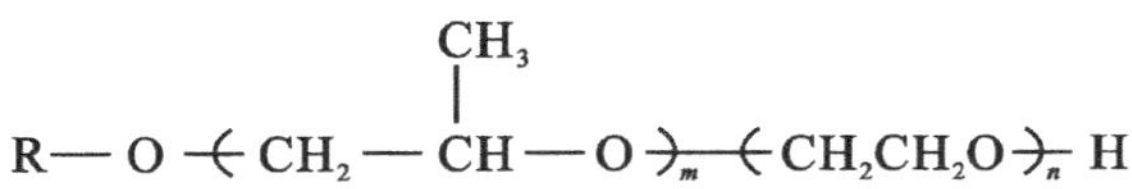

R: C_{16}–C_{30}

(Polyoxyethylene polyoxypropylene alky lether)

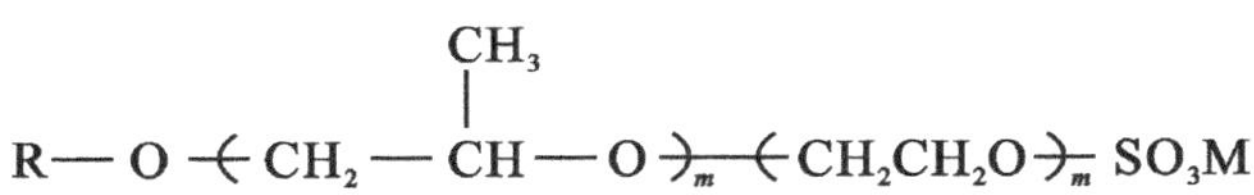

R: C_{16}–C_{30}

(Polyoxyethylene polyoxypropylene alky lether sulfate salt)

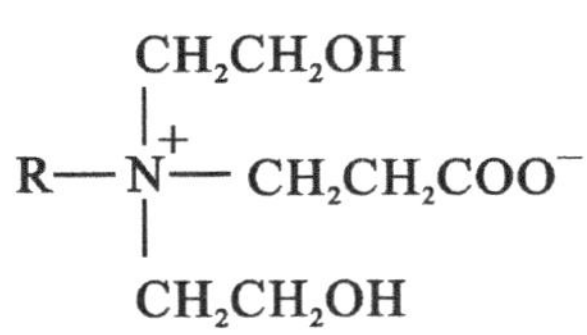

R: C_{16}–C_{30}

[Alkyl di(2-Hydroxyethyl) carboxyethyl quaternary ammonium salt]

Available water-soluble salts include inorganic salt and organic salt such as potassium chloride, potassium nitrate, and sodium salicylate.

Available water-soluble alcohols include ethyl alcohol and isopropanol

When surfactant with long carbon chain is dissolved in certain concentration of salt and/or alcohol solution, the water-solubility of surfactant is inhibited by salt and/or alcohol. Thus, the long carbon chains associate to form linear micelles, which are entangled with each other to generate structure (Fig. 8.1). In this way, the viscosity of fracturing fluid is enhanced so that it can carry sand to fracture the formation.

The linear micelles of surfactants show viscoelasticity under stretching.

Viscoelastic surfactant fracturing fluids have characteristics such as non-residue, low damage, and shear stable.

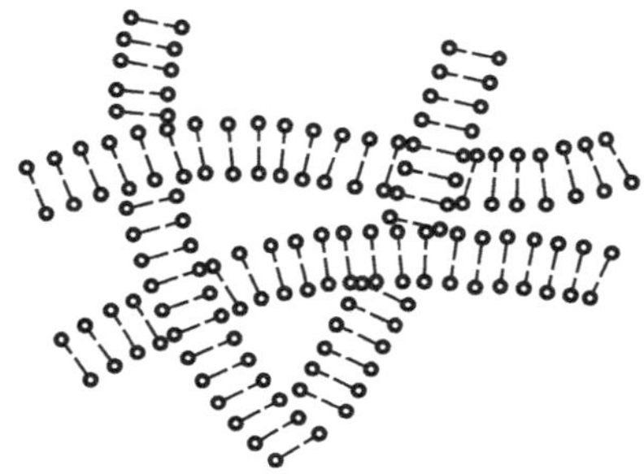

Fig. 8.1 The structure generated by the linear micelles of surfactants

8.1.1.4 Oil-in-Water Fracturing Fluids

Oil-in-water fracturing fluids are prepared mainly by water, oil, and emulsifier. Available water includes freshwater, brine, and thickened water. Available oils include crude oil or its fractions such as kerosene, diesel, and condensate. As shown in Table 8.4, available emulsifiers include ionic surfactant, non-ionic surfactant, and amphoteric surfactant. The HLB value should be in the range of 8–18. The mass

Table 8.4 The emulsifier used in oil-in-water fracturing fluids

Surfactants	Examples
Ionic types	$R—OSO_3Na$ R: C_{10}–C_{18} $R—SO_3Na$ R: C_{10}–C_{18} 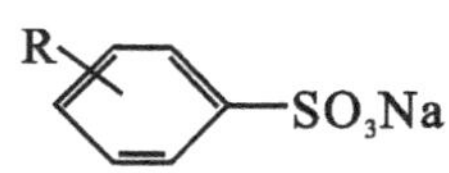 R: C_8–C_{14} $R—COONa$ R: C_9–C_{17} Quaternary ammonium salt surfactant Pyridine salt surfactant
Non-ionic types	$R—O—[CH_2CH_2O]_n—H$ R: C_{10}–C_{18}; $n>2$ 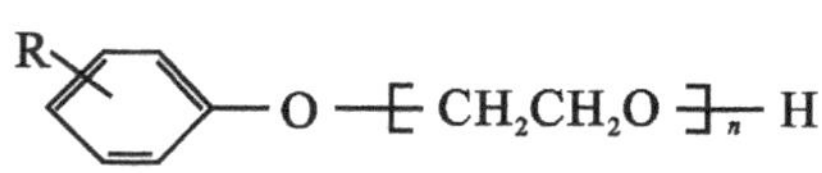 R: C_8–C_{14}; $n>2$ Twain surfactant
Amphoteric types	$R—O—[CH_2CH_2O]_n—SO_3Na$ R: C_{10}–C_{18}; n:3–5 $R—O—[CH_2CH_2O]_n—CH_2SO_3Na$ R: C_{10}–C_{18}; n:3–5 $R—O—[CH_2CH_2O]_n—CH_2COONa$ R: C_{10}–C_{18}; n:3–5

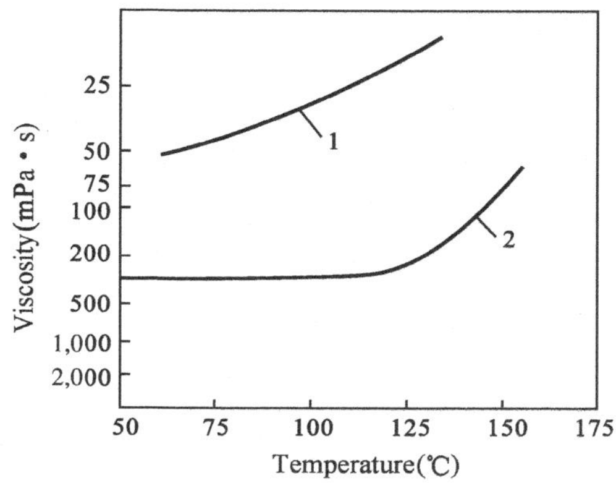

Fig. 8.2 The viscosity–temperature relationships of thickened water and oil-in-water emulsion. 1—thickened water; 2—oil-in-water emulsion

fraction of emulsifier in water phase should be in the range of 1–3%. The volume ratio of oil to water should be in the range of 50:50 to 80:20. Compared with thickened water, oil-in-water emulsion has better viscosity–temperature relationship (Fig. 8.2).

The oil-in-thickened water fracturing fluids (polymer emulsions) are prepared using thickened water as external phase and oil as internal phase. This kind of fracturing fluid can be used in high temperature (160 °C), has good drag reduction performance and can automatically break the emulsion and drain the fluid. With cationic surfactants as emulsifier, surfactants are easily adsorbed on the surface of formation to break the emulsion. If non-ionic surfactants with cloud point below the formation temperature are used as emulsifiers, emulsifiers will separate out and break the emulsion when formation temperature is higher than the cloud point of emulsifier. Then fracturing fluids can be easily discharged from the formation since viscous emulsion transforms into oil and water with low viscosity.

8.1.1.5 Oil-in-Water-Based Gel Fracturing Fluids

When the cross-linker which can cross-link the polymers in thickened water is added into the oil-in-thickened water fracturing fluid, oil-in-water-based gel fracturing fluid can be obtained. Compared with the oil-in-thickened water fracturing fluid, oil-in-water-based gel fracturing fluid has stronger capacities of sand-carrying and reducing filtrate loss. Gel breaker needs to be added into the fracture fluids so that they can be easily discharged from the formation after use (Kakadjian et al. 2001).

8.1.1.6 Water-Based Foam Fracturing Fluids

The fracturing fluids using water as dispersion medium and gas as dispersion phase are called water-based foam fracturing fluids.

Water-based foam fracturing fluids are mainly composed of water, gas, and foaming agent. Available water includes freshwater, brine, and thickened water. Available gases include carbon dioxide, nitrogen, and natural gas. Available foaming agents include alkane sulfonate, alkylbenzene sulfonate, alkane sulfovinate, quaternary ammonium salt, and OP surfactant. The mass fraction of foaming agent in water is generally in the range of 0.5–2%. The foam characteristic value should be in the range of 0.5–0.9.

Water-based foam fracturing fluids have advantages such as low viscosity (yet strong sand-carrying capacity), low friction, low filtration loss, low water content, easy to be discharged and small damage to the formation.

Carbon dioxide and nitrogen are preferably used in liquid form during foam fracturing. When using carbon dioxide, temperature and pressure should be adjusted to make carbon dioxide in liquid form, which can be injected into the formation together with the sand-carrying foaming agent solution. When the temperature of the mixture is over the critical temperature of carbon dioxide (31 °C), foam is produced due to the conversion of liquid carbon dioxide into the gas state. The liquid nitrogen transported to field by tank truck is heated slightly to get high-pressure nitrogen gas at the wellhead, which can be mixed with sand-carrying foaming agent solution to generate foam.

Foam can also be prepared using mixed gases of carbon dioxide and nitrogen. Because nitrogen plays a role in the former stage and carbon dioxide plays a role as dissolved-gas drive in the latter stage during the drainage process, this kind of foam is more conducive to the drainage after fracturing (King 1980; Almond et al. 1984; Harris 1999; Chatterji et al. 2003; Zhou et al. 2004).

8.1.1.7 Water-Based Gel and Foam Fracturing Fluids

The fracturing fluids using water-based gel as dispersion medium and gas as dispersion phase are called water-based gel and foam fracturing fluids.

Water-based gel and foam fracturing fluids are prepared by water, gas, thickening agent, foaming agent, cross-linker, and gel breaker. During fracturing process, thickened solution of foaming agent is prepared by dissolving thickening agent and foaming agent in water. Then gas is injected to generate foam. In the end, cross-linker and gel breaker are injected. The cross-linker can cross-link the thickening agents in foam external phase into gel, producing water-based gel and foam fracturing fluids. The gel breaker makes the fracturing fluids easily discharged from the formation after fracturing.

The characteristics of water-based gel and foam fracturing fluids can be concluded from the characteristics of water-based gel fracturing fluids and water-based foam fracturing fluids.

8.1.2 Oil-Based Fracturing Fluids

The fracturing fluid using oil as solvent or dispersion medium is called oil-based fracturing fluid. The following fracturing fluids are oil-based fracturing fluids which are suitable for fracturing water-sensible formations (containing swelling clay) (Smith et al. 1998; Elgassier et al. 2008).

8.1.2.1 Thickened Oil Fracturing Fluids

Thickened oil fracturing fluid is prepared by dissolving the thickening agent in oil. Available thickening agents are divided into three types.

Fatty Acid Soap

Carbon atom number of fatty acid should be greater than 8 so that the fatty acid soap can dissolve in oil. Surpassing certain concentration, the fatty acid soap can form structure in oil to generate structure viscosity, thickening the oil. For example, after surpassing certain concentration, the polar portions of fatty acid sodium soap and fatty acid calcium soap can combine with each other according to the rules of similar polarity, to form the following structure and generate structure viscosity, thus thickening the oil.

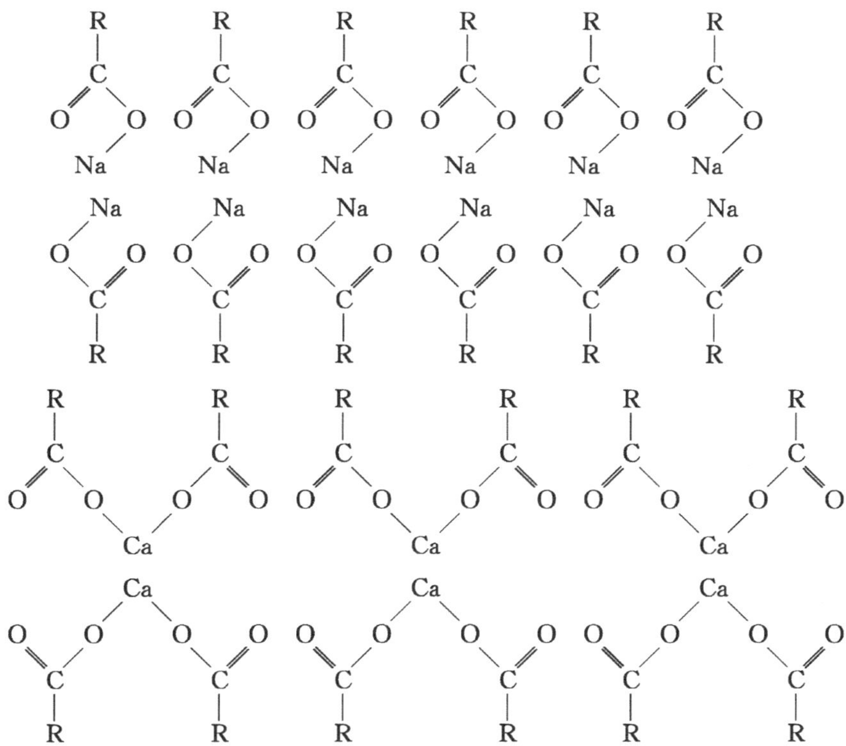

There are two types of fatty acid aluminum soap—single soap and double soap.

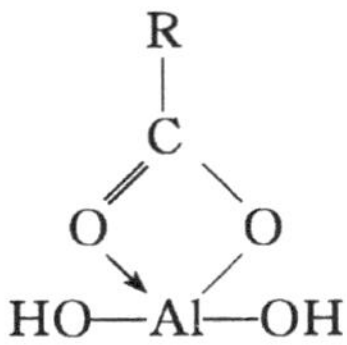

(Single)

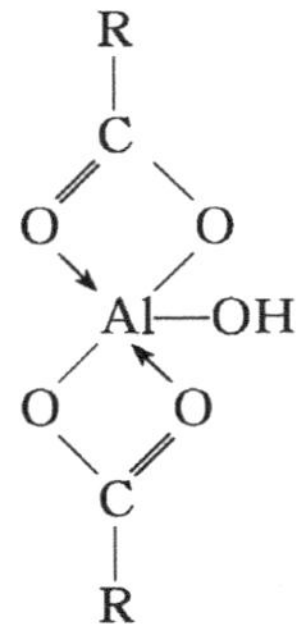

(Double)

They can be linked through ol bridge to form the following structure and thicken the oil.

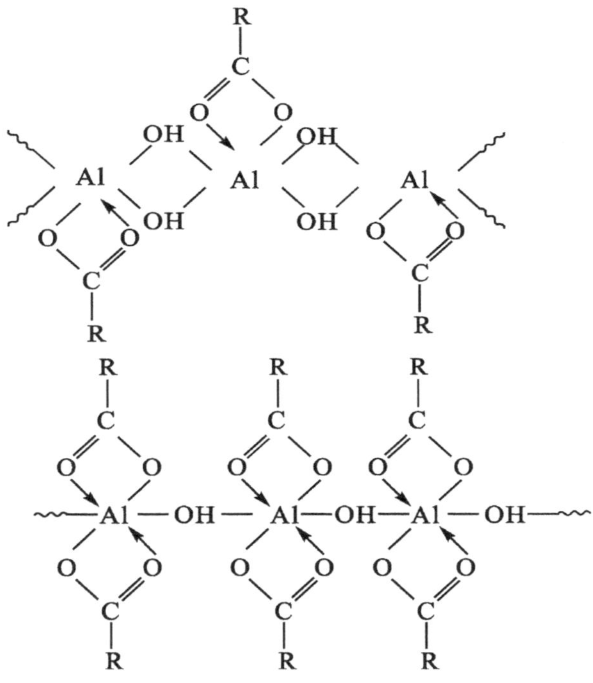

Phosphate

Phosphate is produced by the reaction of alcohol (ROH, R is C_1–C_{20}) with phosphorus pentoxide (Maberry et al. 1997; Syrinek et al. 1988; Maberry et al. 1998).

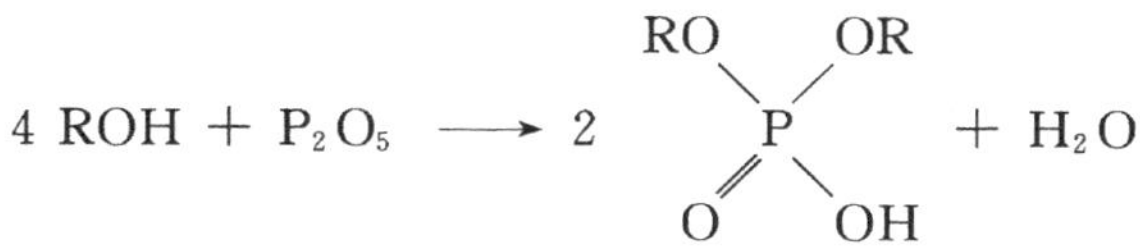

To prepare oil-based fracturing fluids, phosphate is dissolved in oil followed by activation by aluminum salt such as aluminum nitrate and aluminum chloride. The following products are generated.

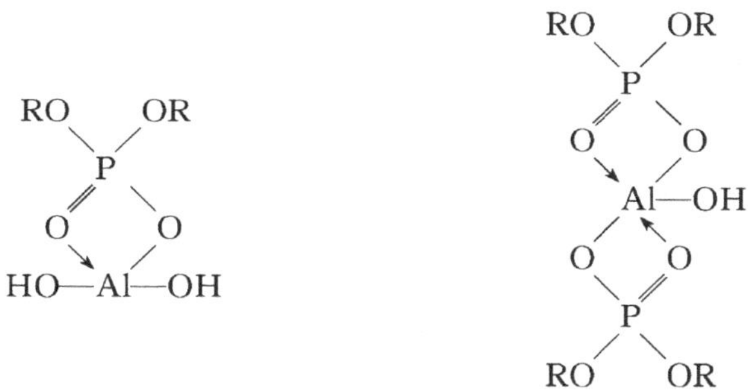

These products can be connected by ol bridge in oil to form the following structure, thickening the oil.

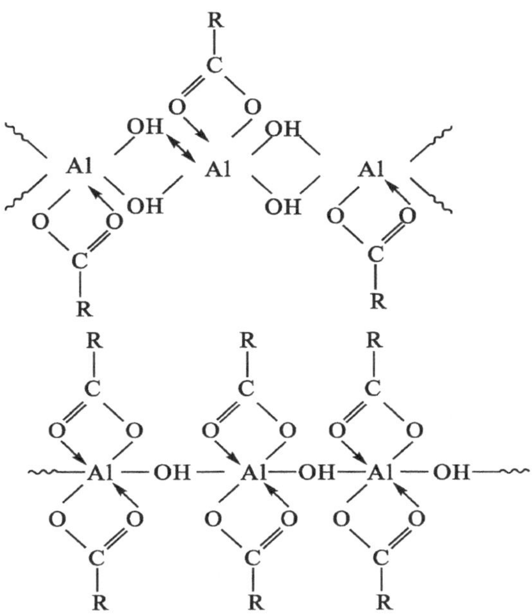

OilSoluble Polymer

The oil-soluble polymer reaching certain concentration in oil can form structure to generate structure viscosity, thickening the oil. The following oil-soluble polymers can be used to thicken the oil.

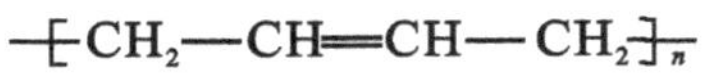

(Cis-polybutadiene)

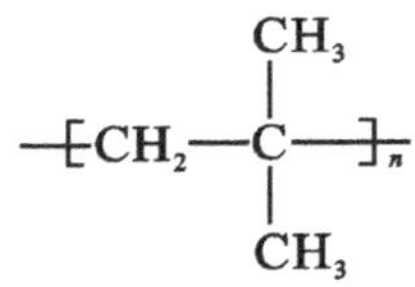

(Polyisobutene)

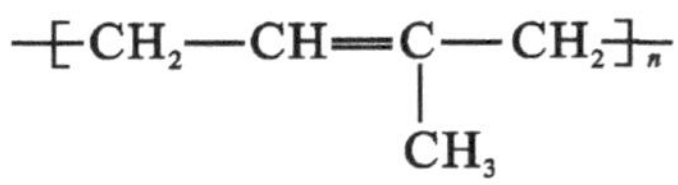

(Polyisoprene)

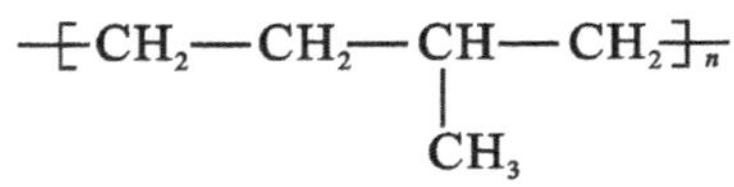

(Hydrogenated polyisoprene)

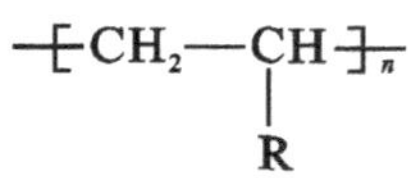

R: C_6–C_{20}

(α-olefin)

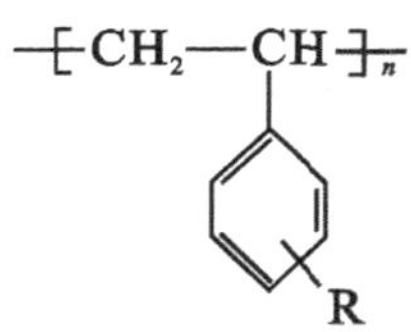

R: C_2–C_{16}

(Polyalkylstyrene)

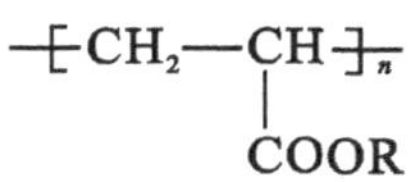

R: C_{14}–C_{40}

(Polyacrylate)

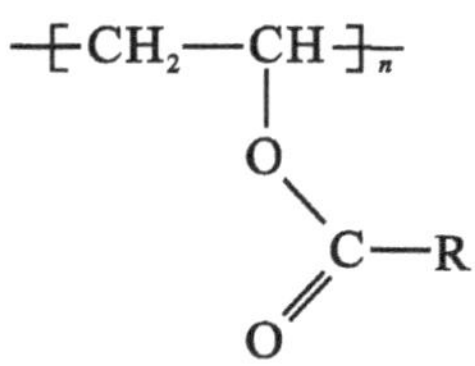

R: C_{15}–C_{35}

(Polyethylated carboxylate)

8.1.2.2 Oil-Based Gel Fracturing Fluids

When the concentration of thickening agent in oil is large enough, thickened oil fracturing fluid will transform into oil-based gel fracturing fluid (Smith et al. 1995). Since the oil-based gel has higher viscosity and stronger sand-carrying capability than that the thickened oil, oil-based gel fracturing fluid can be used for fracturing deeper formation.

8.1.2.3 Water-in-Oil Fracturing Fluids

Water-in-oil fracturing fluid is mainly composed of oil, water, and emulsifier. The available oils include crude oil, diesel, or kerosene. The available water includes freshwater or brine. The emulsifier can be Span 80 or lauric diethanolamide (soluble in oil and water, respectively). With oil as dispersion medium and water as dispersion phase, the advantages of this fracturing fluid include high viscosity, high sand-suspending capacity, low filtration, and small reservoir damage; however, its disadvantage is the high flow friction. The acid-in-oil fracturing fluid can be prepared by replacing the water in water-in-oil fracturing fluid with acid. The acid-in-oil fracturing fluid can not only reduce the corrosion of acid to the pipeline, but also release acid after the demulsification to enlarge the fractures and improve the fracturing performance.

8.1.2.4 Oil-Based Foam Fracturing Fluids

Oil-based foam fracturing fluid is composed of oil, gas, and foaming agent. The available oils are crude oil, diesel oil, and kerosene. The available gases include carbon dioxide and nitrogen. The available foaming agent can be the following fluoropolymers, where R_1 is H or F, R_2 is H or CH_3, R_3 is the alkyl of C_{10}–C_{20}, m and n are values making the mass ratio of polymer segments in the range of 25:75 to 60:40, q is 1 or 2, and r is in the range of 2–10.

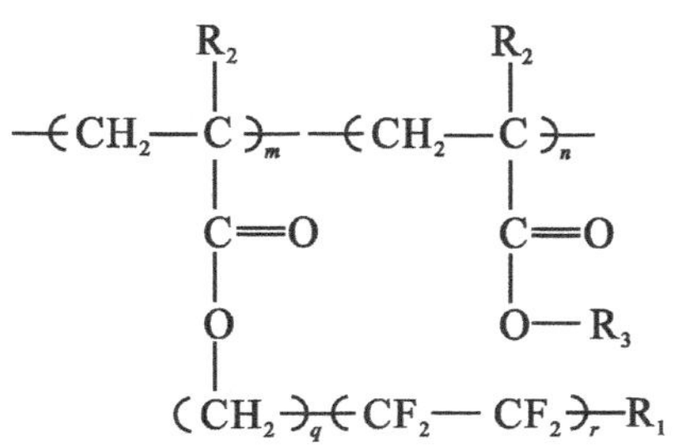

(Alkyl fluoroalkylated acrylate and alkylated acrylate copolymer)

The mass fraction of foaming agent in oil should be in the range of 0.05–5% (Raynolds et al. 1984).

Oil-based foam fracturing fluid can also be prepared into oil-based gel foam fracturing fluid (Vezza et al. 2001).

8.2 Fracturing Additives

To improve the fracturing performance, many additives such as demulsifier, clay stabilizer, cleanup additive, wettability reversal agent, proppant, breaker, friction reducer, and filtrate reducer are added into the fracturing fluids. The former four kinds are the same as those used in acidizing, thus only the latter four kinds of additives are introduced here.

8.2.1 Proppant

Proppant refers to the support material which is carried into the fracture by fracturing fluid to support the fracture when the pressure is released. A good proppant should have low density, high strength and good chemical stability, and also be cost-effective.

The particle size of proppant is generally in the range of 0.4–1.2 mm.

Natural proppants include quartz sand, bauxite, alumina, zircon, and walnut shells.

High-strength proppants include sintered bauxite (sintered ceramsite), aluminum alloy ball, and plastic ball.

Low-density proppants include microporous sintered bauxite (microporous sintered ceramsite) (Rickards et al. 2006) and walnut shells.

Proppants with good chemical stability are the proppants coated by resin (such as phenolic resin) or organosilicone.

If the resin coating of the proppant is completely or partly changed to the special resin which can solidify in the formation whose temperature surpasses 54 °C, the proppant will not only bear good chemical stability, but also be fixed stably within the fracture to support it.

A certain proportion of solid particles with special purpose can also be mixed into proppant. For example, water-swelling particulate, paraffin inhibitor, scale inhibitor, demulsifier, and corrosion inhibitor can be used during oil well fracturing, while clay stabilizer and fungicide can be employed during water well fracturing. After fracturing, these solid particles with special purpose can play corresponding role in the process of oil production and water injection.

8.2.2 *Breaker*

The breaker of fracturing fluid refers to the chemical agents which can significantly reduce the viscosity of fracturing fluid in a certain period of time. Since the destroyed fracturing fluid is easy to be discharged from the stratum, the pollution of fracturing fluid to the stratum can be reduced (Rae et al. 1996).

The breaker of gel fracturing fluid is gel breaker.

The gel breakers of water-based gel fracturing fluid can be classified into the following types.

8.2.2.1 Peroxide

Peroxide is the chemical compound containing peroxy (–O–O–). Some examples are given below (Terracina et al. 1999).

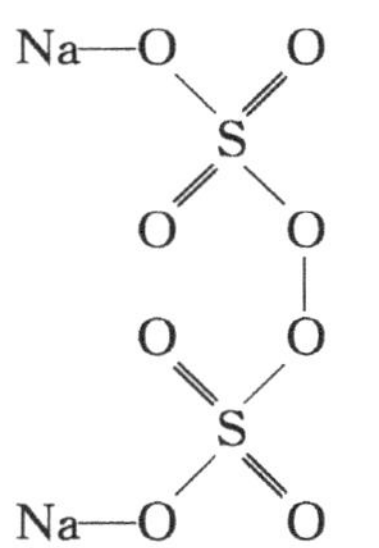

(Sodium persulfate)

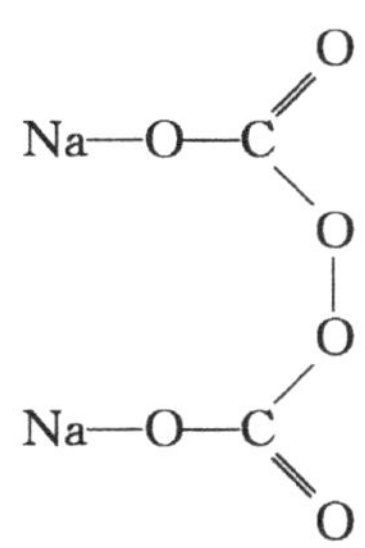

(Sodium percarbonate)

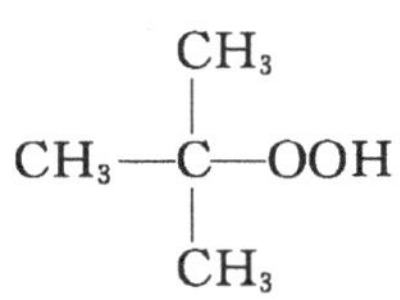

(Isobutanol peroxide)

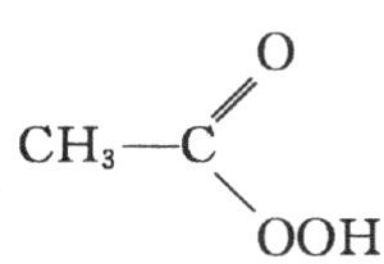

(Peracetic acid)

Peroxides can destroy the gel structure by oxidative degradation of polymer.

8.2.2.2 Enzyme

Enzyme, such as α-amylase, β-amylase, cellulase, hemicellulase, sucrose, and maltase, is a kind of special protein which can destroy the gel structure by catalyzing the hydrolytic degradation of glycan. Enzymes can only be used when temperature is lower than 65 °C and pH is in the range of 3.5–8.0 (Malone et al. 2000).

8.2.2.3 Potential Acid

Potential acid refers to the substance which can change into acid under certain conditions. Some examples are given below.

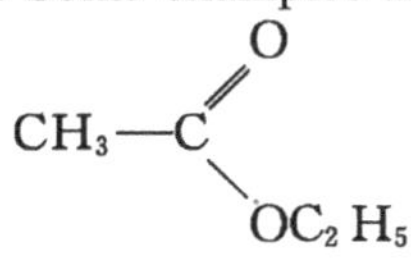

(Ethyl acetate)

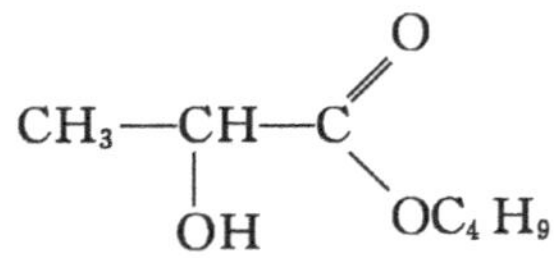

(α-hydroxy butyl propionate)

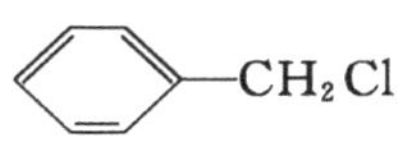

(Benzyl chloride)

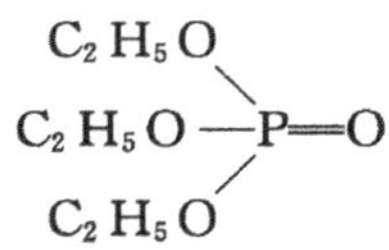

(Triethyl phosphate)

Potential acid functions by destroying the cross-linked structure of gel through condition (pH) changing, instead of destroying the polymer.

8.2.2.4 Potential Chelating Agent

Potential chelating agent refers to chemical agents which can change into chelating cross-linker under certain conditions (Tackett 1992; Dawson et al. 2004).

(Dimethyl oxalate)

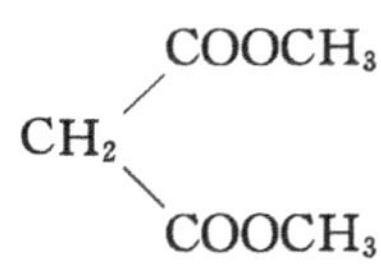

(Dimethyl malonate)

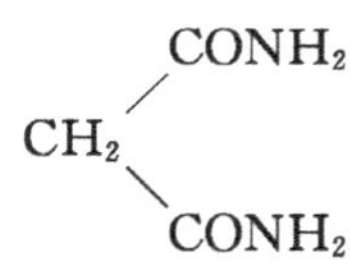

(Malonamide)

The above-mentioned potential chelating agents can hydrolyze to produce oxalic acid or malonic acid under the condition of below 60, 60 °C and above 60 °C, respectively, to chelate metal ions (cross-linker) and destroy the cross-linked gel structure.

Oil-based gel fracturing fluid is prepared by the activation of phosphate using aluminum salt. The available gel breakers include sodium acetate and sodium phenylacetate (Newhouse et al. 2000), whose function mechanism is to break oil-base gel by competitive complexation. For example, the gel breaking of oil-based gel produced by the thickening of aluminum phosphate salt by sodium acetate can be explained by the following reaction.

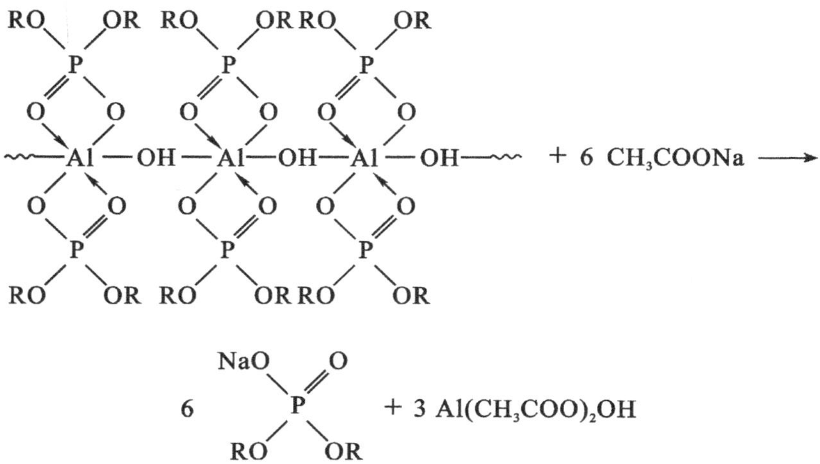

The breakers of viscoelastic surfactant fracturing fluid are formation oil and formation water. These surfactants can be dissolved by formation oil or diluted by formation water, resulting in the destruction of the structure of twisted linear micelles. In addition, other breakers of viscoelastic surfactant fracturing fluid can also be applied, including alcohol (such as ethyl alcohol and isopropyl alcohol) injected after fracturing and ester (such as ethyl acetate, alkyl sulfate, and polyacrylate) injected together with fracturing fluid, which can hydrolyze to produce corresponding alcohol to break the fracturing fluid (Lee et al. 2004; Dantas et al. 2006).

8.2.3 Friction Reducer

Friction reducer of fracturing fluid refers to chemical agents which can reduce the flow resistance of fracturing fluid in turbulent state by storing the turbulence energy. The mechanism will be introduced in the section of friction reduction mechanism of pipeline oil friction reducer.

Friction reducers usually are polymers. A kind of polymer can be thickening agent and friction reducer at the same time. HPAM can be thickening agent when used in high concentration, whereas, it can be friction reducer when used in low concentration (Fig. 8.3).

Water-based fracturing fluid uses water-based friction reducer such as the following polymers (Johnston et al. 1998; Ely 1981).

Fig. 8.3 The influence of mass concentration of HPAM on friction reduction efficiency

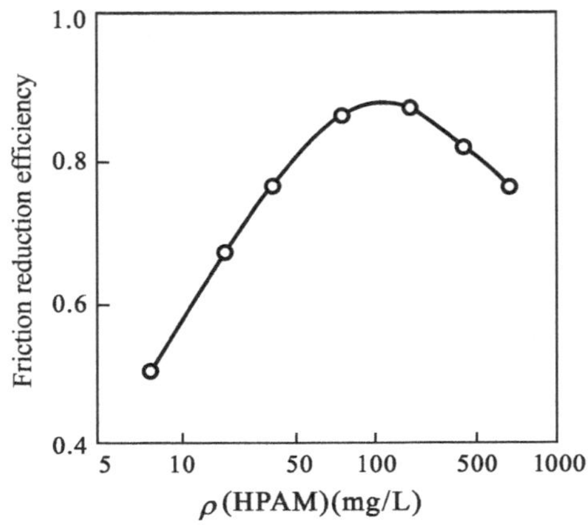

(Polyethylene glycol)

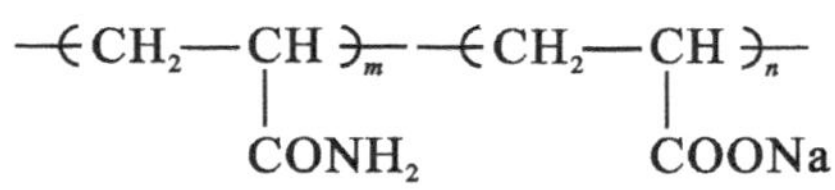

(Acrylamide and sodium acrylate copolymer)

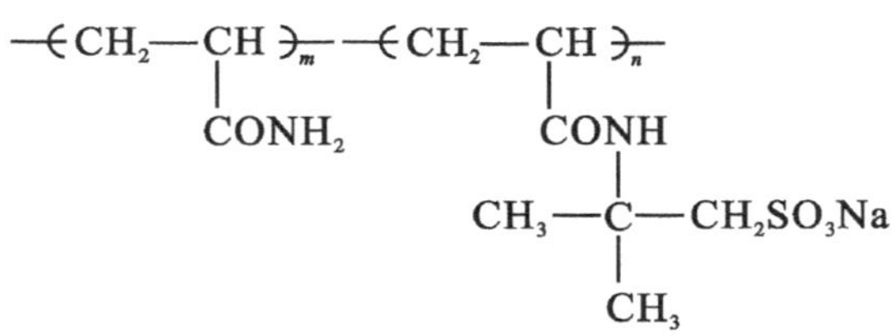

[Acrylamide and (2-acrylamido-2-methyl) propyl sulfonic acid sodium copolymer]

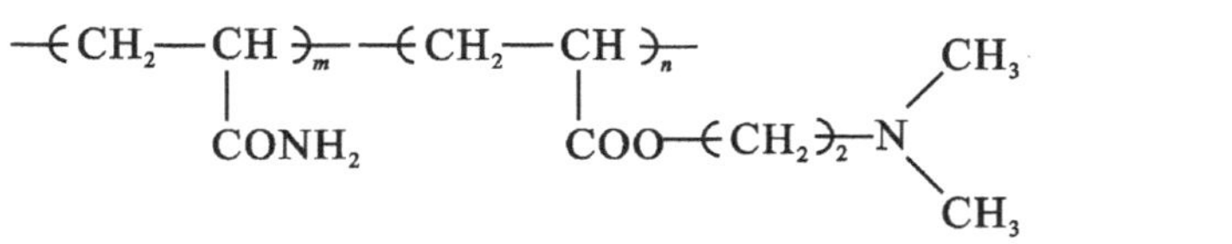

(Acrylamide and crylic acid-1， 2-ethyl dimethyl amine copolymer)

Oil-based fracturing fluid uses oil-based friction reducer such as the following polymers (Joyce et al. 1999).

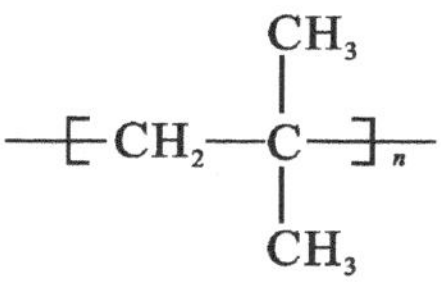

(Polyisobutene)

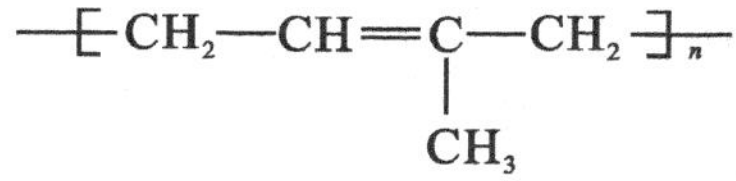

(Polyisoprene)

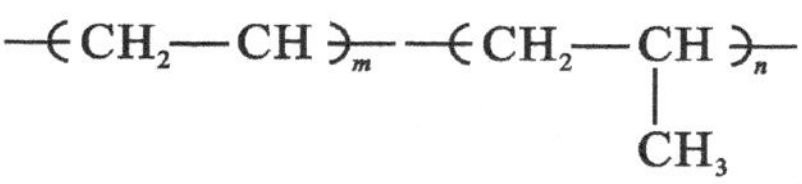

(Ethylene and propylene copolymer)

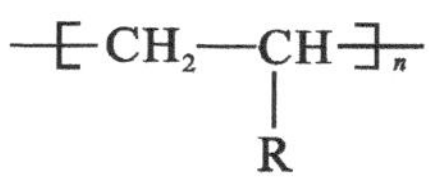

R: C_6– C_{20}

(Poly-α-olefine)

8.2.4 Filtration Reducer

Filtration reducer refers to chemical agents which can reduce the leak-off of fracturing fluid from fractures to formation. It can decrease the pollution of fracturing fluid on formation and can increase pressure sharply during fracturing.

Filtration reducer can be dissolved after fracturing to avoid polluting formation. Filtration reducer includes the following categories.

- Water-soluble filtration reducer such as particles of water-soluble polymer (Sierra et al. 2005; Githens et al. 1994; Gabrysch et al. 1998), water-soluble soap (Blair et al. 2003), and water-soluble salt (Shah et al. 2006).
- Oil-soluble filtration reducer such as particles of wax, naphthalene, anthracene, rosin, rosin dimer, polystyrene, and copolymer of styrene and vinyl toluene, and copolymer of ethylene and vinyl acetate.
- Acid-soluble filtration reducer such as quartz powder and calcium carbonate powder.

References

Almond SW, Harris PC (1984) Fracturing method for stimulation of wells utilizing carbon dioxide based fluids. US Patent 4,480,696, 6 Nov 1984

Blair CC, Chang KT, Treybig DS et al (2003) Use of dispersion polymers as friction reducers in aqueous fracturing fluids. WO Patent 085,232,A1, 16 Oct 2003

Chatterji J, Crook R, King L (2003) Foamed fracturing fluids, additives and methods of fracturing subterranean zones. US Patent 0,004,067, 2 Jan 2003

Colaco A, Marchand JP, Li F et al (2006) Viscoelastic surfactant fluids having enhanced shear recovery, rheology and stability performance. US 2006/0105919A1, 18 May 2006

Dahayanake MS, Yang J, Niu JHY et al (2007) Viscoelastic surfactant fluids and related methods of use. US Patent 0,249,505,A1, 25 Oct 2007
Dantas TNC, Santanna VC, Neto AAD et al (2006) Methodology to break test for surfactant-based fracturing gel. J Petrol Sci Eng 50(3–4):293–298
Dawson JC, Le HV (2005) Fracturing fluids and methods of making and using same. US Patent 6,844,296,B2, 18 Jan 2005
Dawson J, Kesavan S, Le HV (2004) Breaker system for fracturing fluids used in fracturing oil bearing formations. US Patent 6,767,868, 27 July 2004
Dobson JW Jr, Hayden SL, Hinojosa BE (2003) Crosslinked polymer fluids and crosslinking concentrates therefor. EP Patent 1,331,358,A1, 30 July 2003
Elgassier MM, Stolyarov SM (2008) Reasons for oil-based hydraulic fracturing in western Siberia. SPE 112092
Ely JW (1981) Methods of water flooding and fracturing using clean, non-damaging fracturing fluids. US Patent 4,265,311, 5 May 1981
Funkhouser GP, Norman LR (2003) Synthetic polymer fracturing fluid for high-temperature applications. SPE 80236
Gabrysch A, Chesser BG (1998) Use of sized salts as bridging agent for oil based fluids. EP 817818, 14 Jan 1998
Gawiezel KE, Elbel JL (1992) A new system for controlling the crosslinking rate of borate fracturing fluid. SPE Prod Eng 7(3):275–279
Githens CJ, Harrison WG (1994) Method for using soap as a soluble fluid loss additive in the hydraulic fracturing treatment of oil and gas wells. US Patent 5,301,751, 12 Mar 1994
Harris PC (1999) Foamed fracturing fluid. US Patent 5,990,052, 23 Nov 1999
Hughes T, Jones T, Tustin G (2001) Viscoelastic surfactant based gelling composition for wellbore service fluids. US Patent 6,232,274, 15 May 2001
Jennings AR Jr (1996) Fracturing fluids—Then and now. JPT 48(7):604–610
Johnston RL, Lee YN (1998) Nonaqueous drag reducing suspensions. WO Patent 16,586, 23 Mar 1998
Joyce VA, Navarrete R, Constien VG et al (1999) Fluid loss control. US Patent 5,929,002, 27 July 1999
Kakadjian S, Rauseo O, Marquez R et al (2001) Crosslinked emulsion to be used as fracturing fluids. SPE 65038
King GE (1980) Low fluid loss foams. US Patent 4,217,231, 12 Aug 1980
Lee J, Nelson E (2004) Viscosity reduction of viscoelastic surfactant based fluids. WO Patent 007,904,A1, 22 Jan 2004
Liu X, Yi M, Zhao J et al (2001) Viscoelastic surfactant based fracturing fluids. Oilfield Chem 18 (3):273–277
Maberry LJ, Mcconnell SB, Hinkel JJ (1997) New complexation chemistry provides improved continuous mix gelled oil. SPE 37227
Maberry LJ, Mcconnell SB, Tanner KV et al (1998) Chemistry and field application of an improved continuous-mix gelled oil. SPE Prod Facil 13(4):236–242
Malone MR, Nelson SG, Jackson R (2000) Enzyme breaker technology increases production, Grayburg-Jackson field, southeast new Mexico: a case history. SPE 59709
Mathew S, Roger JC, Nelson EB et al (1997) Polymer free fluid for hydraulic fracturing. SPE 38622, pp 554–559
Mitchell TO, Parris MD (2001) High temperature hydraulic fracturing fluid. US Patent 6,277,295, B1, 8 May 2001
Moorhouse R, Cottrell IW (1997) Carboxyalkyl substituted polygalactomannan fracturing fluids. US Patent 5,697,444, 16 Dec 1997
Newhouse DP, Lai QJ (2000) Use of oil-based gel-breaker/inhibitor compounds with polymer gels in well treatments. US Patent 6,133,204, 17 Oct 2000
Nimerik KH (1997) Borate crosslinked fracturing fluid and method. US Patent 5,681,796, 28 Oct 1997

Qu Q, Nelson EB, Willberg DM, et al (2002) Compositions containing aqueous viscosifying surfactants and methods for applying such compositions in subterranean formations. US Patent 6,435,277, 20 Aug 2002
Rae P, Di Lullo G (1996) Fracturing fluid and breaker systems—a review of the state-of-the-art. SPE 37359
Raynolds S, Fournier LB (1984) Hydrocarbon foams. US Patent 4,432,882, 21 Feb 1984
Rickards AR, Brannon HD, Wood WD et al (2006) High strength ultralight weight proppant lends new dimensions to hydraulic fracturing applications. SPE Prod Oper 21(2):212–221
Shah SN, Kamel A, Zhou Y (2006) Drag reduction characteristics in straight and coiled tubing—an experimental study. J Petrol Sci Eng 53(3–4):179–188
Sierra L, Dalrymple ED, Eoff LS et al (2005) Method useful for controlling fluid loss in subterranean treatments. US Patent 0,199,396,A1, 15 Sept 2005
Smith KW, Persinski LJ (1998) Hydrocarbon gels useful in formation fracturing. EP Patent 817,819, 14 Jan 1998
Smith KW, Persinski LJ (1995) Hydrocarbon gels useful in formation fracturing. US Patent 5,417,287, 23 May 1995
Syrinek AR, Huddleleston DA (1988) Hydrocarbon gellant. US Patent 4,781,845, 1 Nov 1988
Tackett JE Jr (1992) In situ reversible crosslinked polymer gel used in hydrocarbon recovery applications. US Patent 5,082,056, 21 Jan 1992
Terracina JM, Mccabe MA, Shuchart CE et al (1999) Novel oxidizing breaker for high-temperature fracturing. SPE Prod Facil 14(2):144–146
Vezza M, Martin M, Thompson JE et al (2001) Morrow production enhanced by new, foamed, oil-based gel flacturing fluid technology. SPE 67209
Westland JA, Lenk DA, Penny GS (1993) Rheological characteristics of reticulated bacterial cellulose as a performance additive to fracturing and drilling fluids. SPE 25204
Willberg D, Nagl M (2004) Method for preparing improved high temperature fracturing fluid. US Patent 6,820,694,B2, 23 Nov 2004
Yang J (2002) Viscoelastic wormlike micelles and their applications. Curr Opin Colloid Interface Sci 7(5–6):276–281
Zhao F, Wang J, Sun Y (1987) Synergism of viscosity in component systems. J East China Petrol Inst 11(2):36–46
Zhao F (1989) Water-base titanium gel fracturing fluid. China Patent 85,105,346.7, 18 Oct 1989
Zhou J, Zhu W, Lu Y et al (2004) Studies and uses of carbon dioxide foamed hydrofracturing fluid. Oilfield Chem 21(4):316–319

Chapter 9
Sand Control in Oil and Water Wells

The production of sand from oil and water wells is called sand production, which will severely affect the regular operation of oil and water wells. The sand production in oil wells will cause blockage in oil production interval, jam or damage in pipeline and equipment. In severe cases, it will cause the collapse of borehole wall, during which the squeezing will result in the deformation and damage of the casing pipe. Though the sand production problem in water wells is not as significant as that in oil wells, the sand production in water wells (usually happens during well washing or fluid flowback after operation) will also lead to blockage in water injection formation, affecting the normal water injection. Therefore, sand control in sand production wells is requisite in order to maintain normal oilfield production.

Five types of chemistry-related sand control methods in oil and water wells are introduced here.

9.1 Sand Control by Chemical Bridging

Loose sand grains can be bridged at their contact points using bridging agent to achieve the purpose of sand control (Fig. 9.1).

Bridging agent is a chemical agent which can bridge the loose grains at their contact points.

There are two types of available bridging agents.

C. Dai and F. Zhao, *Oilfield Chemistry*,
https://doi.org/10.1007/978-981-13-2950-0_9

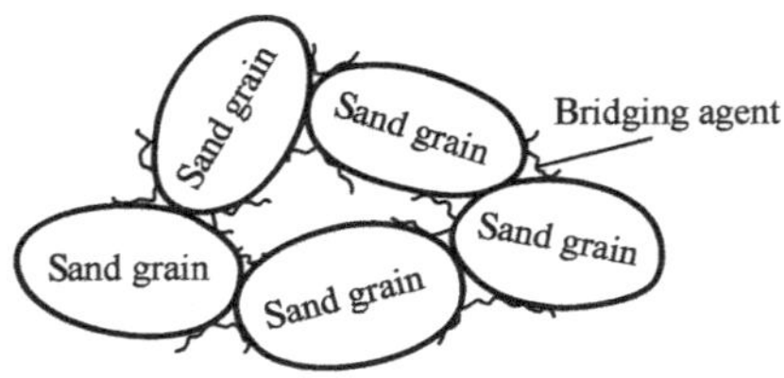

Fig. 9.1 Sand grains are bridged by bridging agent

9.1.1 Non-organic Cationic Polymer

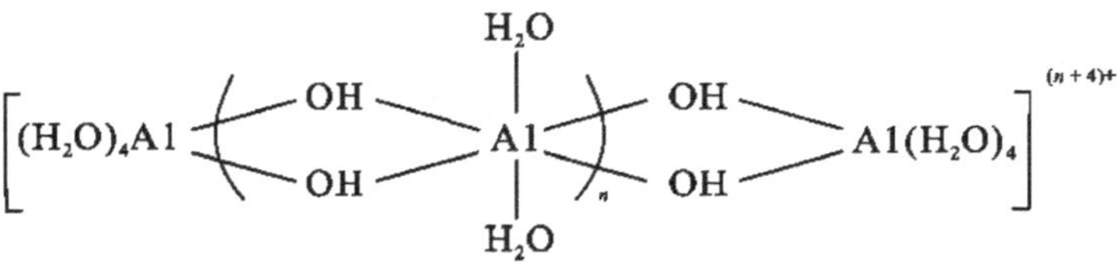

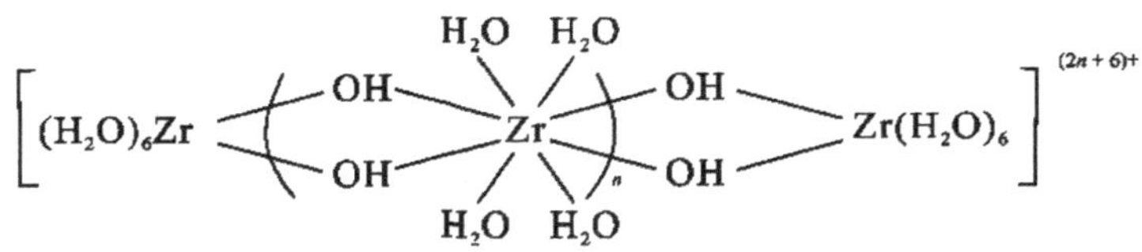

Multinuclear olation complex ions composed of aluminum ions and zirconium ions, together with their corresponding anions are called hydroxy-aluminum and hydroxy-zirconium, respectively. They are typical non-organic cationic polymers which can be used as bridging agents.

9.1.2 Organic Cationic Polymer

Organic cationic polymers with quaternary ammonium salt structures on the branched chain are important bridging agents.

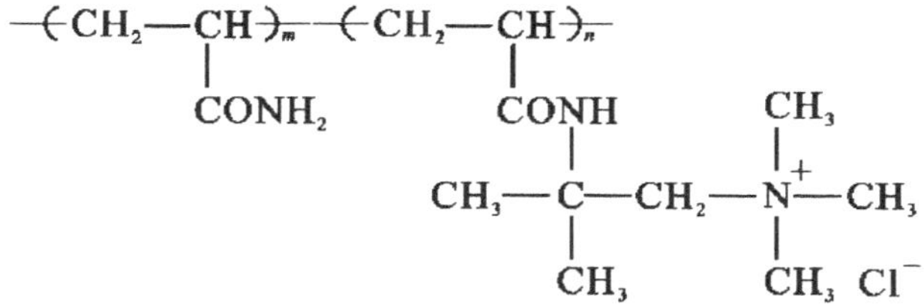

[Copolymer of acrylamide and (2-acrylamide-2-methyl) propyl trimethyl ammonium chloride]

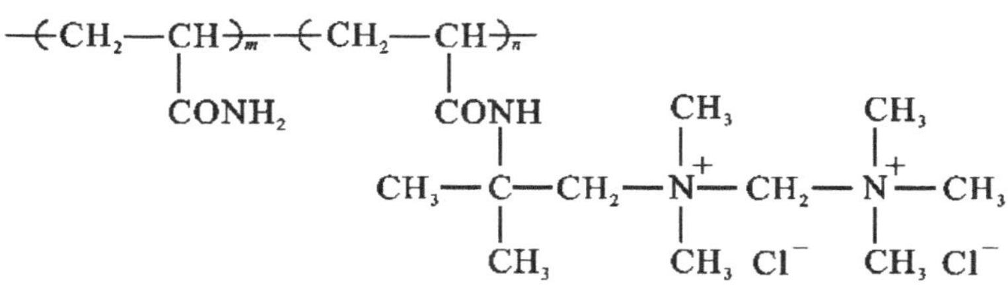

[Copolymer of acrylamide and (2-acrylamide-2-methyl) propyl methylene pentamethyl double ammonium chloride]

By injecting the aqueous solution of bridging agent into the sand producing interval, and shutting the well for certain time, the bridging agent will reach equivalent adsorption among the sand grains, achieving the purpose of sand control.

9.2 Sand Control by Chemical Cementing

Loose sand grains can be cemented at their contact points using cementing agent to achieve the purpose of sand control (Fig. 9.2).

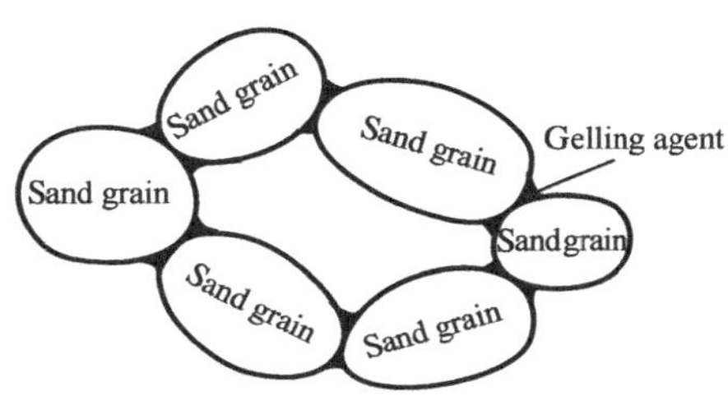

Fig. 9.2 Sand grains are cemented by cementing agent

9.2.1 Sand Control Procedures by Cementing

In order to cement the loose sand grains in sand layer, the following procedures are generally required (Lezzi et al. 2002).

Formation Pre-treatment
Different pre-treatment agents are used according to different purposes.

- If crude oil is to be replaced from sand formation, brine water can be used.
- If oil film on sand grain is to be removed, oil solvent can be used. Oil solvent includes liquefied petroleum gas, gasoline, and kerosene.
- If carbonate which affects the curing of cementing agents is to be removed, hydrochloric acid can be used.
- If the sand surface is required to be cementing agent wetted, alcohol or alcohol ether such as hexyl alcohol and ethylene glycol monobutyl ether can be applied.

Injection of Cementing Agent
Cementing agent is injected into loose sand layer to contact with sand. To have the cementing agent injected into the formation uniformly, a slug of diverting agent can be injected before cementing agent. Diverting agent can reduce the permeability in high-permeability zone, providing a homogeneous permeability in sand layer so that the cementing agent can enter the sand layer uniformly. The mixture of iso-propanol, diesel, and ethyl cellulose is a kind of diverting agent.

Injection of Pore-Forming Fluid
Pore-forming fluid is used to push the extra cementing agent into the in-depth formation. It is required that pore-forming fluid cannot dissolve cementing agent, nor affect the curing of cementing agent.

Curing of Cementing Agent
If the curing agent is added during the injection of cementing agent, this procedure is well shut-in for cementation. If the curing agent is not added during the injection of cementing agent, this procedure is injecting curing agent and then shutting the well for cementation. Different curing agents are required for different cementing agents.

9.2.2 Cementing Agent

Cementing agent refers to the chemical agent which can cement the loose sand grains at their contact points.

There are two types of cementing agents.

Non-organic Cementing Agent

- Silicic Acid

If sodium silicate, pore-forming oil and hydrochloric acid are injected in sequence into the sand layer, silicic acid will be generated at the contact points of sand grains, cementing the sand grains.

- Calcium Silicate

If sodium silicate, pore-forming oil and calcium chloride are injected in sequence into the sand layer, calcium silicate will be generated at the contact points of sand grains, cementing the sand grains.

Organic Cementing Agent

- Gel Type Cementing Agent (Sydansk 1992; James et al. 2002)

The gel with certain gelation time at reservoir temperature can be used for cementing loose sand layer, such as chromium gel. During cementing of loose sand layer with chromium gel, cross-linker (such as chromium acetate) can be added into polyacrylamide solution and injected into loose sand layer. After injection of pore-forming fluid (such as kerosene and diesel), the well is shut down for certain period for gelation, during which the loose sand grains will be cemented. The gel which gelates instantly at reservoir temperature can also be used for cementing loose sand layer, such as zirconium gel. During cementing of loose sand layer with zirconium gel, polyacrylamide solution can be injected into loose sand layer before the pore-forming oil. Then cross-linker solution (such as zirconium oxychloride) is injected, making the polyacrylamide retained at the contact points of sand grains cross-linking into gel to cement the loose sand grains. Therefore, various gels can be used as cementing agent.

- Resin Type Cementing Agent (Xu 2001; Song et al. 2002; Nguyen et al. 2006; Nguyen et al. 2006)

Important resin type cementing agents include phenolic resin, urea resin, epoxy resin, and furan resin, among which phenolic resin is the most common one.

There are two forms for the usage of phenolic resin. One is ground pre-condensed thermoset phenolic resin. Hydrochloride with a concentration of 10% is used as curing agent and injected into the formation after phenolic resin and pore-forming fluid. Another one is phenolic resin synthesized underground. Tin chloride is used as curing agent because it can generate hydrochloride gradually during hydrolysis to cure the phenolic resin slowly. Therefore, tin chloride can be injected into formation together with phenol and methanol before the injection of pore-forming fluid. For phenolic resin synthesized underground, the mass ratio among phenol, methanol, and tin chloride is 1:2:0.24. Since this kind of phenolic resin needs to undergo polycondensation underground, it is only applicable for sand layer with temperature no less than 60 °C.

Urea resin and epoxy resin are mainly pre-condensed resin. Urea resin is similar to phenolic resin; the curing agent (such as hydrochloric and oxalic acid) is injected into the formation after resin and pore-forming fluid. While for the epoxy resin, cross-linker (such as ethidene diamine and phthalic anhydride) is added into resin before injected into formation.

Furan resin is a kind of resin containing furan ring (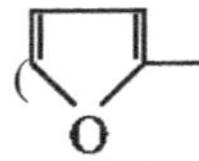). Furfuryl alcohol resin belongs to furan resin, which is a thermoset obtained from the polycondensation of furfuryl alcohol (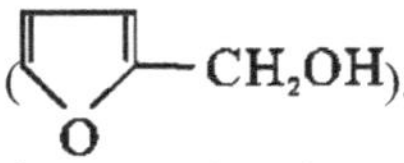). During operation, it is injected into the formation first. After the pore forming, it is cured by the injection of curing agent (such as hydrochloric acid). Furfuryl alcohol resin is resistant to high temperature, acid, alkali, and organic solvent. It is a good cementing agent (Friedman 1984; Friedman 1985).

Coupling agent can be used to enhance the bonding between the resins discussed above and sand grain surface (Yang et al. 1995). γ-aminopropyl triethoxy monosilane is a typical coupling agent whose molecular formula is shown below.

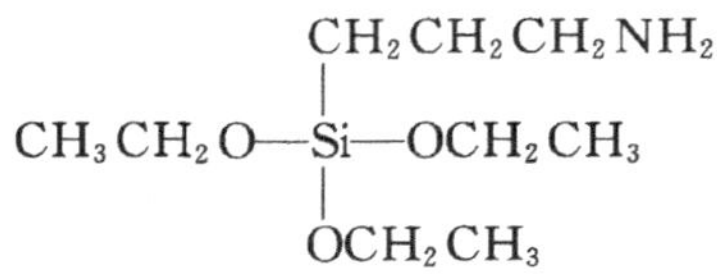

It can generate silanol by hydrolysis.

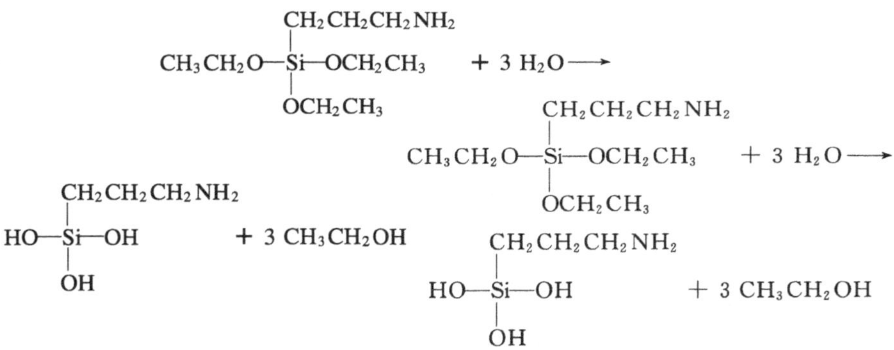

Polysilanol can be generated from the condensation and dehydration of silanol.

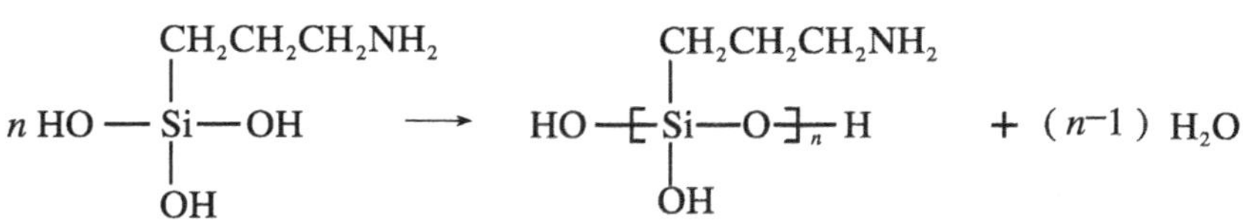

The hydroxyl part in polysilanol can be bonded with the hydroxyl part in sand grain surface through hydrogen bonding, while another part in polysilanol can be

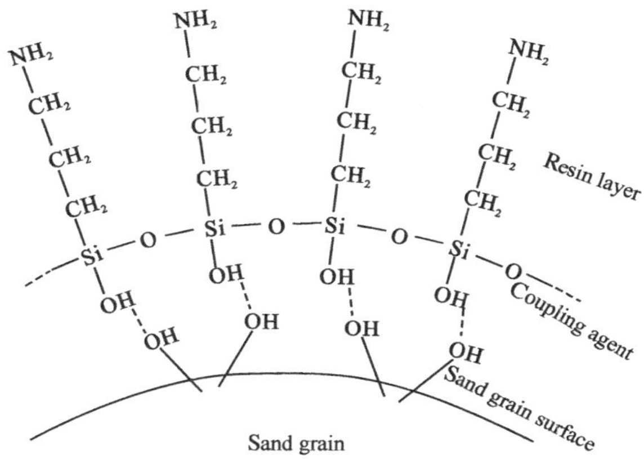

Fig. 9.3 Function of coupling agent (— is hydrogen bond)

bonded with the resin type cementing agent (Fig. 9.3). Therefore, the cementing performance of cementing agent can be improved.

- Polycarbamate Type Cementing Agent

The aforementioned selective water plugging agent polycarbamate can also be used as cementing agent. During operation, water is used to flush the sand layer first before the pore forming using oil, while oil solution of polycarbamate is injected in the end. Water at the contact points of sand grains can induce a series of reactions of polycarbamate to make it solidify, thus cementing the loose sand grains.

- Coke Type Cementing Agent

In order to cement the sand grains using coke, heavy oil (oil with high content of colloid and asphaltene) is injected into the sand layer, and water is injected to enlarge pores. Then heavy oil is processed with one of the following methods.

- Heating the sand layer to vaporize the light component in heavy oil, colloid, and asphaltene are left. Continue heating the sand layer until the partial carbonization of colloid and asphaltene, which will cement the loose sand grains (Nichols 1985).
- Solvent such as trichloroethane is used to precipitate the asphaltene in heavy oil, which will be heated to partial carbonization to cement the loose sand grains (Wiechel et al. 1985).
- The injection of heated air can not only lead to the vaporization of the light component in heavy oil and the partial carbonization of the colloid and asphaltene, but also result in the oxidation of the colloid and asphaltene, therefore increasing the cementation strength, cementing the loose sand grains (Redford 1976; Aggour et al. 2002).

9.3 Sand Control by Artificial Borehole Wall

This method is applicable for the sand control when sand is already produced. A sand barrier composed of cemented particles with sufficient permeability is constructed at the void space of sand layer, which is called artificial borehole wall.

Artificial borehole wall can be constructed by the following methods.

9.3.1 Sand Pack Cementing Method

The void space of sand producing layer is filled with sand first. Then cementing agent is used to cement the sands according to the aforementioned steps. In this way, artificial borehole wall is created.

9.3.2 Resin-Coated Sand Method

Resin-coated sand is the sand whose surface is pre-coated with a layer of resin (cementing agent). The aforementioned resin cementing agents all can be used to coat sand; however, the most commonly used are phenolic resin and epoxy resin. The void space of the sand layer is filled with resin-coated sand first. Then the resin is cross-linked into three-dimensional structure under the effect of heat and (or) curing agent. Sand grains are thus cemented, forming artificial borehole wall.

To increase the permeability of artificial borehole wall, resin-coated fiber with a mass fraction of 8–12% can be added into the resin-coated sand. Phenolic resin and epoxy resin can be used to coat fiber. The fiber to be coated can be polyester fiber, polyamide fiber, polypropylene fibers, carbon fiber, glass fiber, etc. (Daccord et al. 2004; Zhao et al. 2005).

9.3.3 Cement Slurry Method

Cement slurry is prepared by water, cement, and silica sand in mass fraction ratio of 0.5:1:4. When the cement slurries solidify at the void space of the sand layer, it can function as artificial borehole wall.

9.3.4 Cement Clinker Method

Cement clinker is prepared by the sintering of limestone and clay with certain proportion. Cement clinker blocks grinded to certain granularity (e.g. 0.3–1 mm) can be used to fill the void space of the sand layer. Cement clinker is cured under the effect of water, forming artificial borehole wall.

9.4 Sand Control by Screen Pipe

To prevent sand production, the sand screen pipe in Fig. 9.4 can be positioned in the sand producing layer.

The outer layer of sand screen pipe is protection shield, the intermediate layer is sand filter, and the inside layer is central pipe.

Protection shield is used to protect the sand filter. The holes around the shield can direct the formation fluid to enter the pipe in tangential direction, thus avoiding its direct flushing to the screen pipe.

Sand filter is generally made of metal wool, metal felt, or metal net whose material is stainless steel such as chrome alloy steel (chrome content $\geq 12\%$) and chrome–nickel alloy steel (usually contains 18% chrome and 8% nickel). Sand filter can also be made of cemented silica sands, which are cemented by cementing agents such as epoxy resin and concrete. When using epoxy resin to cement silica sands, epoxy, ethidene diamine (cross-linker), and dibutyl phthalate (flexibilizer) are prepared with the mass ratio of 100:8:10. Then sand filter is made of cemented molding with 20 kg silica sand per 1 kg epoxy resin. When using concrete to cement silica sands, water, concrete, and silica sand can be prepared with the mass ratio of 1:2.4:12, before molded into sand filter.

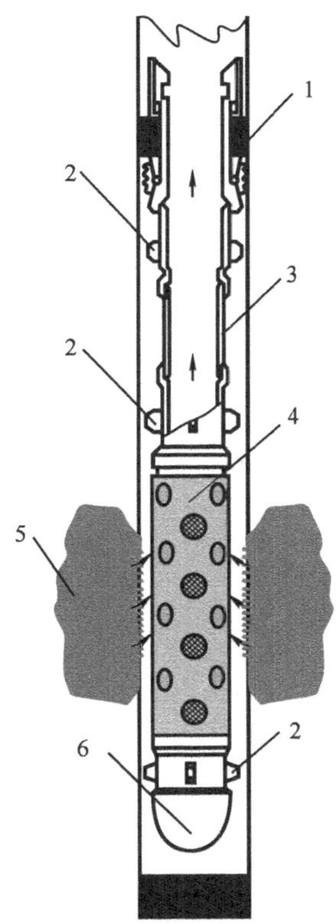

Fig. 9.4 Sand control with screen pipe. 1—suspension packer; 2—centralizer; 3—oil pipe joint; 4—sand screen pipe; 5—sand producing section; 6—screw plug

Through the central pipe with multiple holes, the protection shield and sand filter are connected with the screw plug downside, and the centralizer, oil pipe joint and centralizer upside. Then it is linked to the suspension packer, so that the sand filter is positioned in the sand producing layer for sand control.

9.5 Sand Control by Wire Wrapped Screen Pipe with Gravel Packing

A screen drum is made by welding stainless steel on the longitudinal bar (the spacing between the stainless steel wires is 0.2–0.4 mm). Then the screen drum is covered on the central pipe. The wire wrapped screen pipe is made by welding the two ends of screen drum onto the central pipe.

Wire wrapped screen pipe is placed into the sand production layer, then gravel (quartz sand with medium granularity 5–6 times of the sand from sand production layer) is carried to fill the annulus between the sand production layer and wire wrapped screen pipe by proppant-carrying fluid. In this way sand barrier is formed (wire wrapped screen pipe is used to block the packed gravel, and packed gravel is used to block formation sand), making wire wrapped screen pipe with gravel packing an effective method for sand control.

The gravel packing requires proppant-carrying fluid. There are two types of proppant-carrying fluids for gravel packing. One type is thickened water proppant-carrying fluid, which is prepared by dissolving water-soluble polymer in water. Available water-soluble polymers include hydrolyzed polyacrylamide, sodium carboxymethyl cellulose, hydroxypropyl galactomannan, and xanthan gum. In order to carry gravel, the content of water-soluble polymer in water should be adjusted so that the viscosity of thickened water proppant-carrying fluid lies in the range of 30–60 mPa · s under 50 °C and 170 s^{-1}. Appropriate amount of viscosity reducer (mainly peroxide such as ammonium persulfate and isobutanol peroxide) should be added into the thickened water proppant-carrying fluid so that its viscosity can be reduced below 10 mPa · s after gravel packing (about 5 hours). Another type is water-based gel proppant-carrying fluid which is prepared by polymer, cross-linker, and gel breaker. The water-soluble polymers used for preparing thickened water proppant-carrying fluid all can be used to prepare the water-based gel proppant-carrying fluid. Cross-linker is chosen based on cross-linkable groups in the polymer chains. For example, carboxyl can be cross-linked by multinuclear olation complex ions composed of high-valence metal ions (such as Cr^{3+} and Zr^{4+}), *cis*-ortho hydroxyl can be cross-linked by boracic acid radical (such as BO_3^{3-}). The viscosity of water-based gel proppant-carrying fluid should be in the range of 300–500 mPa · s at 50 °C and 170 s^{-1}. Gel breaker functions the same as the viscosity reducer in the thickened water proppant-carrying fluid. The viscosity of water-based gel proppant-carrying fluid is required to be reduced below 10 mPa · s after gravel packing (about 5 hours).

References

Aggour MA, Abu-Khamsin SA, Osman EA (2002) Method for sand control in oil, gas and water wells. US Patent 6,364,019,B1, 2 Apr 2002
Daccord G, Bedel JP, Ermel M (2004) A well treatment fluid containing fibres and a method of use. GB Patent 2,392,682,A, 3 Oct 2004
Friedman RH (1984) Sand consolidation method. US Patent 4,427,069, 24 Jan 1984
Friedman RH (1985) Sand consolidation methods using adsorbable catalysts. US Patent 4,512,407, 23 Apr 1985
James SG, Nelson EB, Guinot FJ (2002) Sand consolidation with flexible gel systems. World Patent 04,784,A2, 17 Jan 2002
Lezzi A, Galbaariggi G, Pitoni E (2002) Process for the consolidation of sand. EP Patent 829,616, 25 Sept 2002
Nguyen PD, Dusterhoft RG, Ali SA et al (2006) Stabilizing wellbores in unconsolidated, clay-laden formations. SPE Prod & Oper 21(3):339–348
Nguyen PD, Rickman RD, Dusterhoft RG (2006) Method of stabilizing unconsolidated formation for sand control. US Patent 0,219,405,A1, 5 Oct 2006
Nichols DP (1985) Method of consolidation of oil-bearing sand. US 4,501,328, 26 Feb 1985
Redford DA (1976) Sand control method employing low temperature oxidation. US Patent 3,974,877, 17 Aug 1976
Song X, Sun C, Zhang L et al (2002) Sand control uses phenolic resin and sand consolidation uses modified resin. Oil Drill Prod Technol 24(6):57–60
Sydansk RD (1992) Sand consolidation treatment for a hydrocarbon production well bore using an overdisplacement fluid. US 5,082,057, 21 Jan 1992
Wiechel JF, French CR, Hall WL (1985) Sand control employing halogenated, oil soluble hydrocarbons. US 4,494,605, 22 Jan 1985
Xu Y (2001) Urea-formaldehyde aqueous cementing fluid for sand consolidation and its application in Wenliu oilfields. Oilfield Chem 18(3):196–198
Yang Y, He S (1995) Organosilane coupling agent and its application. Chem Eng 40–42
Zhao X, Li P, Qiu G et al (2005) A research and evaluation of fibred as sand control in Gudong oilfield. Drill Prod Technol 28(6):93–95

Chapter 10
Wax Control and Removal in Oil Well

Wax deposition severely affects the normal production of oil wells. Wax control and cleanup are also problems to be solved during oil recovery.

The inner reason for wax deposition is the existence of wax in crude oil. The higher the content of wax in crude oil, the higher its condensation point is (as shown in Table 10.1). Thus, oil wells with high wax content in crude oil are all troubled seriously by wax deposition.

Wax is straight-chain alkane (C_{15}–C_{70}) which is solid at ambient condition. Wax is dissolved in crude oil at reservoir conditions. When crude oil flows from oil reservoir to wellbore bottom, and then lifted to wellhead, wax deposition occurs due to the reduction of pressure and temperature.

Wax deposition can be separated into three stages—wax precipitation stage, wax crystal-growing stage, and deposition stage. If wax is separated out from an active spot on certain solid surface (e.g., steel surface), wax will grow up on that active spot, resulting in wax deposition. In this case, only the former two stages exist during wax deposition.

Table 10.1 Properties of some crude oil

Well No.	Density (g/cm^3)	Viscosity (mPa · s)	Condensation point (°C)	*w* (max) (%)
1	0.9505	1,685	4	7.31
2	0.9293	312	10	10.48
3	0.9232	240	18	12.20
4	0.8861	26	23	15.40
5	0.8657	15	25	21.90

C. Dai and F. Zhao, *Oilfield Chemistry*,
https://doi.org/10.1007/978-981-13-2950-0_10

10.1 Wax Control in Oil Well

The purpose of wax control can be achieved by terminating the wax deposition process at any stage. Here, two methods to prevent the wax deposition of oil wells are introduced.

10.1.1 Wax Control with Wax Inhibitors

Wax inhibitors are chemicals which can inhibit the wax crystals in crude oil from precipitation, growing up, and/or deposition on solid surface.

There are three types of wax inhibitors (Zhao 1988; Tung et al. 2001; Becker 2001).

10.1.1.1 Polycyclic Aromatic Hydrocarbon Wax Inhibitors

Polycyclic aromatic hydrocarbon refers to the kind of aromatic hydrocarbon composed of two or more fused benzene rings.

(Naphthalene)

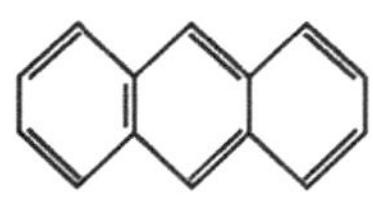

(Anthracene)

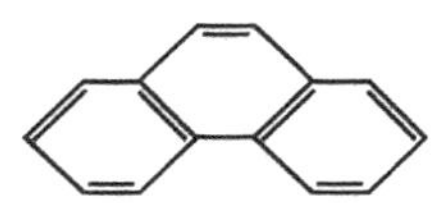

(Phenanthrene)

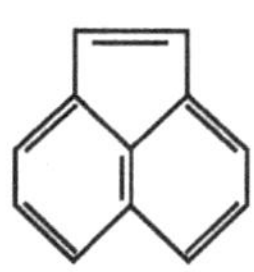

(Acenaphthene)

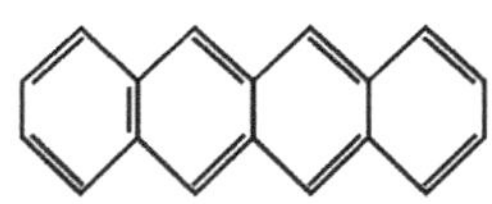

(Naphthacene)

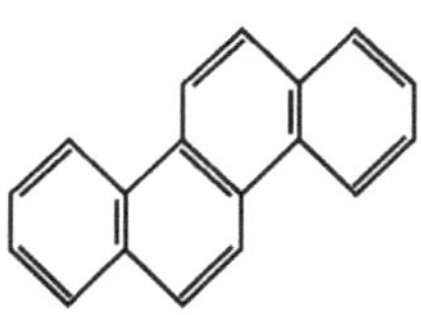

(Chrysene)

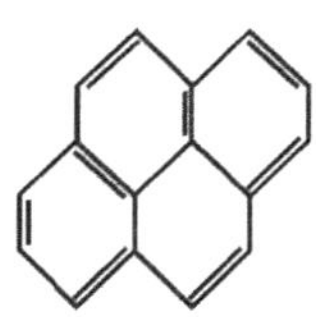

(Pyrene)

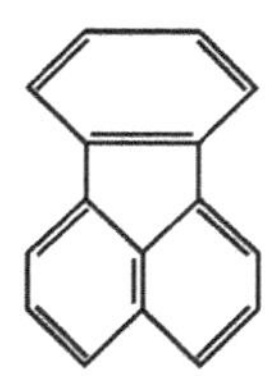

(Benzoacenaphthene)

The above-listed are all polycyclic aromatic hydrocarbons, mainly derived from coal tar. The derivatives of polycyclic aromatic hydrocarbons (some examples are listed below) possess similar properties as polycyclic aromatic hydrocarbons.

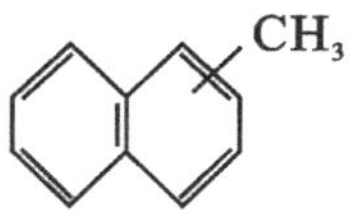

(Methylnaphthalene)

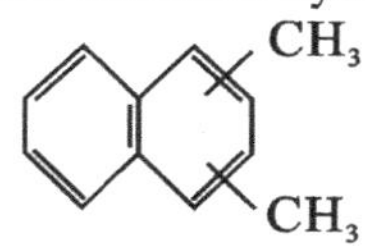

(Dimethylnaphthalene)

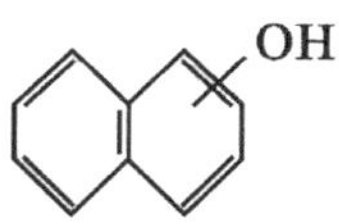

(Naphthol)

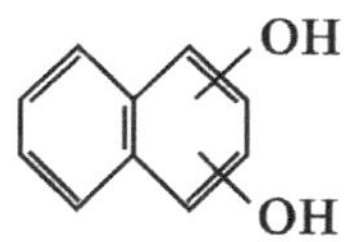

(Naphthalenediol)

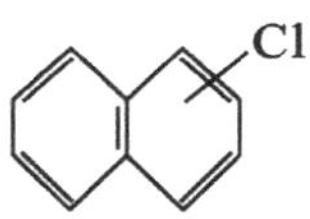

(Chloronaphthalene)

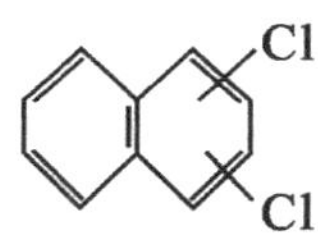

(Dichloronaphtalene)

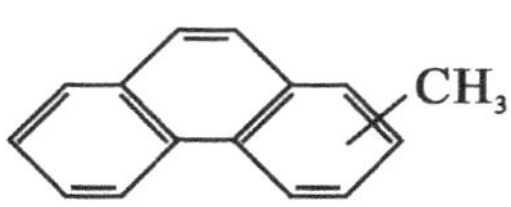

(Methylphenanthrene)

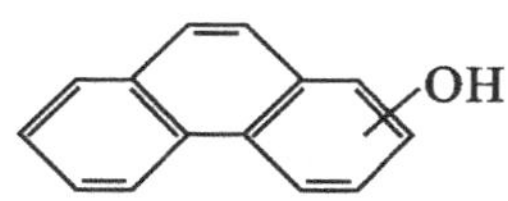

(Phenanthrenol)

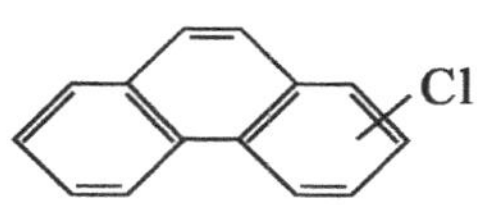

(Chlorophenanthrene)

Polycyclic aromatic hydrocarbon wax inhibitors can participate in the formation of wax crystal nuclei, which will distort the crystal nuclei and prevent the growth of wax crystals.

Polycyclic aromatic hydrocarbons can be dissolved in solvent before added into crude oil, or they can be molded before placed into wells. For molded wax inhibitor, to control its dissolving rate in crude oil, it can be appropriately compounded with its derivatives. For example, naphthalene dissolves fast, whereas α-naphthol dissolves slowly in crude oil. The wax inhibitor compounded by them can maintain effective wax control for a long time.

10.1.1.2 Surfactant Wax Inhibitors

This type of wax inhibitors can be further divided into oil-soluble surfactants and water-soluble surfactants.

Oil-soluble surfactants work in wax inhibition by altering the surface properties of wax crystals. Due to the adsorption of surfactants on wax crystal surface, the crystal surface is changed into polar one which is adverse to the growth of wax crystal (Fig. 10.1). Available oil-soluble surfactants are mainly petroleum sulfonates and amine surfactants. Some examples are given below.

$$\mathrm{RArSO_3M} \qquad \mathrm{M{:}1/2Ca, Na, K, NH_4}$$

(Petroleum sulfonates)

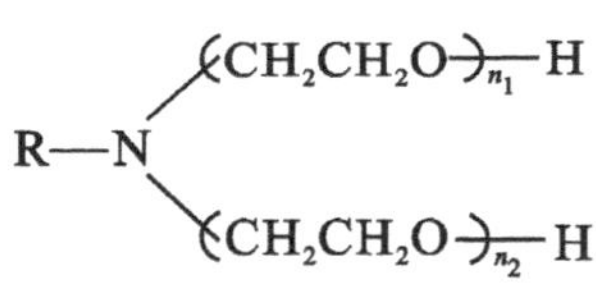

$\mathrm{R{:}C_{16}-C_{22}}$; $n_1+n_2{:}2-4$

(Amine surfactants)

Water-soluble surfactants work in wax inhibition by altering the properties of the surfaces where wax deposits (such as oil pipe, sucker rod, and instrument surface). Water-soluble surfactants can adsorb on the surfaces where wax deposits to turn them into polar surfaces, which are covered by a layer of water film and are not conducive for the wax deposition. Some available water-soluble surfactants are shown below.

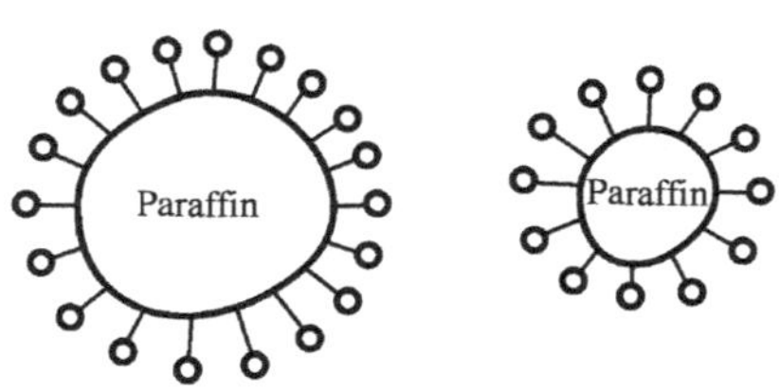

Fig. 10.1 Surfactants change wax surface to polar surface

$R—SO_3Na$ R: C_{12}–C_{18}

(Alkyl sodium sulfonate)

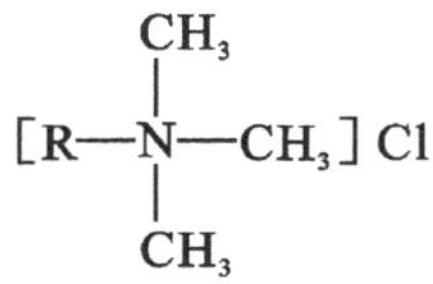

R: C_{12}–C_{18}

(Alkyl trimethyl ammonium chloride)

$R—O—[CH_2CH_2O]_n—H$ R: C_{12}–C_{18}; n: 5–100

(Polyoxyethylene alkyl alcohol ether)

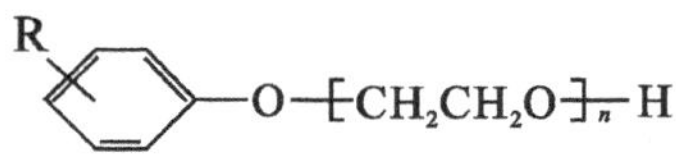

R: C_8–C_{14}; n: 5–100

(Polyoxyethylene alkyl phenol ether)

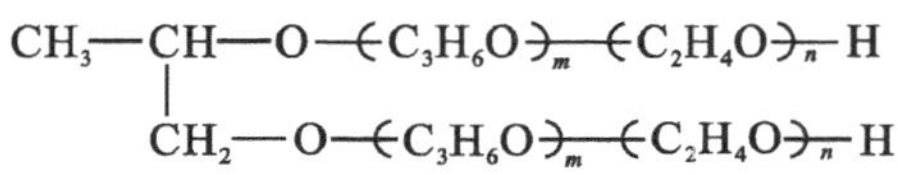

m: 17; n: 15–53

(Polyoxyethylene polyoxypropylene propylene alcohol ether)

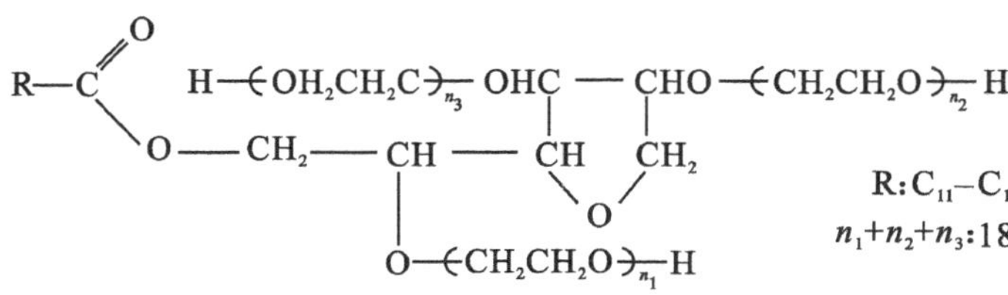

R: C_{11}–C_{17};
$n_1+n_2+n_3$: 18–26

(Sorbide monocarboxylic ester polyoxyethylene ether)

$R—O—[CH_2CH_2O]_n—SO_3Na$ R: C_{12}–C_{18}; n: 1–10

(Polyoxyethylene alkyl alcohol ether sodium sulfate)

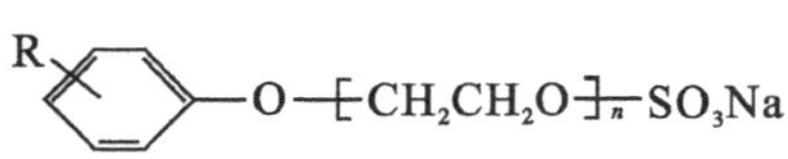

R: C_8–C_{14}; n: 1–10

(Polyoxyethylene alkyl alcohol ether sodium carboxylate)

$R—O—[CH_2CH_2O]_n—R'SO_3Na$ R: C_{12}–C_{18}; R′: C_1–C_3;

(Polyoxyethylene alkyl phenol ether sodium carboxylate) n: 1–10

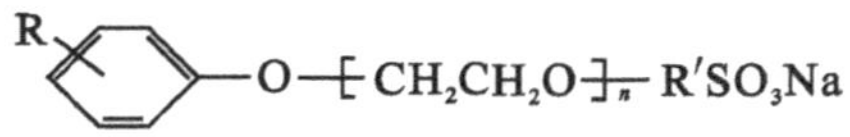

R: C_8–C_{14}; R′: C_1–C_3; n:1–10

(Polyoxyethylene alkyl phenol ether sodium sulfonate)

$$R{-}O{+}CH_2CH_2O{+}_n R'COONa$$

R: C_{12}–C_{18}; R′: C_1–C_3; n:1–10

(Polyoxyethylene alkyl alcohol ether sodium carboxylate)

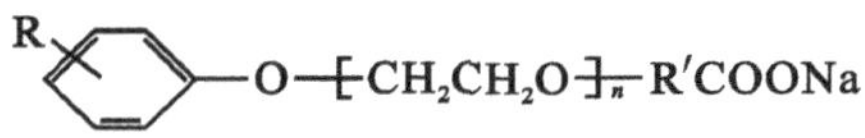

R: C_8–C_{14}; R′: C_1–C_3; n:1–10

(Polyoxyethylene alkyl phenol ether sodium carboxylate)

10.1.1.3 Polymer Wax Inhibitors

The nonpolar chains or nonpolar part in polar chains of this type of wax inhibitor can crystallize together with wax. The polar chains of this type of wax inhibitor can cause the distortion of the wax crystal, which is not conducive to the growth of wax crystal to a network structure. Therefore, polymer wax inhibitors possess excellent performance of wax control. The followings are some important polymer wax inhibitors.

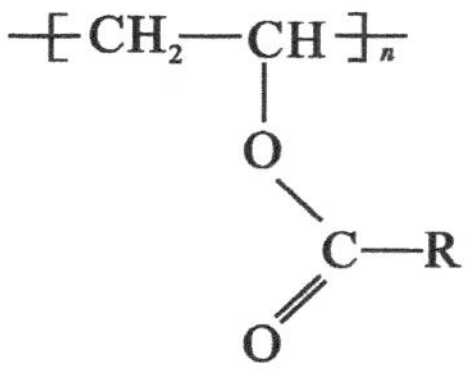

R: C_{15}–C_{35}

(Polyoxyethylene carboxylic ester)
(Wang et al. 2003)

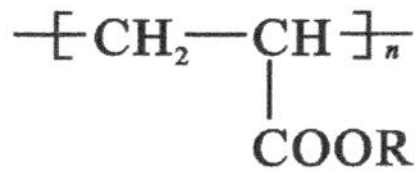

R: C_{14}–C_{40}

(Polyacrylate)
(Duncum et al. 1999)

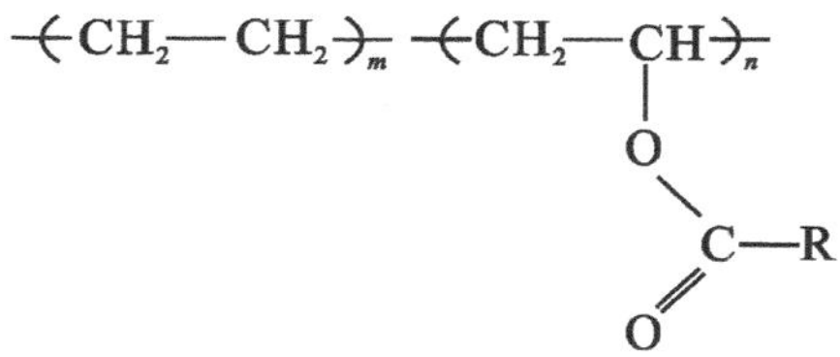

R: C_1–C_{25}

(Copolymer of ethylene and ethylene carboxylic ester)
(Machado et al. 2001)

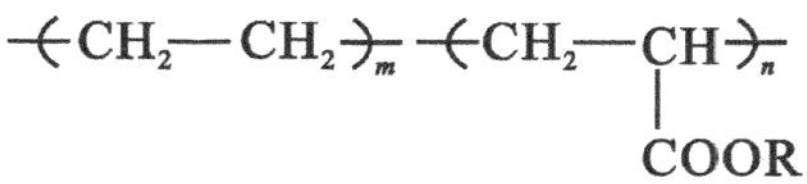

R: C_1–C_{26}

(Copolymer of ethylene and acrylate)

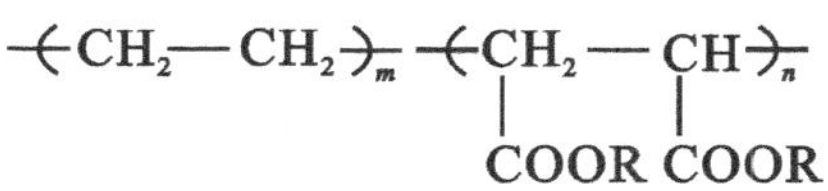

R: C_1–C_{26}

(Copolymer of ethylene and Maleic acid ester)
(Newberry 1985)

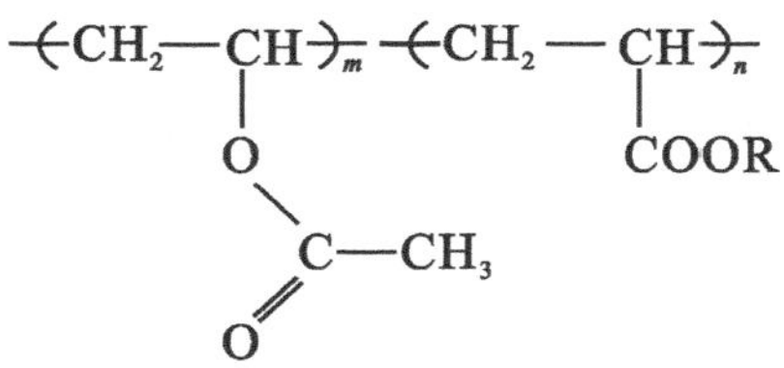

R: C_{14}–C_{40}

(Copolymer of vinyl acetate and acrylate)

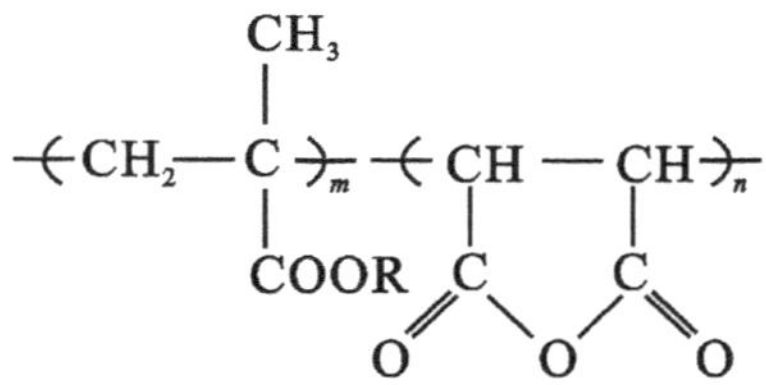

R: $C_{16}-C_{30}$

(Copolymer of methacrylate and maleic anhydride)

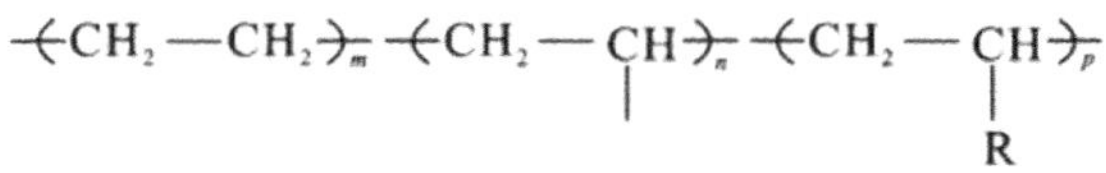

R: C_1, C_2, C_3…

(High pressure polyethylene)

(Li et al. 2001)

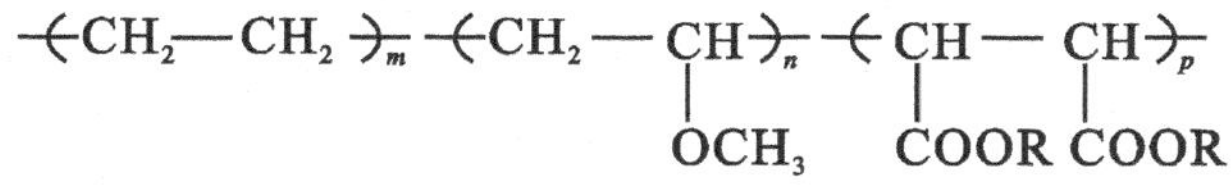

R: C_1–C_{26}

(Copolymer of ethylene, vinyl methyl ether and maleic acid ester)

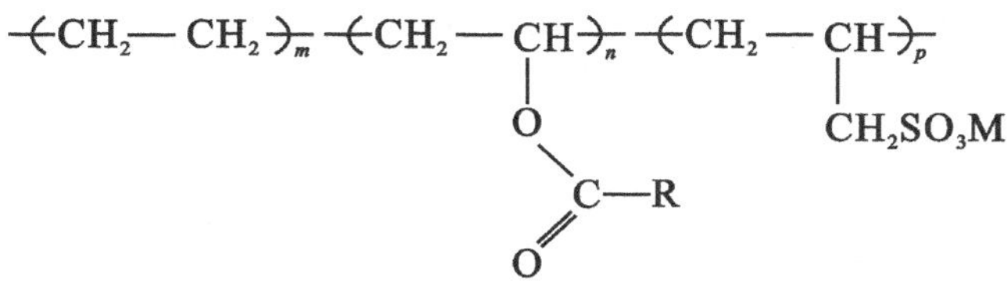

R: C_1–C_{25};
M: Na, K, NH_4

(Copolymer of ethylene, ethylene carboxylic ester and allyl sulfonate)

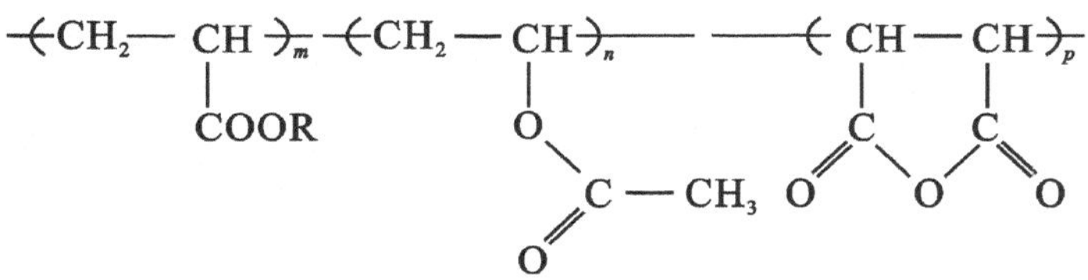

R: C_{16}–C_{30}

(Copolymer of acrylate, vinyl acetate and maleic anhydride)
(Shmakova-Lindeman 2005)

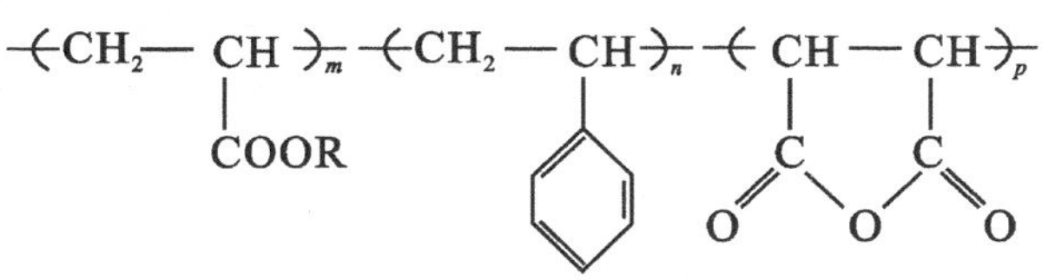

R: C_{16}–C_{30}

(Copolymer of acrylate, styrene and maleic anhydride)

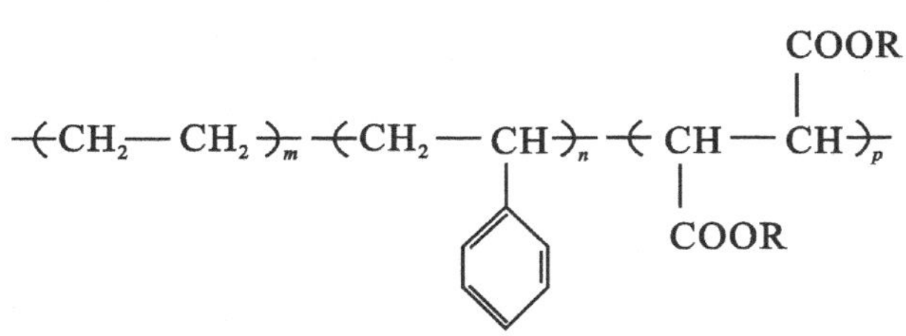

R: C_1–C_{26}

(Copolymer of ethylene, styrene and fumaric acid ester)
(Patil et al. 2002)

Polymer wax inhibitors can be dissolved in solvent before added to the crude oil, or they can be molded and then placed into wells.

The above-mentioned three types of wax inhibitors are all additional. In fact, the colloid and asphaltene in crude oil are natural wax inhibitors because the polycyclic aromatic hydrocarbons account for a considerable proportion in their polycyclic parts. Since there is always a certain amount of colloid and asphaltene in crude oil, all the additional wax inhibitors should be considered to work cooperatively with the colloid and asphaltene in crude oil.

10.1.2 Wax Control by Altering Surface Properties of Oil Pipe

Wax control using glass oil pipe and coated pipe belongs to this method.

Glass oil pipe can be produced by coating 0.4–1.5 mm glass on the inner wall of the oil pipe. Since glass surface is polar, smooth, and thermally insulated, it can prevent the wax deposition on it. The glass oil pipe is especially suited to the wax deposition wells whose water content surpassing 5%. The higher the oil production is, the better the wax inhibition performance will be. However, for those wells with low production and low water content, the glass oil pipe is not so effective.

Coated oil pipe is obtained by coating the inner wall of the oil pipe with wax inhibitors. The coating material is mainly polyurethane (Yue 2008), and other alternatives such as furfuryl alcohol resin, urushiol furfuryl resin, and imidazole epoxy resin can also be employed. Coated oil pipes possess the same wax inhibition mechanism with glass oil pipes. The surface of coated oil pipe is not only smooth and thermally insulated, but also contain chemical structures which are not conducive to the wax deposition. For example, polyurethane has urethane bond in its structure.

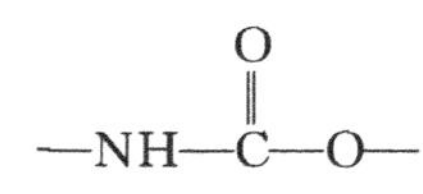

Furfuryl alcohol resin has the heterocyclic polymeric structure.

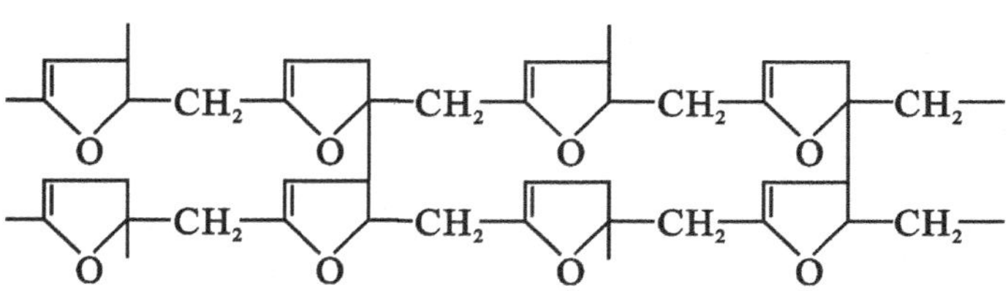

Urushiol furfural resin has the heterocyclic–non-heterocyclic polymeric structure.

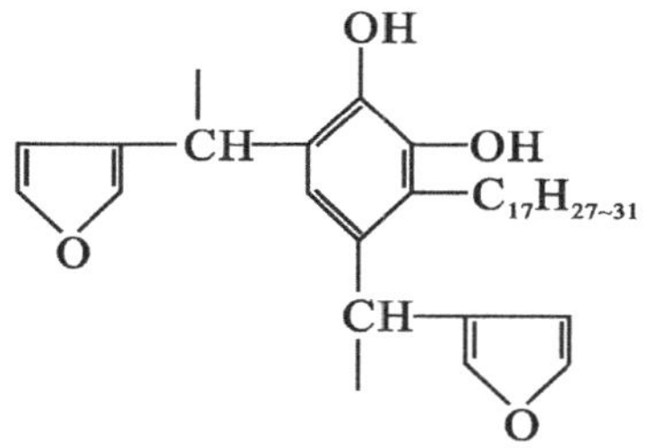

Imidazole epoxy resin is a kind of epoxy resin which is cured by 2-methylimidazole isooctyl glycidyl ether, thus it has the following structure.

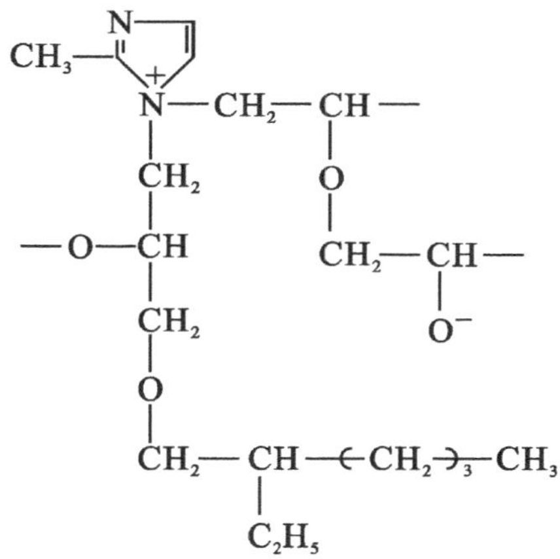

Those structures are all polarized, so all wax inhibition coatings possess polar surfaces which are not conducive to the deposition of wax.

10.2 Wax Removal in Oil Well

Oil wells already with wax deposition can be cleanup through mechanical methods (such as wax scraper) or thermal methods (such as circulation of hot-oil and circulation of hot-water). Wax remover can also be utilized to remove the wax.

Wax remover is a chemical agent which can clean up wax deposition. It can be classified into three types.

10.2.1 Oil-Based Wax Remover

Oil-based wax remover is a kind of solvent which can dissolve a large amount of wax such as gasoline, kerosene, diesel, benzene, methylbenzene, dimethylbenzene, phenylethane, phenylpropane, cyclopentane, cyclohexane, and decalin (decahydronaphthalene) (Jennings 2007). The major disadvantage of this type of wax remover is that they are toxic and flammable, which makes it very unsafe during utilization. Carbon disulfide, trichloromethane, and carbon tetrafluoride have superior wax cleanup performance; however, they can cause severe corrosion and catalyst poisoning during post-processing of crude oil. Therefore, they have been banned from using. Some vegetable oils (such as pine oil, camphor oil, oleum menthae, and citronella oil), which are extracted or stream-distilled from stems and leaves of some woody plants (such as pine trees and camphor trees) and herbaceous plants (such as mint and citronella), can dissolve wax and be used as wax remover. Their major component is terpene, whose molecular formula can be described by an integer multiple of isoprene as follows. Due to the hypotoxicity, low flammability, and biodegradability (Curtis 2003), vegetable oil has attracted increased attention.

$$\left(CH_2{=}\underset{\displaystyle CH_3}{\underset{|}{C}}{-}CH{=}CH_2\right)_n$$

To improve the wax removal performance of wax remover, various kinds of wax removers can be compounded for use. In addition, mutual solvent (such as alcohol and alcohol ether) can be added to increase the solubility of polar compounds (such as asphaltene) of wax in wax remover. The formula of a kind of compound oil-based wax remover is shown in Table 10.2.

10.2.2 Water-Based Wax Remover

Using water as dispersion medium, water-based wax remover is prepared with surfactants, mutual solvents, and/or alkaline matters. The function of surfactant is wettability alteration. They can convert the wax-depositing surface to hydrophilic surface, facilitating the falling off of wax from the surface and inhibiting the wax deposition. Available surfactants include water-soluble sulfate, quaternary ammonium, polyether, tween, peregal, OP surfactants, and peregal or OP surfactants which are sulfated or sulfo-alkylated.

The effect of mutual solvent is to increase the mutual solubility of oil (including wax) and water. Available mutual solvents are alcohol and alcohol ether such as methanol, ethanol, isopropanol, isobutanol, ethylene glycol butyl ether, and diethylene glycol ether. Alkali can react with the polar matter such as asphaltene in wax. As their product is easy to disperse in water, it can be feasibly removed from the surface with water-based wax remover. Available alkali includes sodium

Table 10.2 A kind of compound oil-based wax remover (Maly et al. 1976)

Component	φ (component) (%)
Kerosene	45–85
Benzene	5–45
Ethylene glycol butyl ether	0.6–6
Isopropanol	1–15

hydroxide, potassium hydroxide, and some salts which can dissolve in water to make the water alkaline such as sodium silicate, sodium orthosilicate, sodium phosphate, sodium pyrophosphate, and sodium hexametaphosphate.

Here are some examples of water-based wax removers. The numbers in the following examples are mass fractions of corresponding components.

Example 1 Water-based wax remover prepared by surfactants and alkali (Martin 1983).

$R-O-[CH_2CH_2O]_n-H$	10%
R: C_{12}–C_{18}, n: 8–20	
$Na_2O \cdot mSiO_2$	2%
H_2O	88%

Example 2 Water-based wax remover prepared by surfactants and mutual solvents (Mcclaflin et al. 1987).

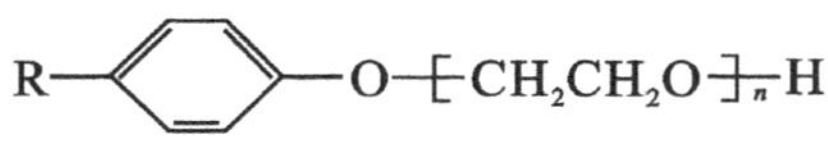

(structure above)	20%
R: C_6–C_{18}, n: 30–40	
CH_3OH	20%
H_2O	60%

Example 3 Water-based wax remover prepared by surfactants, mutual solvents and alkali (Mccormick 1977).

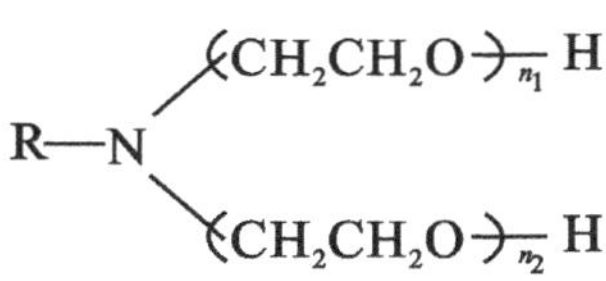

15%–65%

R: C_{12}–C_{18}, n_1+n_2: 6–12

$R-O-[CH_2CH_2O]_n-SO_3Na$ 15%–50%

R: C_{12}–C_{18}, n: 1–10

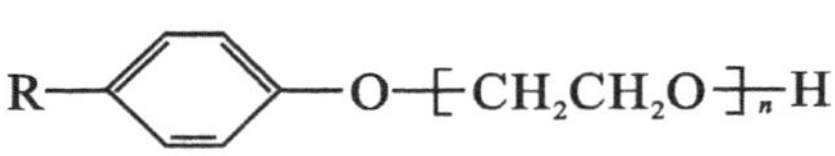

15%–50%

R: C_8–C_{18}, n: 4–20

$C_4H_9-O-CH_2CH_2OH$ 5%–30%

After water-based wax remover is prepared, alkali is utilized to adjust the water to alkaline.

Example 4 Water-based wax remover prepared by surfactants and compound mutual solvents (alcohol + alcohol ether).

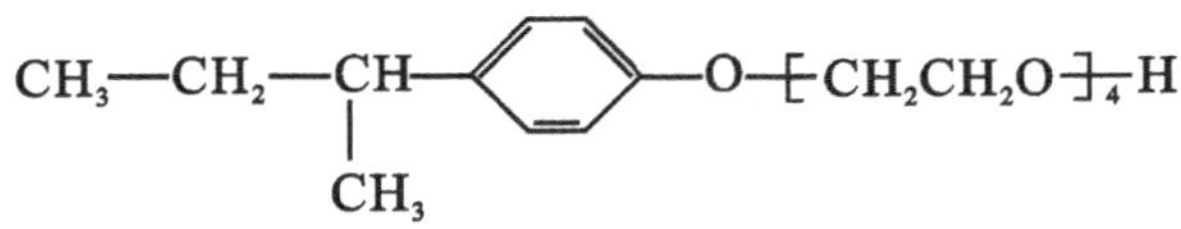
10%

$C_4H_9—O—[CH_2CH_2O]_2—H$ 25%

CH_3OH 25%

H_2O 40%

Example 5 Water-based wax remover prepared by surfactants and compound mutual solvents (two kinds of alcohol ethers).

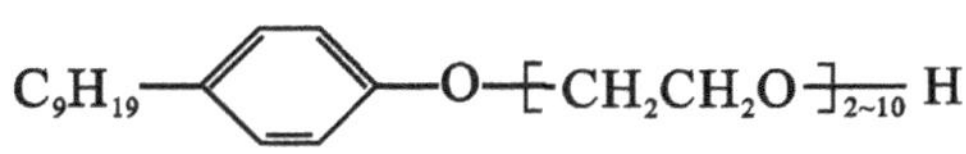
6.63%

$C_4H_9—O—CH_2CH_2OH$ 3.26%

$C_2H_5—O—[CH_2CH_2O]_2—H$ 6.63%

H_2O 83.48%

10.2.3 Oil-in-Water Wax Remover

Oil-in-water wax remover is the combination of oil-based and water-based wax removers, in which the oil phase is oil-based wax remover while the water phase is aqueous solution of surfactants and mutual solvents.

Oil-based wax removers used for oil phase include kerosene, toluene, xylene, cyclopentane, cyclohexane, decalin, pine oil, and camphorated oil; Surfactants used for water phase include alkyl sulfate, alkyl sulfonate, polyoxyethylene alkyl alcohol ether, polyoxyethylene alkyl phenol ether, polyoxyethylene alkyl alcohol ether sulfate, and polyoxyethylene alkyl alcohol ether sulfonate. Mutual solvents used for water phase include methanol, ethanol, propyl alcohol, glycol, diethylene glycol, ethylene glycol butyl ether, and diethylene glycol ether.

Table 10.3 A kind of typical oil-in-water wax remover (Matta 1985)

Component	*w* (component) (%)
Limonene	10–60
Surfactants	10–30
Water[a]	20–70

Note [a]There are 2–10% mutual solvents in water

The formula of a typical oil-in-water wax remover is shown in Table 10.3. When this kind of wax remover is delivered to the wax deposition intervals across the annulus, demulsification of the wax remover will occur due to the changes of condition (such as temperature and water salinity) to precipitate out the oil phase and water phase, which will respectively clean up wax through their own function.

Pseudo-tricomponent phase diagram, as shown in Fig. 10.2 can be used to demonstrate the content variation range of each component in oil-in-water wax remover in Table 10.3.

Limonene in Table 10.3 is a kind of terpene, whose molecular formula is shown as following. It exists in pine oil, camphor oil, oleum menthae, and citronella oil.

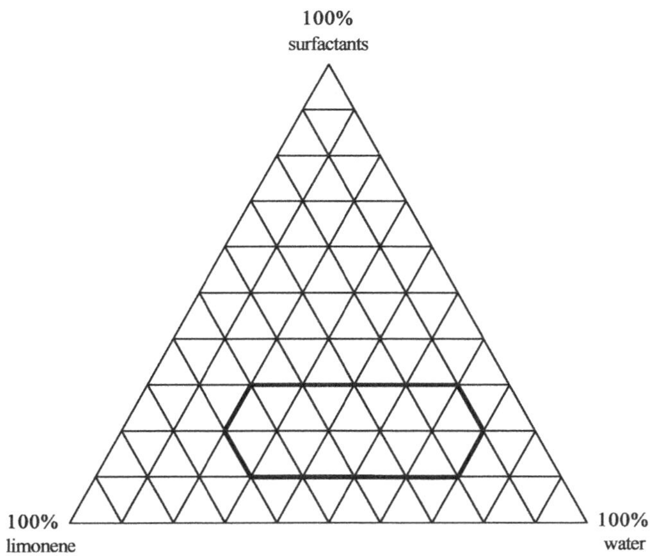

Fig. 10.2 Content variation range of each component in oil-in-water wax remover

Oil-in-water wax remover which is made of limonene, surfactants and mutual solvents represents the development trend of wax remover (Hamilton et al. 1993; Kaplan et al. 1994; Mehta et al. 1995; Hansen et al. 1995; Blunk 2001).

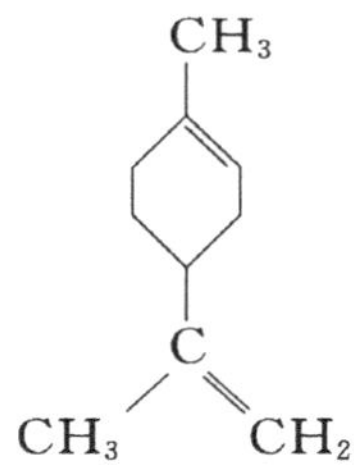

References

Becker JR (2001) Wax-crystal-modifier studies in field and laboratory. SPE 70030

Blunk JA (2001) Composition for wax removal from oilfield equipment. US Patent 6,176,243,B1, 23 Jan 2001

Curtis J (2003) Environmentally favorable terpene solvents find diverse application in stimulation, sand control and cementing operations. SPE 84124

Duncum SN, Hodgson DKG, James K et al (1999) Inhibitors and their uses in oils. WO Patent 9,821,446, 2 July 1999

Hamilton CR, Gustafson RM (1993) Process for cleaning articles with an aqueous solution of terpene and recycle water after separation. US Patent 5,271,773, 21 Dec 1993

Hansen KE, Requejo LP (1995) Dusting and cleaning composition. US Patent 5,399,282, 21 Mar 1995

Jennings DW (2007) Wax inhibitor compositions and their use in oil and gas production. US Patent 0,213,231,A1, 13 Sept 2007

Kaplan RI, Jerry EP (1994) Offshore rig cleaner. US Patent 5,336,428, 9 Aug 1994

Li K, Zhao X, Zheng Y et al (2001) Synthesis of copolymer of stearyl methacrylate and maleic anhydride and its pour point depressing performance. Adv Fine Petrochem 2(2):22–23

Machado ALC, Lucas EF (2001) The influence of vinyl acetate content of the poly (ethylene-co-vinyl acetate) (EVA) additive on the viscosity of Brazilian crude oil. Petrol Sci Technol 19(1–2):197–204

Maly GP, Beach N, Landess JA et al (1976) Method and solvent composition for stimulating the production of oil from a producing well. US Patent 3,998,743, 21 Dec 1976

Martin KR (1983) Petroleum and gas well enhancement agent. US 4,380,268, 19 Apr 1983

Matta GB (1985) D-limonene based aqueous cleaning compositions. US Patent 4,511,488, 16 Apr 1985

Mcclaflin GG, Yang K (1987) Composition and method for treatment of well bores and well formation containing wax. US Patent 4,668,408, 26 May 1987

Mccormick TH (1977) Wax removing compositions. US Patent 4,005,020, 25 Jan 1977

Mehta NK, Krajicek RW (1995) Decontamination of hydrocarbon process equipment. US Patent 5,389,156, 14 Feb 1995

Newberry ME (1985) Particulate compositions. US Patent 4,518,509, 21 May 1985

Patil AO, Zushma S, Berluche E et al (2002) Wax crystal modifiers [LAW657]. US Patent 0,040,123,A1, 4 Apr 2002
Shmakova-Lindeman OE (2005) Wax inhibitor. US Patent 0215437A1, 29 Sept 2005
Tung NP, Phong NTP, Long BQK et al (2001) Studying the mechanisms of crude oil pour point and viscosity reductions when developing chemical additives with the use of advanced analytical tools. SPE 65024
Wang K, Wu C, Creek JL et al (2003) Evaluation of effects of selected wax inhibitor on wax appearance and disappearance temperatures. Petrol Sci & Technol 21(3–4):359–368
Yue D (2008) Polyurethane coated pipeline for wax control. Oilfield Chem 25:207–209
Zhao F (1988) Paraffin inhibitor and remover. Oil Drill Prod & Technol 10(1):81–86

Part III
Gathering and Transportation Chemistry

Oil gathering and transportation chemistry is part of oilfield chemistry.

Oil gathering and transportation chemistry is an edge discipline between gathering engineering and chemistry.

Oil gathering and transportation chemistry studies how to use chemical methods to solve problems encountered during the process of oil gathering and transportation.

There are many chemical issues in the process of oil gathering and transportation, such as the corrosion and anti-corrosion of buried pipeline, the demulsification of the emulsified crude oil and defoaming of the foaming crude oil, the pour point depression transportation and drag reduction transportation of crude oil, the dehydration and deacidification of natural gas, the oil and oxygen removal, flocculation of suspended solids, scale inhibition, corrosion inhibition, and sterilization of the oilfield wastewater.

Various chemical agents are applied to address these issues, such as the chemical agents in the anti-corrosive coating of buried pipeline, demulsifier of the emulsified crude oil, defoamer of the foaming crude oil, pour point depressant, drag reducer, various natural gas treating agents, and oilfield wastewater treating agents.

In Part III, there are five chapters elaborating the chemical issues in the process of oil gathering and transportation and their solutions.

Chapter 11
Corrosion and Anti-corrosion of Buried Pipeline

The materials used for the buried pipelines are mostly metals. The phenomenon that the surfaces of the metal materials are destroyed through chemical and electrochemical reactions after contact with soil is called metal corrosion. To reduce the corrosion of buried pipelines, effective protective methods are needed (Xiao 1994).

11.1 Corrosion of the Buried Pipeline

The corrosion of buried pipeline is generated by its contact with soil. Therefore, the soil must be understood in order to understand and control the corrosion of buried pipeline.

11.1.1 Soil

Soil consists of four parts, which are mineral, organic matter, water, and air.

The mineral in soil comes from the weathered rock. As rock is composed by one kind or several kinds of minerals, mineral particles with different particle sizes can be generated after the weathering of rock, which can construct the framework of the soil, with certain porosity and permeability.

The organic matter in soil is generated by the plant and animal residues under chemical and microbial reactions. The organic matter in soil is mainly humic acid, whose production process is illustrated in Fig. 11.1.

The humic acid in the soil is a kind of mixture of complex matters, whose molecule is composed of nucleus, substituent group on the nucleus, and bridge bonds. The nucleuses of humic acids are mainly aromatic ring (such as benzene,

C. Dai and F. Zhao, *Oilfield Chemistry*,
https://doi.org/10.1007/978-981-13-2950-0_11

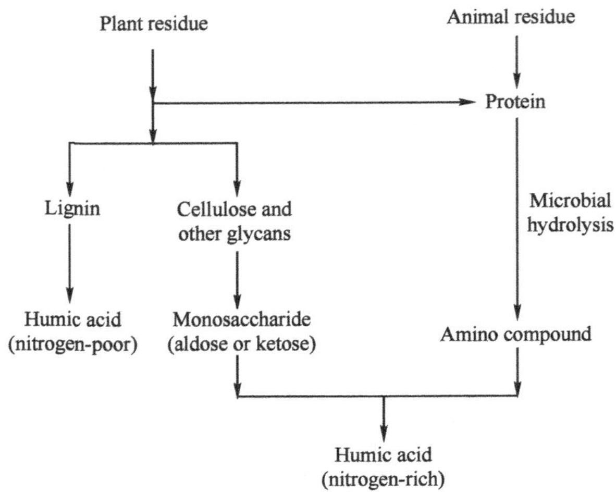

Fig. 11.1 Production process of humic acid

naphthalene, and anthracene) and aromatic ring (such as pyridine). The substituent groups on the nucleus include hydroxyl (such as alcoholic hydroxyl group and phenolic hydroxyl group), carboxyl, methoxyl, etc. The bridge bonds of humic acid include alkylene, imidogen, carbonyl, and ether bond, which can link the nucleuses of substituent groups into humic acid molecule. These humic acid molecules can form supermolecular structure by hydrogen bond and metal ion interaction. Figure 11.2 shows a hypothetical structural formula illustrating the concept of molecular structure and supermolecular structure of humic acid.

The water in soil is called soil water, in which various soluble organic matters, inorganic salts, and gases are dissolved. Soil water exists and flows in the pores of mineral framework.

The air in the soil exists in the pores of the soil, which is connected with the atmospheric air for exchange of substance. The oxygen in the soil water mainly comes from the air.

If analyzed by phase composition, soil is composed of the solid phase, liquid phase, and gas phase, among which the solid phase consists of mineral particles with different grain sizes and organic matters with different degrees of decomposition, the liquid phase is soil water, and the gas phase is air.

11.1.2 Soil Corrosion

The corrosion using soil as the corrosion medium is called soil corrosion (Wang et al. 1993; Liu et al. 2006). Soil corrosion is mainly related to the properties of the soil water.

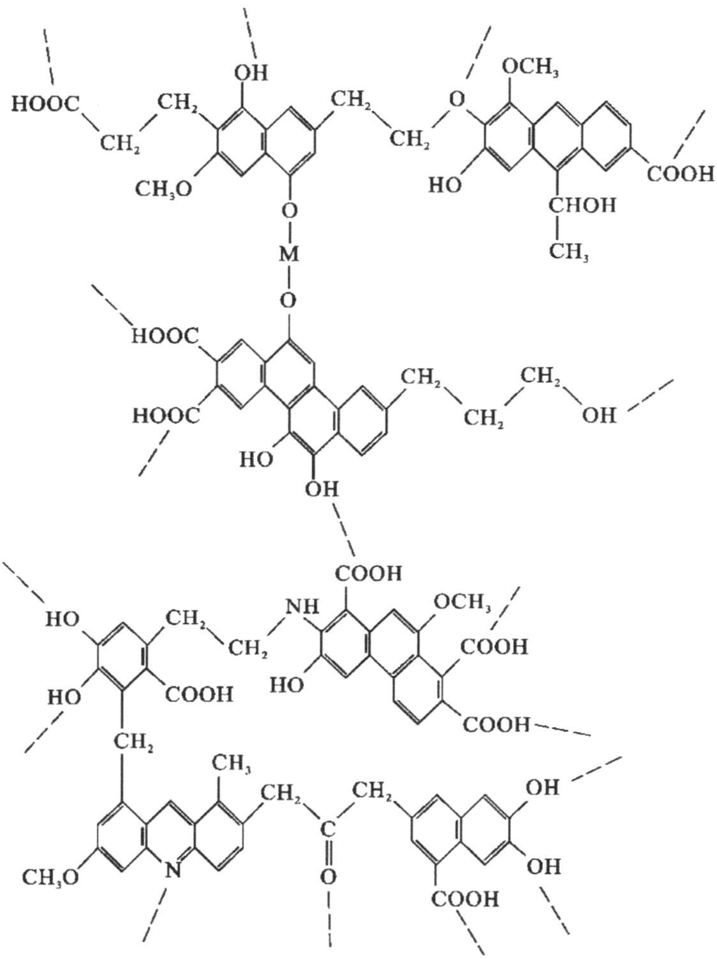

Fig. 11.2 Hypothetical structural formula illustrating the concept of molecular structure and supermolecular structure of humic acid

The Corrosion Generated When Soil Water Contains Oxygen

If soil water contains oxygen, the microbattery in the uneven part of the buried pipeline can generate corrosion on the anode part, through the following electrode reaction and cell reaction.

Anodic reaction

$$2\,Fe \rightarrow 2\,Fe^{2+} + 4e^{-}$$

Cathodic reaction

$$O_2 + 2\,H_2O + 4\,e^- \rightarrow 4\,OH^-$$

Cell reaction

$$2\,Fe + O_2 + 2\,H_2O \rightarrow 2\,Fe^{2+} + 4\,OH^-$$

The Corrosion Generated When Soil Water Contains Acid Gas

If soil water contains hydrogen sulfide and carbon dioxide, they can generate H^+ in water by dissociation.

$$H_2S \rightarrow H^+ + HS^-$$
$$CO_2 + H_2O \rightarrow H^+ + HCO_3^-$$

The H^+ in water can have the microbattery in the uneven part of the buried pipeline generate corrosion on the anode part, through the following electrode reaction and cell reaction.

Anodic reaction

$$Fe \rightarrow Fe^{2+} + 2\,e^-$$

Cathodic reaction

$$2\,H^+ + 2\,e^- \rightarrow H_2$$

Cell reaction

$$Fe + 2\,H^+ \rightarrow Fe^{2+} + H_2$$

The Corrosion Generated by Bacteria in Soil

If soil water contains sulfate-reducing bacteria (anaerobic bacteria) which can reduce sulfate into sulfide, the microbattery in the uneven part of the buried pipeline can generate corrosion on the anode part, through the following electrode reaction and cell reaction.

$$SO_4^{2-} \xrightarrow{SRB} S^{2-} + 4[O]$$

Anodic reaction

$$Fe \rightarrow Fe^{2+} + 2\,e^-$$

Cathodic reaction

$$[O] + H_2O + 2e^- \rightarrow 2\ OH^-$$

Cell reaction

$$Fe + [O] + H_2O \rightarrow Fe^{2+} + 2\ OH^-$$

There also exist iron bacteria (aerobic bacteria) in the soil water, which can reduce Fe^{3+} to Fe^{2+}.

$$4\ Fe^{2+} + O_2 + 2\ H_2O \xrightarrow{IB} 4\ Fe^{3+} + 4\ OH^-$$

The substances produced from the above-mentioned various corrosion cell reactions need to undergo further reaction to generate different kinds of solid corrosion products.

$$\begin{aligned} Fe^{2+} + 2\ OH^- &\rightarrow \underset{(White)}{Fe(OH)_2} \downarrow \\ 3\ Fe(OH)_2 + 2\ OH^- &\rightarrow \underset{(Black)}{Fe_3O_4} + 4\ H_2O + 2\ e^- \\ 2\ Fe_3O_4 + 2\ OH^- + (3n - 1)H_2O &\rightarrow 3\ \underset{(Red)}{Fe_2O_3 \cdot nH_2O} \downarrow + 2\ e^- \\ Fe^{2+} + S^{2-} &\rightarrow \underset{(Black)}{FeS} \downarrow \\ Fe^{3+} + 3\ OH^{2-} &\rightarrow \underset{(Brown)}{Fe(OH)_3} \downarrow \end{aligned}$$

Furthermore, soil corrosion is closely related to the soil porosity, permeability, and the water and gas saturation degrees. When differences in the aforementioned factors exist in the soil contacting the buried pipeline, differential concentration corrosion is generated. The soil with small porosity, low permeability, and/or low gas saturation degree is the anode, generating anodic reaction. While the soil with large porosity, high permeability, and/or high gas saturation degree is the cathode, generating cathodic reaction. Therefore, differential concentration corrosion is an important corrosion form in the soil corrosion of buried pipeline.

If a cubic soil is cut off and the resistance between parallel sections is measured, the electrical resistivity can be defined using the following equation, where ρ is the electrical resistivity of the soil ($\Omega \cdot m$), R is the resistance between parallel sections of soil (Ω), A is the sectional area of the parallel section (m^2), and L is the vertical distance between parallel sections (m).

$$\rho = R \cdot A/L \tag{11.1}$$

As the soil electrical resistivity can comprehensively reflect the soil corrosivity and is easy to measure, it can be used as standard for dividing the soil corrosivity.

Table 11.1 Soil corrosivity level classified according to the soil electrical resistivity

Corrosivity level	Strong corrosivity	Medium corrosivity	Weak corrosivity
Soil electrical resistivity (Ω•m)	<20	20–50	>50

According to the soil electrical resistivity, the soil corrosivity can be divided into three levels, which are strong, medium, and weak corrosivity (Table 11.1).

Table 11.1 shows that the lower the soil electrical resistivity, the more serious the soil corrosivity will be.

11.2 Anti-corrosion of the Buried Pipeline

Generally, anti-corrosion coating method and cathodic protection method are employed to reduce the corrosion of buried steel pipeline in the soil.

11.2.1 Anti-corrosion Coating Method

The protective layer covered on the metal surface to separate the metal surface from the corrosive medium is called covering layer.

The method to inhibit metal corrosion using covering layer is called anti-corrosion coating method. The covering layer which inhibits metal corrosion is called anti-corrosion layer (Liu et al. 2008; Wang et al. 2007; Yang et al. 2005; Guo et al. 2009; Yang 2005).

Good anti-corrosion layer should bear properties such as thermal stability, chemical stability, biological stability, high mechanical strength, high resistivity, and low permeability.

In the structure of anti-corrosion layer, the coating is an important component.

Coating refers to the material which can form a tough protective film on the surface (commonly known as paint (Chen 1995; Ni 1994)).

The coatings used for buried pipeline are mainly petroleum asphalt coating, coal-tar asphalt coating, polyethylene coating, epoxy resin coating, and polyurethane coating. Among these coatings, some are tough protective films generated through melting and then cooling (such as petroleum asphalt coating, coal-tar asphalt coating, and polyethylene coating), while others are tough protective films generated through chemical reactions (such as epoxy resin coating and polyurethane coating).

In the structure of anti-corrosion layer, besides surface coating, other components such as the primer, intermediate coat, finishing coat, inner tape, and outer tape are also used when necessary.

The following are commonly used buried steel pipeline coatings.

11.2.1.1 Petroleum Asphalt Anti-corrosion Coating

This is a kind of coating using petroleum asphalt as the main material.

Petroleum asphalt comes from crude oil. The residual oil on tower bottoms after reduced pressure distillation of crude oil or the asphalt eluted (with oxidation or without oxidation) using solvent (such as propane) are all petroleum asphalt mentioned here.

Petroleum asphalt is mainly composed of oil, colloid, and asphaltene. There are two types of available petroleum asphalts. One is petroleum asphalt I with a softening point of 95–110 °C, while the other is petroleum asphalt II with a softening point of 125–140 °C. The former one is used in buried pipeline transporting liquid with temperature below 50 °C, while the latter one is used in buried pipeline transporting liquid with temperature in the range of 50–80 °C.

Petroleum asphalt anti-corrosion coatings with different structures can be selected according to the corrosivity of soil (Table 11.2).

From Table 11.2, we can see that the primer is produced by dissolving petroleum asphalt in industrial gasoline. The inner tape is glass cloth (woven glass fiber), and the outer tape is polyvinyl chloride industrial film which is coated by adhesive prepared by mixing neoprene and rosin.

Petroleum asphalt coating has advantages such as wide source of raw materials, low cost, and simple construction technology.

11.2.1.2 Coal-Tar Enamel Anti-corrosion Coating

Coal-tar enamel is composed of coal-tar, coal-tar asphalt, and coal powder. There are three types of available coal-tar enamels whose softening points are >100 °C, >105 °C, >120 °C, respectively. The flow coating temperature of these coal-tar enamels is all in the range of 230–260 °C.

Coal-tar enamel anti-corrosion coatings with different structures can be selected according to the corrosivity of soil (Table 11.3).

Table 11.2 Level and structure of petroleum asphalt coating

Coating level	Coating structure	The total thickness of the coating (mm)
Ordinary level	Primer–petroleum asphalt–inner tapes–petroleum asphalt–inner tapes–petroleum asphalt–outer tapes	≥ 4.0
Strengthen level	Primer–petroleum asphalt–inner tapes–petroleum asphalt–inner tapes–petroleum asphalt–inner tapes–petroleum asphalt–outer tapes	≥ 5.5
Extra-strong level	Primer–petroleum asphalt–inner tapes–petroleum asphalt–inner tapes–petroleum asphalt–inner tapes–petroleum asphalt–inner tapes–petroleum asphalt–outer tapes	≥ 7.7

Table 11.3 Level and structure of the coal-tar enamel coating

Coating level	Coating structure	The total thickness of the coating (mm)
Ordinary level	Primer–coal-tar enamel–outer tapes	≥3.0
Strengthen level	Primer–coal-tar enamel–inner tapes–coal-tar enamel–outer tapes	≥4.0
Extra-strong level	Primer–coal-tar enamel–inner tapes–coal-tar enamel–inner tapes–coal-tar enamel–outer tapes	≥5.0

The primer listed in Table 11.3 is coal-tar primer, which is prepared by dissolving coal-tar and coal-tar asphalt in xylene. The inner tapes and outer tapes are both felt strips woven by fiberglass; the former is soaked in adhesives (such as copolymer of ethylene and vinyl acetate) while the latter is soaked in coal-tar enamel.

Coal-tar enamels have many advantages such as waterproof, high mechanical strength, chemical stability, strong abilities of anti-bacterial and anti-penetration of plant roots, wide raw material sources, and low cost.

11.2.1.3 Polyethylene Anti-corrosion Coating

Two kinds of polyethylene are applied in the polyethylene anti-corrosion coating. One is the high-density polyethylene with a density of 0.935–0.950 g/cm^3, while the other is the low-density polyethylene with a density of 0.900–0.935 g/cm^3. Their structures are shown as follows.

$$\left[CH_2-CH_2 \right]_n$$

(High density polyethylene)

$$\left(CH_2-CH_2 \right)_m \left(CH_2-\underset{|}{CH} \right)_n \left(CH_2-\underset{\underset{R}{|}}{CH} \right)_p \qquad R: C_1, C_2, C_3 \cdots$$

(Low density polyethylene)

There are two methods for forming polyethylene coating (Feehan et al. 1977). One method is extrusion coating method while the other is adhesive tape winding method.

When using extrusion coating method, polyethylene can be first heated and melted, then extruded on the outer wall of primer-coated pipeline to form the polyethylene coating. The primer used here is a kind of adhesive working as primer, whose molecule contains nonpolar part which can be closely bonded with the surface of the polyethylene and also polar portion which can bond tightly with the pipe surface (the surface of the pipe is polar due to the air oxidation).

The following copolymer can be used to do primer:

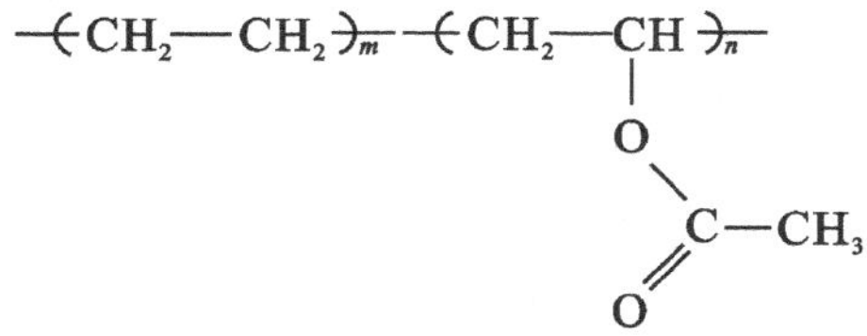

(Copolymer of ethylene and vinyl acetate)

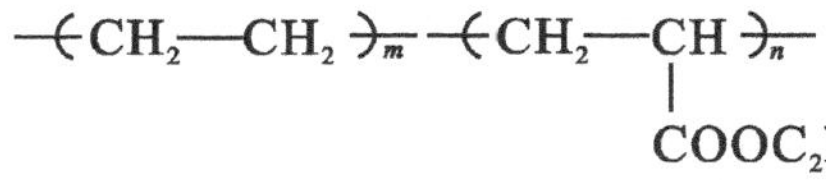

(Copolymer of ethylene and ethyl acrylate)

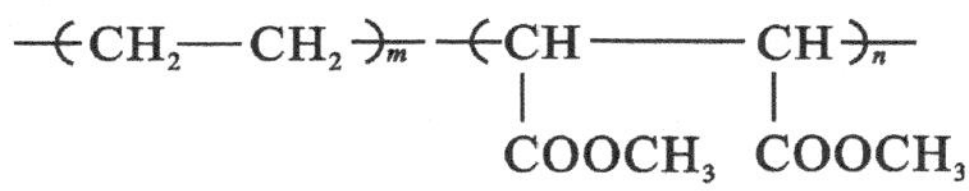

(Copolymer of ethylene and dimethyl maleate)

When using adhesive tape winding method, adhesives are coated on the surface of polyethylene tape to prepare polyethylene adhesive tape, which is winded on the outer wall of primer-coated pipeline to form the polyethylene anti-corrosion coating (Hielema 1974). The adhesives coated on the polyethylene tape contain two major components. One is the adhesive such as polyisoprene which can provide adhesive viscosity.

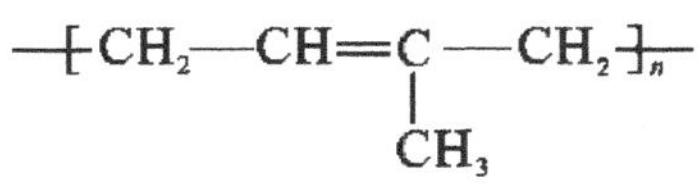

The other is wetting agent such as poly-limonene which can be dissolved in adhesive to improve its wetting ability, reduce the contact angle of the adhesive on the polyethylene surface and the primer surface to improve the adhesive effect of adhesive.

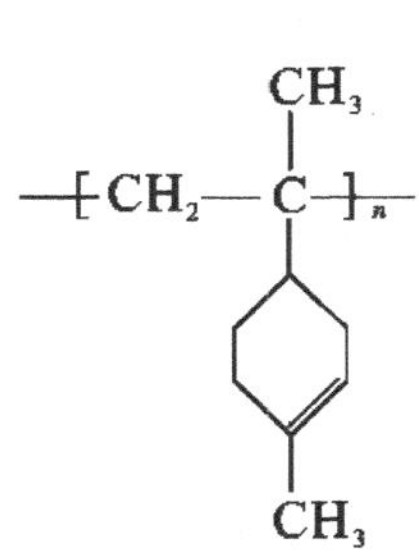

In the adhesive tape winding method, the primer on metal surface can be prepared by dissolving the following rubber type polymers into solvents.

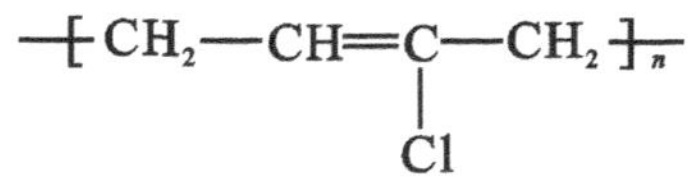

(Neoprene)

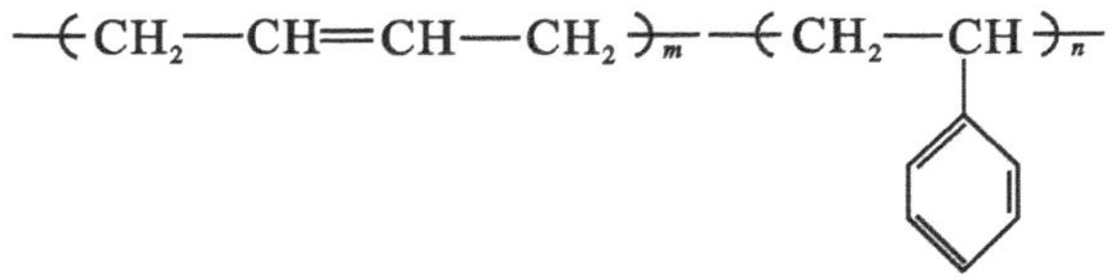

(Styrene-butadiene rubber)

Available solvents include xylene, ethyl acetate, methyl ethyl ketone, methyl isobutyl ketone, etc. After the evaporation of the solvent, the primer can form a film of rubber type polymer on the metal surface, which can increase the bonding force between the polyethylene adhesive tape with the metal surface.

11.2.1.4 Polyethylene Polyurethane Foam Insulation Anti-corrosion Coating

This is a kind of composite insulation anti-corrosion coating using polyurethane foam as the inner thermal insulation layer, and polyethylene as the outer protective layer.

Polyurethane foam is generated by the foaming and curing of polyurethane through different methods. Polyurethane is synthesized by polyisocyanates and polyhydroxy compounds. During the synthesis, the amount of isocyanate groups should be maintained excess compared with that of hydroxyl groups.

Available isocyanates are shown as follows.

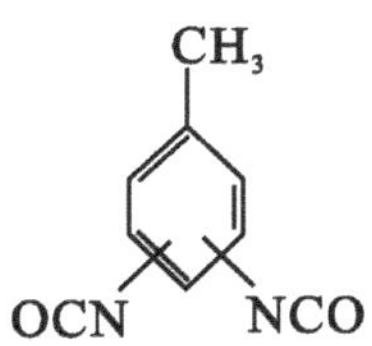

(Toluene diisocyanate, TDI)

$$OCN-(CH_2)_6-NCO$$

(Hexamethylene diisocyanate, HDI)

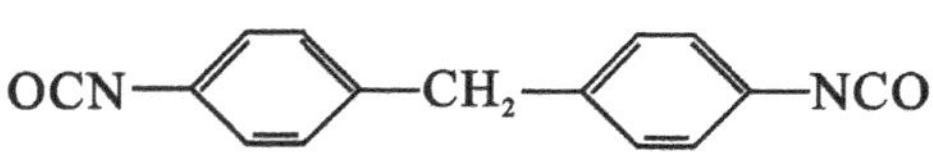

(Diphenyl methane diisocyanate, MDI)

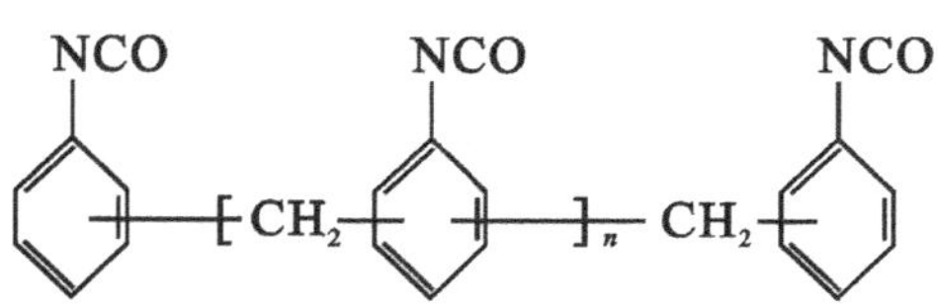

(Polymethylenepolyphenylisocyanate)

Available polyhydroxy compounds are shown as follows.

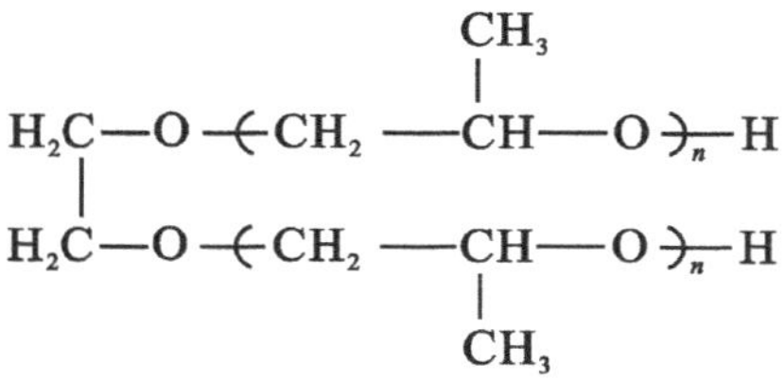

(Polyoxypropylene glycol ether)

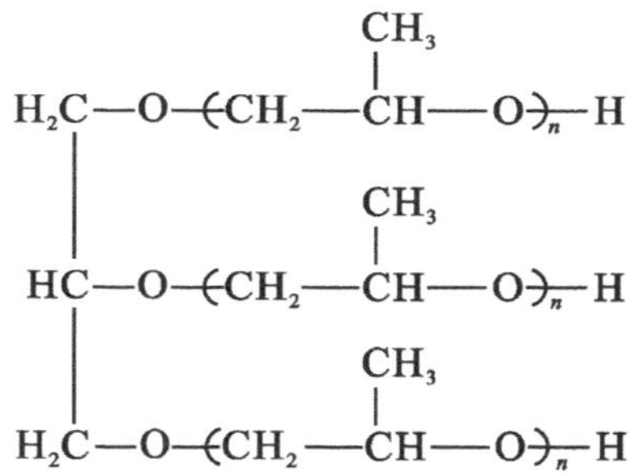

(Polyoxypropylene glycerol ether)

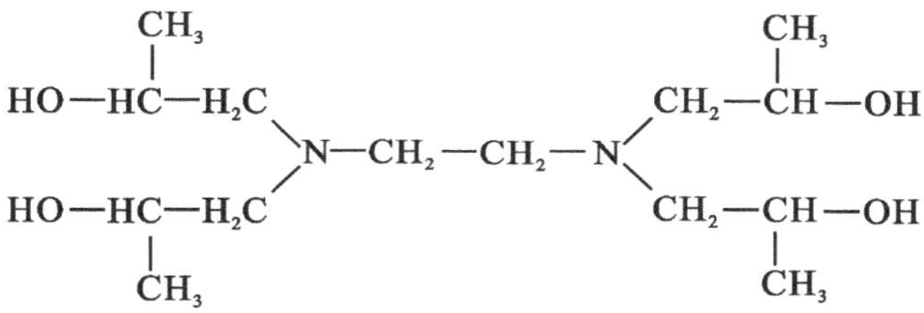

(Tetrahydroxyethyl isopropyl diamine)

Because of the excess amount of isocyanates compared with that of hydroxyl groups during the synthesis, polyurethane can undergo the following reaction after contact with water, causing foaming and curing of the polyurethane to generate polyurethane foam.

$$-NCO + H_2O \longrightarrow -NH_2 + CO_2\uparrow$$

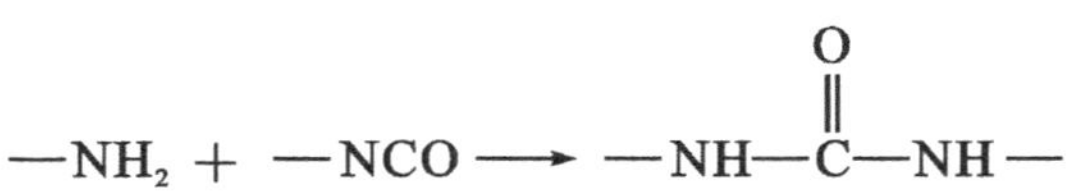

(Urea bond)

In addition, bubbles can also be generated through the vaporization of low molecular alkane (such as butane) or low molecular halogenated hydrocarbon (such as trichloromethane) by injecting air or taking advantage of the reaction heat. Meanwhile, amine (some examples are given below) can be added to solidify the polyurethane to generate polyurethane foam (Oertel 1985).

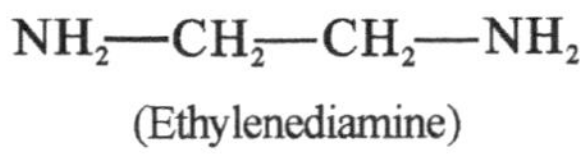

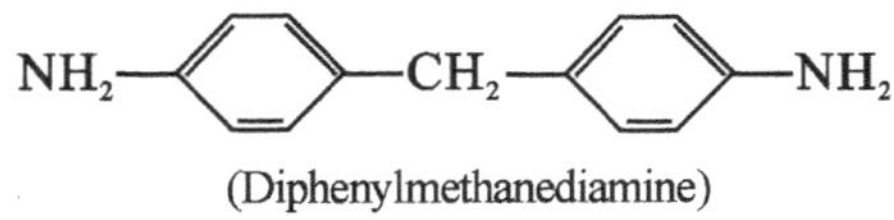

The polyethylene used in external protective coating is mostly high-density polyethylene.

Polyethylene polyurethane foam insulation anti-corrosion coating has wide application in oilfield.

11.2.1.5 Fusion-Bonded Epoxy Powder Anti-corrosion Coating

By spraying the epoxy resin powders added with curing agent onto the metal surface before drying under 150–180 °C for 15 min, tough fusion-bonded epoxy powder coating can be obtained (Torossian et al. 1986; Tran et al. 1996).

Epoxy resins with the following structures can be applied in this kind of anti-corrosion coating.

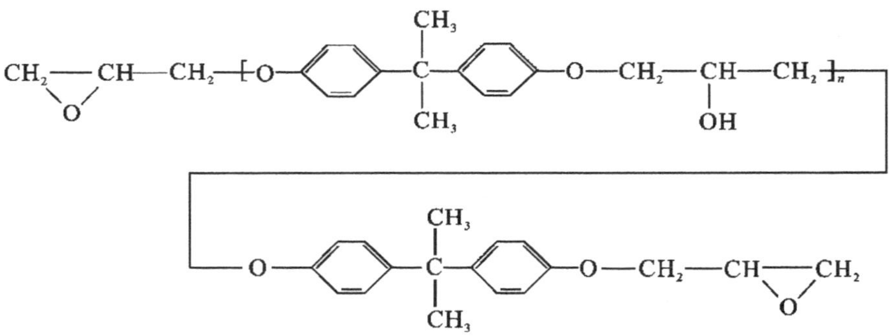

Softening point is one of the most important properties of epoxy resin, which refers to the softening temperature of epoxy resin in prescribed conditions. Table 11.4 shows the relationship between the softening point of epoxy resin and its degree of polymerization (*n* in above formula). Figure 11.3 shows the relationship between the softening point of epoxy resin and its molecular weight.

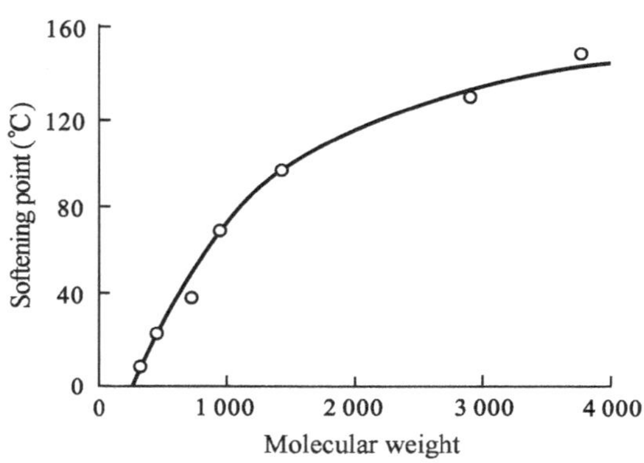

Fig. 11.3 Relationship between the softening point of epoxy resin and its molecular weight

Table 11.4 Relationship between the softening point of epoxy resin and its degree of polymerization

Softening point (°C)	Degree of polymerization
<50	<2
50–100	2–5
>100	<5

The epoxy resin with softening point of 95 °C can be applied to prepare epoxy powder.

There are mainly two types of the curing agents for epoxy resin.

Dicyandiamide

Dicyandiamide is produced by the adduct reaction between two cyanogen amines.

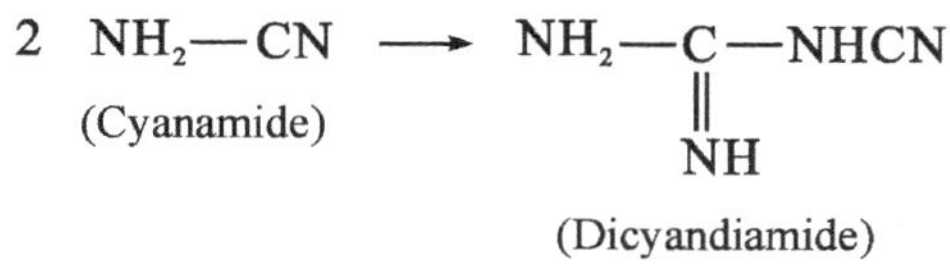

The primary amino and secondary amino of dicyandiamide can cross-link through the epoxy groups of epoxy resin.

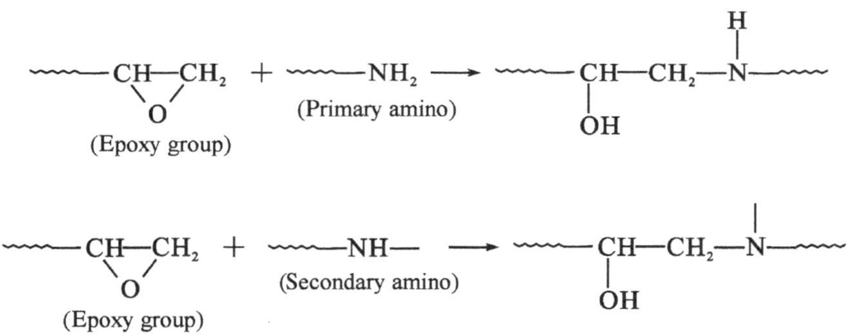

Table 11.5 shows the formula of epoxy powder using dicyandiamide as curing agent. The polyacrylic ester in the formula can reduce the surface tension of the coating material and improve the wettability of the coating to the metal surface so that the coating can easily spread on the metal surface. The titanium dioxide in the formula is a kind of pigment, playing the role of coloring and enhancing the strength of coating.

Table 11.5 Formula of epoxy powder using dicyandiamide as curing agent

Composition	*w* (%)
Epoxy resin (with a softening point of 95 °C)	66.0
Polyacrylic ester	3.4
Titanium dioxide	27.5
Dicyandiamide	3.1
Sum	100.0

Phenolic Resin

Phenolic resin is produced from the polycondensation of phenol (or cresol) and formaldehyde. The phenolic groups in phenolic resin can be cross-linked through the epoxy groups in epoxy resin.

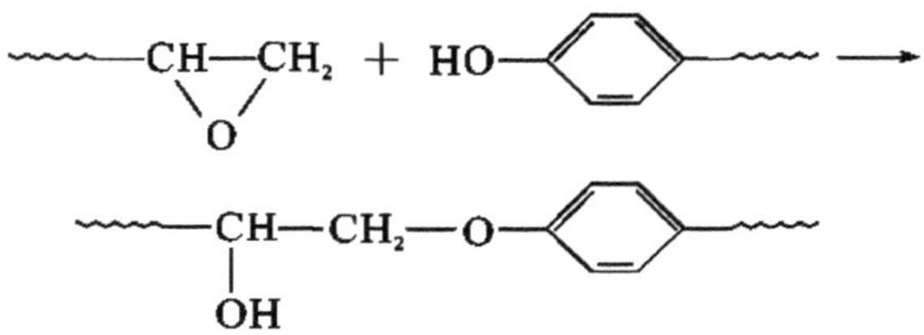

Table 11.6 shows the formula of epoxy powder using phenolic resin as curing agent. The fumed silica in the formula is generated by the atomization of molten silicon dioxide in the gas phase. As the surface is hydroxylated and can form structure through hydrogen bonding, the loss of coating edge during high-temperature drying is prevented. The ferric oxide and titanium dioxide in the formula are both pigments, playing the role of coloring and enhancing the strength of coating.

The epoxy powder using phenolic resin as curing agent can be applied as the anti-corrosive coating for high-temperature pipeline (Norsworthy 1998).

Table 11.6 Formula of epoxy powder using phenolic resin as curing agent

Composition	*w* (%)
Epoxy resin (with a softening point of 95 °C)	64.0
Phenolic resin	16.0
Fumed silica	0.5
Titanium dioxide	18.0
Ferric oxide	1.5
Sum	100.0

Table 11.7 Formula of various coats in the epoxy coal asphalt coating

Ingredient	Composition	w (%)		
		Primer	Intermediate coat	Finishing coat
The first component	Epoxy resin	11.3	11.2	19.6
	Coal-tar asphalt	6.7	14.0	24.5
	Light calcium carbonate	30.2	31.5	15.8
	Iron red	11.3	10.5	5.2
	Zink yellow	7.5	–	–
	Mixed solvent	27.4	27.2	25.1
The second component	Polyamide	2.8	2.8	4.9
	Xylene	2.8	2.8	4.9
	Sum	100.0	100.0	100.0

11.2.1.6 Epoxy Coal Asphalt Anti-corrosion Coating

Epoxy coal asphalt coating is mainly composed of epoxy coal asphalt primer, intermediate coat, and finishing coat. Table 11.7 shows the formula of various coats in the epoxy coal asphalt coating (Yu 1996).

In the first component in Table 11.7, epoxy resin and coal-tar asphalt are used as the main agents, light calcium carbonate is used as the filler, iron red (ferric oxide) and zinc yellow (double salt formed by alkaline zinc chromate and potassium chromate) are used as pigment. The mixed solvent is prepared by toluene, cyclohexanone, xylene, and butyl acetate with a mass ratio of 4:3:2:1. In the second component, polyamide is prepared by the heating polymerization of unsaturated fatty acid into dimer acid, followed by further condensation with diethylenetriamine. The structure of polyamide is shown as follows. Xylene is the solvent.

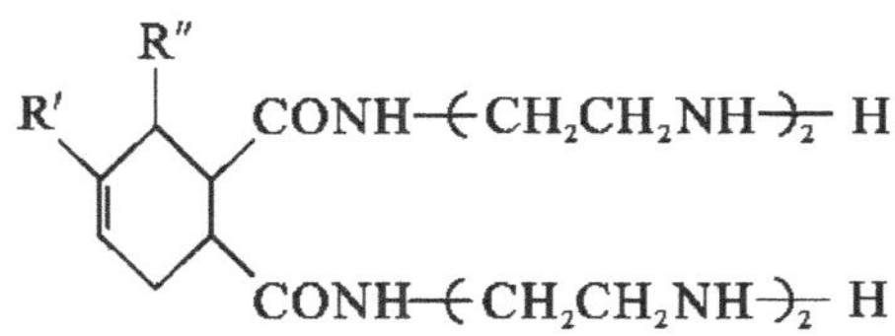

The various coats in the epoxy coal asphalt anti-corrosion coating can be prepared by mixing the first component and the second component in Table 11.7.

Table 11.8 shows the level and structure of epoxy coal asphalt coating. Inner tapes (glass cloth) are added in the strengthen level and extra-strong level anti-corrosion coating.

Because epoxy resin can cure at room temperature, epoxy coal asphalt coating can be coated at room temperature.

Table 11.8 Level and structure of epoxy coal asphalt coating

The level of the anti-corrosion coating	The structure of the anti-corrosion coating	The thickness of the anti-corrosion coating
Ordinary level	Primer–intermediate coat–finishing coat	≥0.2
Strengthen level	Primer–intermediate coat–inner tapes–intermediate coat–finishing coat	≥0.4
Extra-strong level	Primer–intermediate coat–inner tapes–intermediate coat–inner tapes–intermediate coat–finishing coat	≥0.6

11.2.1.7 Trilaminar Composite Anti-corrosion Coating

As various anti-corrosion coatings have their own advantages, composite coating with superior operational performance can be generated by the compound use of different anti-corrosion coatings. The trilaminar composite anti-corrosion coating represents this development trend. In this anti-corrosion coating, the primer is epoxy resin which has properties of good anti-corrosion, adhesion, and thermal stability. The intermediate coat contains various copolymers of vinyl monomers such as copolymer of ethylene and vinyl acetate, copolymer of ethylene and ethyl acrylate, and copolymer of ethylene and maleic acid methyl ester. These copolymers contain not only polar groups which can bond with the primer, but also the nonpolar groups which can bond with the outer layer; therefore they have excellent cohesive performance. The outer layer is high-density polyethylene which has good mechanical strength. If polypropylene is used instead of polyethylene as the outer layer, the anti-corrosion coating can be used in the high temperature up to 93 °C.

The main disadvantage of this kind of trilaminar composite coating is high cost, thus restricting the application.

11.2.2 Cathodic Protection Method

Anti-corrosion coating method is an important method to prevent corrosion of the buried pipeline; however, it must be in combined use with the cathodic protection method to effectively control the corrosion of buried pipelines. As holiday points inevitably exist during the coating process, which would result in phenomena such as peeling, perforation, and cracking under various factors in the application period, the cathodic protection method is the supplement to the anti-corrosion coating method (Gay 1975; Moreland et al. 1995; Huang et al. 2009; Liu 2007).

There are two kinds of cathodic protection methods, which are impressed current method and sacrificial anode method.

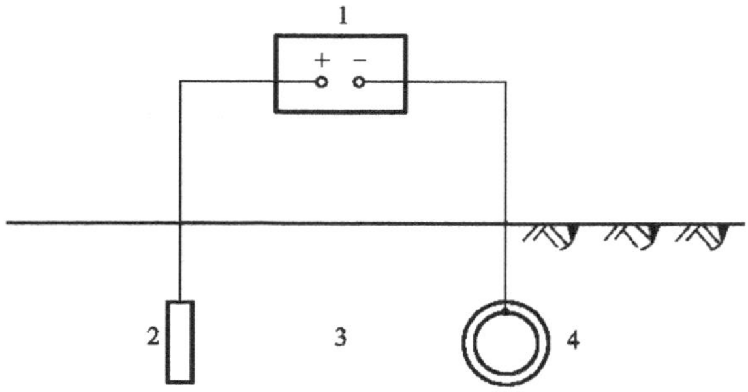

Fig. 11.4 Impressed current cathodic protection of buried pipeline. 1—DC power (potentiostat); 2—auxiliary anode; 3—soil; 4—the protected metal (buried pipeline)

11.2.2.1 The Impressed Current Cathodic Protection Method

In the corrosion cell, anode is the corroded electrode while cathode is not.

A circuit shown in Fig. 11.4 is formed by connecting the negative electrode of DC power to the metal to be protected (buried pipelines), and connecting the positive electrode to the auxiliary electrode (such as high-silicon cast iron). Then voltage is applied, changing the whole protected metal (including a large number of microbatteries generated because the metal is inhomogeneous or in different conditions) into the cathode, generating protection current. After the generation of protection current, electrode reactions occur at the electrode surface. Anodic reaction (oxidation) happens on the anode surface while cathodic reaction (reduction) happens on the cathode surface. Since the metal to be protected is connected to the negative electrode of DC power, cathodic reaction happens on the metal to protect it from corrosion.

Current cathodic protection refers to the way that the protected metal is connected to the negative electrode of DC power, protection current is provided by the impressed current to reduce the corrosion rate.

In cathodic protection, the natural potential and protect potential of the protected metal need to be measured, by connecting the protected metal with the reference electrode.

The reference electrode has stable and reproducible potential. Common reference electrode is copper/saturated copper sulfate electrode (abbreviated as CSE). Figure 11.5 gives the structure of a kind of copper/saturated copper sulfate reference electrode, whose electrode reaction is shown as follows.

$$Cu \underset{\text{Reduction}}{\overset{\text{Oxidation}}{\rightleftarrows}} Cu^{2+} + 2e^{-}$$

If the protected metal is connected with the copper sulfate reference electrode inserted in the soil (Fig. 11.6), the measured potential is the natural potential of the protected metal. The natural potential of the buried steel pipeline is about −0.55 V (with respect to the CSE) (Wu 1991).

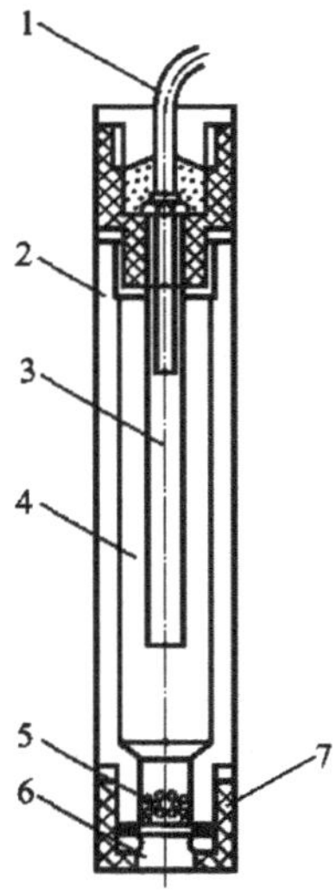

Fig. 11.5 Copper/saturated copper sulfate reference electrode. 1—guide wire; 2—sheath; 3—copper electrode; 4—saturated copper sulfate solution; 5—copper sulfate; 6—microporous material; 7—tightening cap

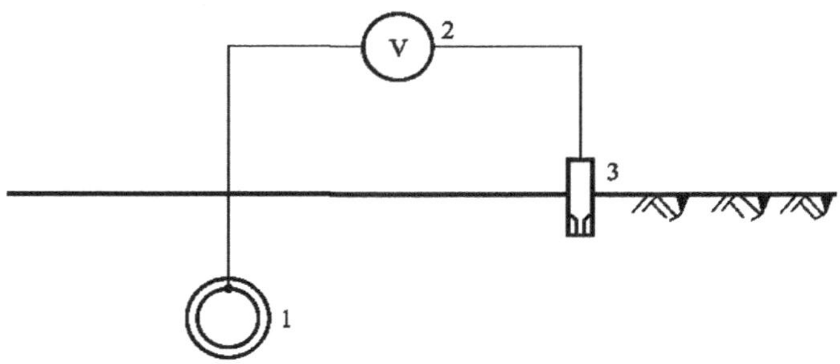

Fig. 11.6 Measurement of the natural potential of the protected metal. 1—the protected metal (buried pipeline); 2—high resistance potentiometer; 3—copper sulfate reference electrode

In the circuit shown in Fig. 11.4, if the impressed current cathodic protection method is applied to protect the metal, protection current is generated in the circuit. In this situation, by connecting the metal to be protected with the copper sulfate reference electrode inserted in the soil (Fig. 11.7), the measured potential is the protection potential of the metal. Under effective cathodic protection, the protective potential is generally controlled in the range of −0.85 to −1.20 V, with respect to CSE. The negative shift of the protection potential relative to the natural potential is due to the blockage of the cathode reaction. The reaction occurs in the cathode surface is reduction reaction, which requires electron-accepting ions (such as H^+) in water contacting the cathode surface. The rate of diffusion, reaction, and the product

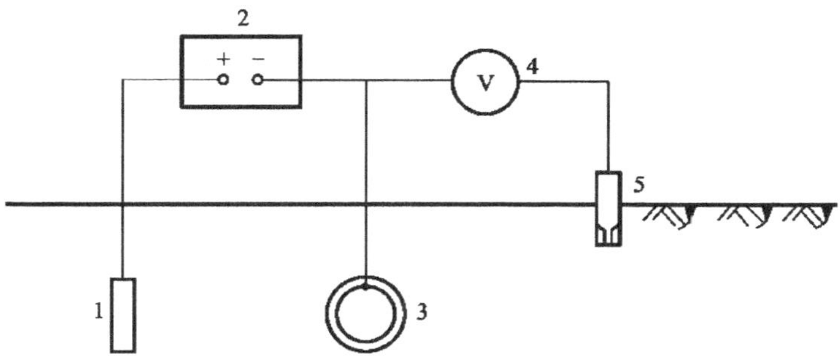

Fig. 11.7 Measurement of the protection potential of the protected metal. 1—auxiliary anode; 2—DC power; 3—the protected metal; 4—high resistance potentiometer; 5—copper sulfate reference electrode

leaving the cathode surface of these ions is lower than movement rate of electrons in metal conductor, causing the accumulation of electrons on the cathode surface. In this case, the measured potential (protection potential) is definitely more negative than the natural potential.

11.2.2.2 The Sacrificial Anode Cathodic Protection Method

The way that the protected metal is connected to the metal or alloy which can provide cathodic protection current (i.e., sacrificial anode) to lower the corrosion rate of protected metal is called sacrificial anode cathodic protection (Jost et al. 1986; Bagnulo 1985; Cao 2006; Wang et al. 2009).

Figure 11.8 shows the sacrificial anode cathodic protection system and monitoring system.

The materials which can be used as sacrificial anode are metal or metal alloy whose potential is more negative than that of the protected metal.

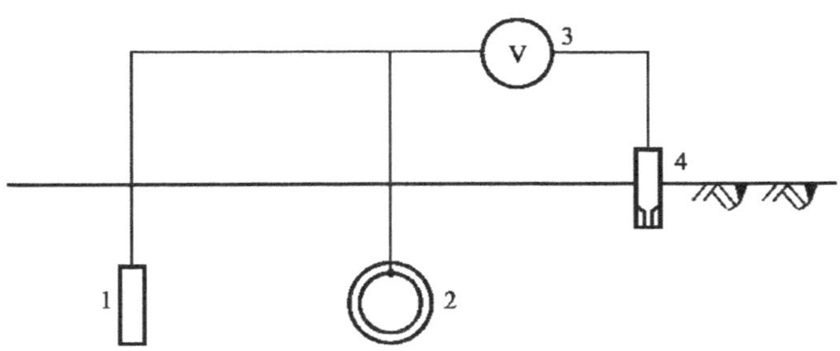

Fig. 11.8 Sacrificial anode cathodic protection system and monitoring system. 1—the sacrificial anode; 2—metal to be protected; 3—high resistance potentiometer; 4—copper sulfate reference electrode

A good sacrificial anode should meet various requirements such as enough negative potential, large electric capacity, high current efficiency, uniform dissolution, easy falling off of the corrosion product, easy preparation, wide source, and low cost. Here the electric capacity refers to the electric quantity of the protection current provided by dissolving unit mass of sacrificial anode. The current efficiency refers to the ratio of the actual electric capacity to the theoretical electric capacity of the sacrificial anode, expressed as a percentage.

Important sacrificial anodes are alloys. Available alloys have two types, one is magnesium-based alloy with magnesium as the main component (Hu et al. 1984), while the other is zinc-based alloy with zinc as the main component. The mass fractions of their chemical elements are shown in Table 11.9, and their electrochemical properties are shown in Table 11.10.

When using the sacrificial anode, filler composed of sodium sulfate, bentonite, and gypsum powder must be added around it. Filler can reduce the grounding resistance of the sacrificial anode, increase the output current, and facilitate the falling off of the corrosion products.

Metal corrosion quantity is calculated according to the following Faraday formula.

$$m = \frac{tM}{nF} \cdot I \tag{11.2}$$

Table 11.9 Mass fraction of chemical elements in the sacrificial anode

(1) Magnesium-based alloy								
Chemical element	Mg	Al	Zn	Mn	Maximum mass fraction of impurities (%)			
					Fe	Ni	Cu	Si
w (%)	Margin	5.3–6.7	2.5–3.5	0.15–0.60	0.005	0.003	0.02	0.1
(2) Zinc-based alloy								
Chemical element	Zn	Al	Cd	Maximum mass fraction of impurities (%)				
				Fe	Cu	Pb	Si	
w (%)	Margin	0.3–0.6	0.05–0.12	0.005	0.005	0.006	0.125	

Table 11.10 Electrochemical performance of the sacrificial anode

Sacrificial anodes	OCP (V)[a]	Working potential (V)[a]	Actual capacity (A•h/kg)	Current efficiency (%)
Magnesium-based alloy	≤ 1.50	≤ 1.40	1,100	55
Zinc-based Alloy	1.05–1.09	1.00–1.05	780	95

Note [a]With respect to the potential of CSE

where,

M—Metal corrosion quantity in the corrosion time, g;
t—Corrosion time, s;
M—Metal molar mass, g/mol;
n—Valence number of metal ions;
F—Faraday constant, equal to 96,500 C/mol;
I—Corrosion current, A.

Apparently, Eq. 11.2 can be used to calculate the theoretical electric capacity of the sacrificial anode.

References

Bagnulo L (1985) Method of corrosion protection. US Patent 4,496,444, 29 Jan 1985
Cao B (2006) Discussion on the selection of sacrificial anode sets in buried steel pipeline. Oil Gas Storage Transp 25(6):24–26, 30
Chen S (1995) Paint process (updated version) first fascicle. Chemical Industry Press, Beijing, pp 1–20
Feehan CE, Wagner EF (1977) Polyethylene coating for ferrous metals. US Patent 4,007,298, 8 Feb 1977
Gay PJ (1975) Cathodic protection. US Patent 3,868,313, 25 Feb 1975
Guo J, Wei C, Du X (2009) Development and new technology of pipeline coating. Pipeline Tech Equip 5:49–51
Hielema R (1974) Pipe coating method. US Patent 3,823,045, 19 July 1974
Hu S, Xu K, Wang W (1984) The make-in-research and application of Magnesium-base sacrifice anode. Oil Gas Storage Transp 3(4):37–41
Huang H, Liu Y (2009) Cathodic protection technology and its application to buried steel pipeline. Pipeline Tech Equip (2):46–48, 51
Jost JW, Ghandehari MH (1986) Corrosion protection with sacrificial anodes. US Patent 4,626,329, 2 Dec 1986
Liu G (2007) Application of cathodic protection technology. Corros Prot Petrochem Ind 23(6):58–61
Liu J, Wang W, Ma J (2006) Corrosion mechanisms of buried pipelines and countermeasures. Corros & Prot Petrochem Ind 23(6):20–22
Liu P, Chen B, Guo D et al (2008) The origin and protect-ion research of the buried oil pipeline to corrode. Pipeline Tech Equip 4:47–48
Moreland PJ, De Peuter FLJ (1995) Cathodic protection system and a coating and coating composition therefore. US Patent 5,431,795, 11 July 1995
Ni Y (1994) Practical anti-corrosion techniques. Chemical Industry Press, Beijing, pp 195–209
Norsworthy R (1998) Coating proves effective on hot pipelines. Pipeline Gas J 225(3):44–45, 48
Oertel G (1985) Polyurethane handbook. Hanser Publishers, New York
Torossian KA, Markovitz M, Cox FE (1986) Unique epoxy resin compositions and composite molded bodies filled therewith. US Patent 4,631,230, 23 Dec 1986
Tran BT, Toerner TJ, Titus PE (1996) Polymer concrete coating for pipe, tubular shapes, other metal members and metal structures. US Patent 5,573,855, 12 Nov 1996
Wang Y, Sun J (2007) External anticorrosive techniques on buried pipelines//China Petroleum and Natural Gas Co., Ltd. Xinjiang Oil Field Branch. Petroleum Industry Press, Beijing, pp 326–329

Wang K, Wu J (1993) The development of soil corrosion mechanism and research technology. Prog Soil Sci 21(4):10–14

Wang C, Pang Y, Li B (2009) Selection and installation of sacrificial anodes materials. Pipeline Tech Equip 1:59–60

Wu J (1991) Effects of soil properties on electrode potential of steel in soils. Acta Pedologica Sinica 28(2):117–123

Xiao J (1994) Corrosion introduction—corrosion of materials and control methods. Chemical Industry Press, Beijing, pp 284–296

Yang Y (2005) Anti-corrosion progress of steel pipes. Petrochem Pipeline 28(3):39–40

Yang J, Ma G, Zhang Y et al (2005) Selection and application on external coating of long-distance pipelines. Xinjiang Pet Sci & Technol 15(2):46–49

Yu Z (1996) Paint process (updated version) second fascicle. Chemical Industry Press, Beijing, pp 496–497

Chapter 12
Demulsification of the Emulsified Crude Oil and Defoaming of the Foaming Crude Oil

There are a variety of surfactants included in the crude oil, such as naphthenic acid, fatty acid, colloid, and asphaltene, which can adsorb on the oil–water interface or vapor–liquid interface to stabilize the droplet and foam. Thus, the problems of emulsion and foaming of crude oil come up. This chapter is focusing on topics on the demulsification of emulsified crude oil and defoaming of the foaming crude oil.

12.1 Demulsification of the Emulsified Crude Oil

Emulsified crude oil refers to the emulsion with crude oil as dispersed phase or dispersed media. The water content in the emulsified crude oil will increase the load of pump, pipeline, and storage tank, leading to the erosion and scaling on metal surface. Therefore, the emulsified crude oil must be demulsified for dehydration before transportation.

12.1.1 Types of Emulsified Crude Oil

There are mainly two types of emulsified crude oil.

Water-in-Oil (W/O) Emulsified Crude Oil
This is a kind of emulsified crude oil using oil as the dispersed medium and water as the dispersed phase. The emulsified crude oil produced from the first and secondary oil recovery is mainly W/O emulsified crude oil. The emulsifiers for stabilizing such kind of emulsified crude oil are mainly active oil acid (such as naphthenic acid and colloid acid) and oil-wet solid particles (such as wax particles and asphaltene particles).

C. Dai and F. Zhao, *Oilfield Chemistry*,
https://doi.org/10.1007/978-981-13-2950-0_12

Oil-in-Water (O/W) Emulsified Crude Oil
This is a kind of emulsified crude oil using water as the dispersed medium and crude oil as the dispersed phase. The emulsified crude oil produced from the ternary oil recovery (especially by alkali flooding and surfactant flooding) is mainly O/W emulsified crude oil. The emulsifiers suitable for stabilization of this kind of emulsified crude oil are the alkali metal salts of active oil acid, water-soluble surfactant, or water-wet solid particles (such as clay particles).

These two kinds of emulsified crude oil mentioned above are the basic kinds of emulsified crude oil. However, based on the observations under microscope, a certain amount of oil-in-water-in-oil (O/W/O) or water-in-oil-in-water (W/O/W) emulsified crude oil is contained in these two kinds of emulsified crude oil, which is called multiple emulsion crude oil. The complexity of emulsified crude oil might be one of the reasons for the difficulties in the thoroughly demulsification of emulsified crude oil.

12.1.2 Demulsification of W/O Emulsified Crude Oil

12.1.2.1 Methods of Demulsification of W/O Emulsified Crude Oil

Heating method, electronic method, and chemical method are commonly used for the demulsification of W/O emulsified crude oil (Luo et al. 1997; Quan 1996; He et al. 2005; Meng 2005). Those methods are always used in combination, named as heat–electronic–chemical method.

Heating Method
Heating method refers to the destruction of the W/O emulsified crude oil by increasing temperature. The increase of temperature can reduce the adsorption quantity of emulsifier, lower the solvation degree of emulsifier and reduce the viscosity of the dispersed medium. Therefore, it is beneficial to the coalescence and stratification of the dispersed phase.

Electronic Method
Electronic method refers to the destruction of the W/O emulsified crude oil under high voltage (1.5×10^4–3.2×10^4 V) direct-current or alternating-current electric field. Under the action of electric field, the water droplets are polarized into spindle shape. The surface-active substances concentrate at the end of the deformed water droplet to weaken the interface protection perpendicular to the power line, causing the coalescence of water droplets along the direction perpendicular to the power line and resulting in the demulsification.

Chemical Method
Chemical method refers to the destruction of the W/O emulsified crude oil using demulsifier.

12.1.2.2 Demulsifiers of the W/O Emulsified Crude Oil

Although low molecular demulsifiers such as fatty acid salt, alkyl sulfate salt, alkyl sulfonate, alkyl benzene sulfonate, OP-type surfactant, peregal surfactant, and Twain-type surfactant can be used for the emulsion breaking, macromolecular demulsifiers are more efficient for the demulsification of W/O emulsified crude oil. These demulsifiers can be produced by the reaction between initiator (such as propylene glycol, glycerin, diethylenetriamine, triethylenetetramine, tetraethylenepentamine, phenolic resin, and phenolic amine resin) and epoxy compound (such as ethylene oxide and propylene oxide). In order to improve the relative molecular mass of the demulsifier, chain extender (such as diisocyanate and binary carboxylic acid) can be used. In order to change the hydrophilic–lipophilic balance of the demulsifier, endcapping agent (such as rosin acid and carboxylic acid) can be used. In the following demulsifiers for W/O emulsified crude oil, the components of initiator and epoxy compound can be found. The components of chain extender and endcapping agent can also be found in some demulsifiers.

$$R{-}O{-}(C_3H_6O)_m{-}(C_2H_4O)_n{-}H$$

(Polyoxyethylene polyoxypropylene alkyl alcohol ether)

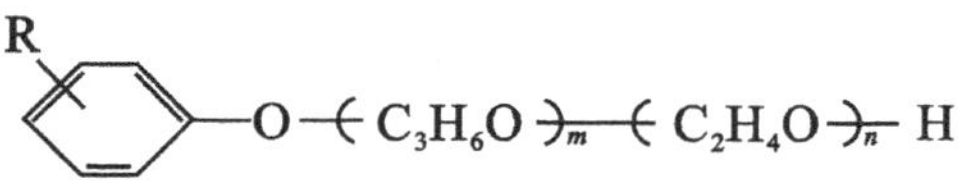

(Polyoxyethylene polyoxypropylene alkylphenol ether)

$$\begin{array}{l} CH_3{-}CH{-}O{-}(C_3H_6O)_m{-}(C_2H_4O)_n{-}H \\ \quad\;\; | \\ \quad\;\; CH_2{-}O{-}(C_3H_6O)_m{-}(C_2H_4O)_n{-}H \end{array}$$

(Polyoxyethylene polyoxypropylene propylene glycol ether, BE type demulsifier)
(Langdon et al.1980)

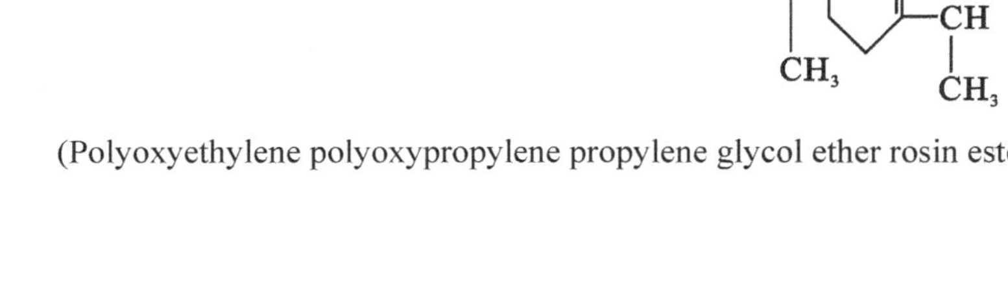

(Polyoxyethylene polyoxypropylene propylene glycol ether rosin ester)

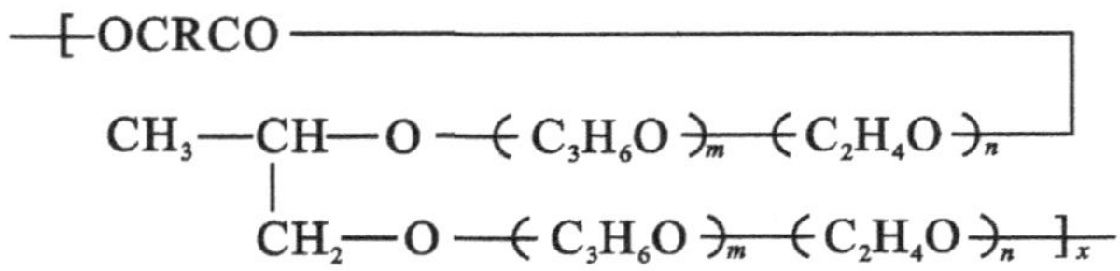

(Dicarboxylic chain-extending products of polyoxyethylene polyoxypropylene propylene glycol ether) (Oppenlaender et al.1974)

$$R-O-(C_3H_6O)_m-(C_2H_4O)_n-(C_3H_6O)_p-H$$

(Polyoxypropylene polyoxyethylene polyoxypropylene alkyl alcohol ether)

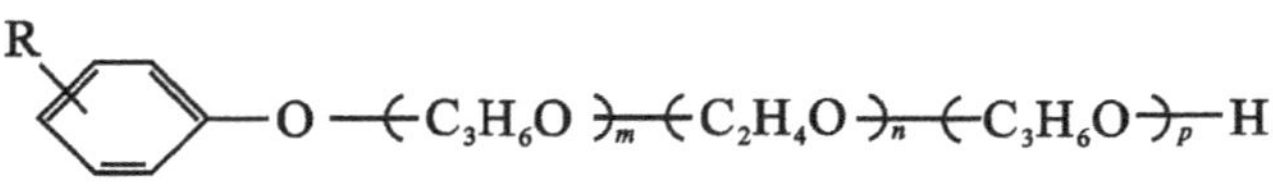

(Polyoxypropylene polyoxyethylene polyoxypropylene alkyl phenol ether)

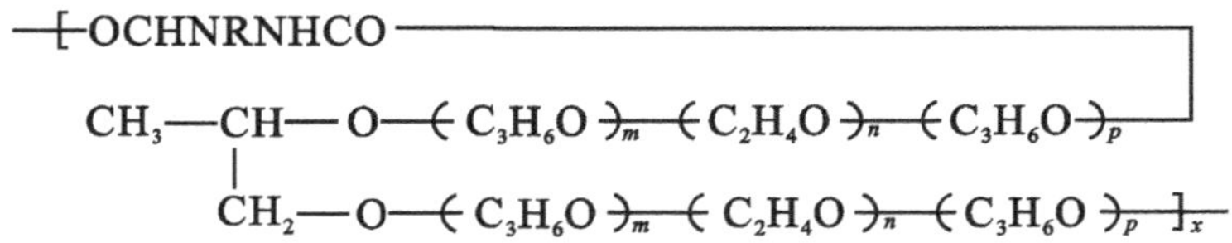

(Diisocyanate chain-extending products of polyoxypropylene polyoxyethylene polyoxypropylene propylene glycol ether, POI type demulsifier)

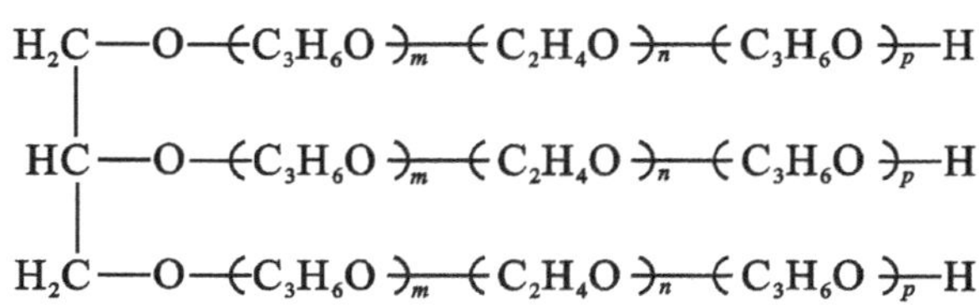

(Polyoxypropylene polyoxyethylene polyoxypropylene glycerol ether, GP type demulsifier)

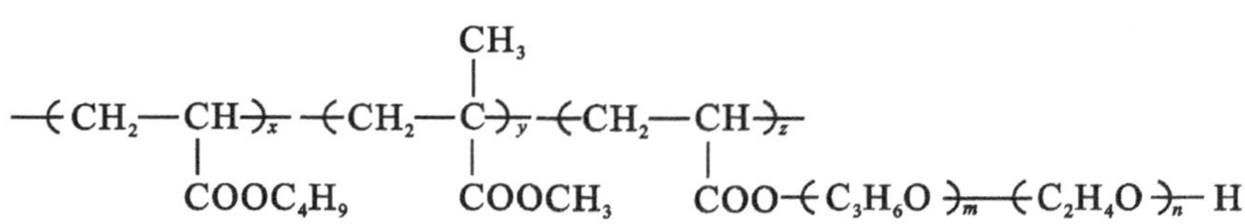

(Copolymer of butyl acrylate, methyl methacrylate and polyoxyethylene polyoxypropylene acrylate) (Stephenson 1990)

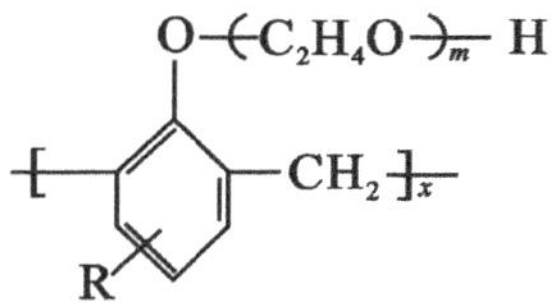

(Polyoxyethylene alkylphenol formaldehyde resin) (Merchant et al.1988)

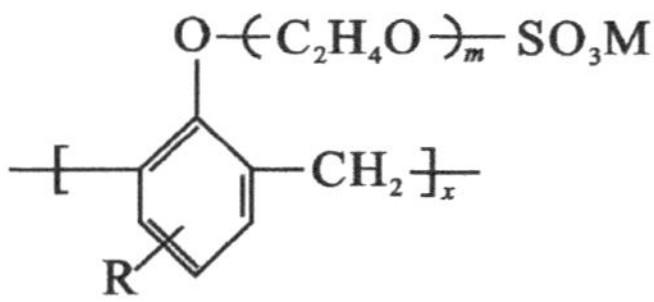

(Polyoxyethylene alkylphenol formaldehyde resin sulfate)

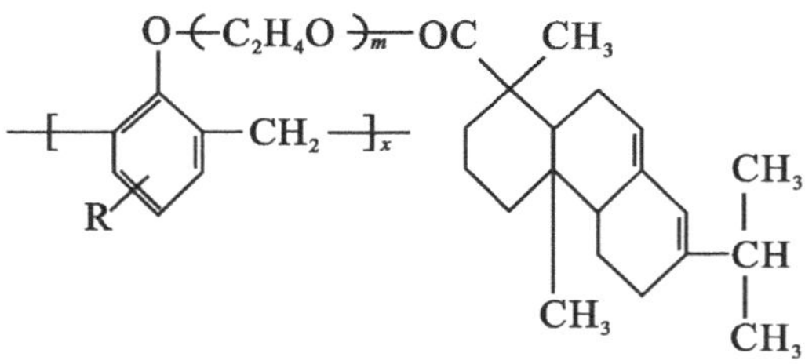

(Polyoxyethylene alkylphenol formaldehyde resin rosin ester)

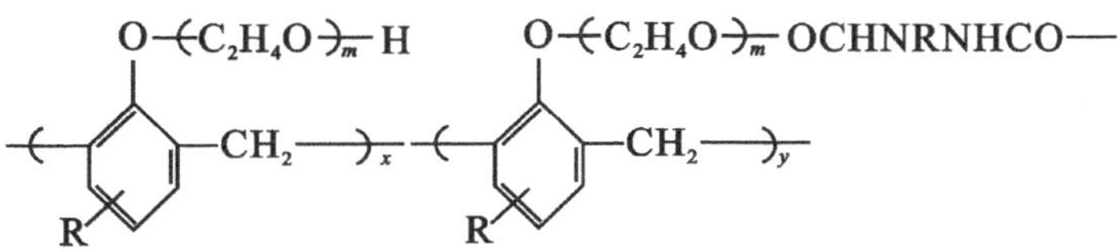

(Diisocyanate chain-extending products of polyoxyethylene alkylphenol formaldehyde resin)

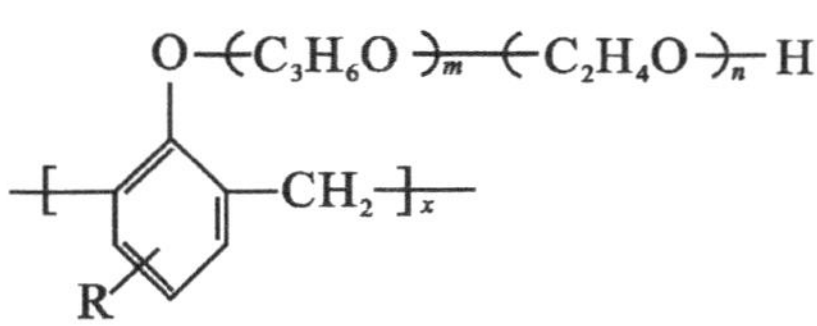

(Polyoxyethylene polyoxypropylene alkylphenol formaldehyde resin, AR type demulsifier)

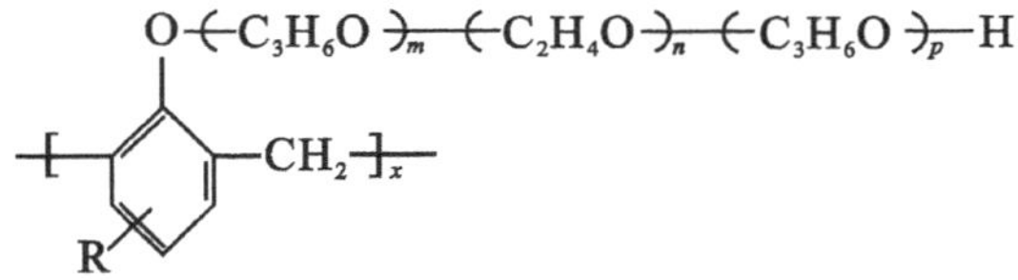

(Polyoxypropylene polyoxyethylene polyoxypropylene alkylphenol formaldehyde resin, AF type demulsifier)

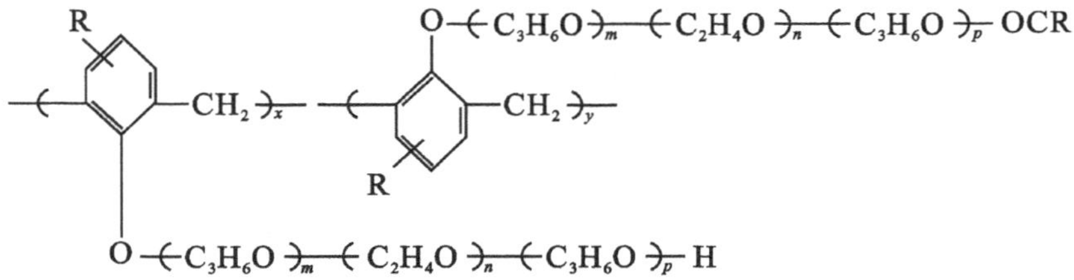

(Polyoxypropylene polyoxyethylene polyoxypropylene alkylphenol formaldehyde resin carboxylic ester)

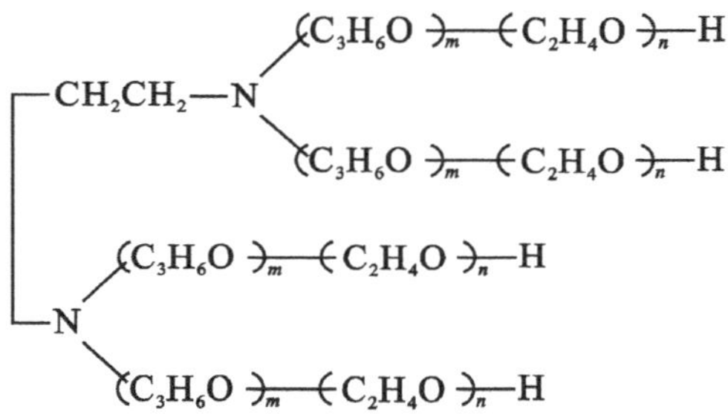

(Polyoxyethylene polyoxypropylene ethylenediamine, AE type demulsifier)

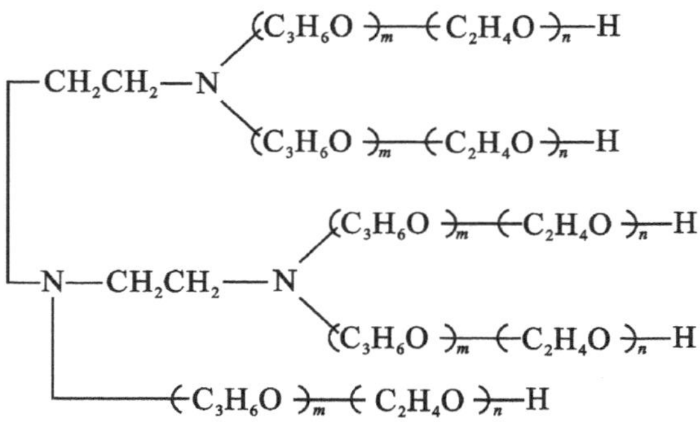

(Polyoxyethylene polyoxypropylene diethylenetriamine, AE type demulsifier) (Liebold et al.1975)

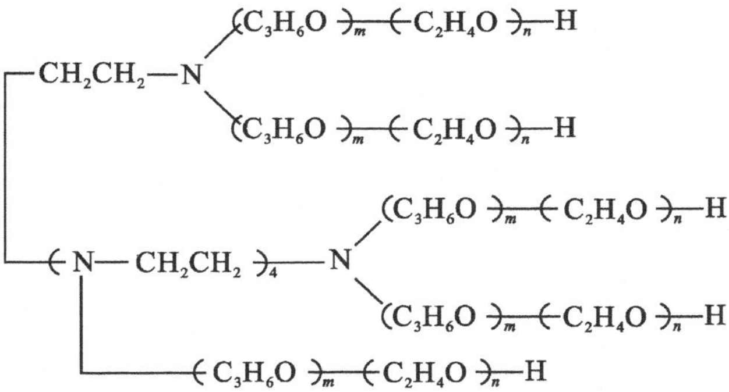

(Polyoxyethylene polyoxypropylene pentaethylenehexamine, AE type demulsifier)

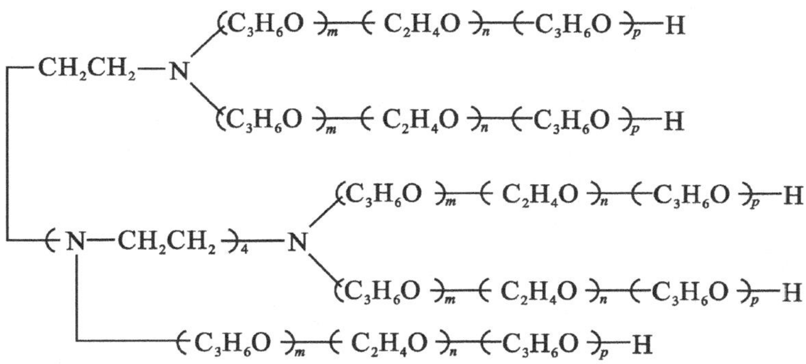

(Polyoxypropylene polyoxyethylene polyoxypropylene pentaethylenehexamine, AP type demulsifier)

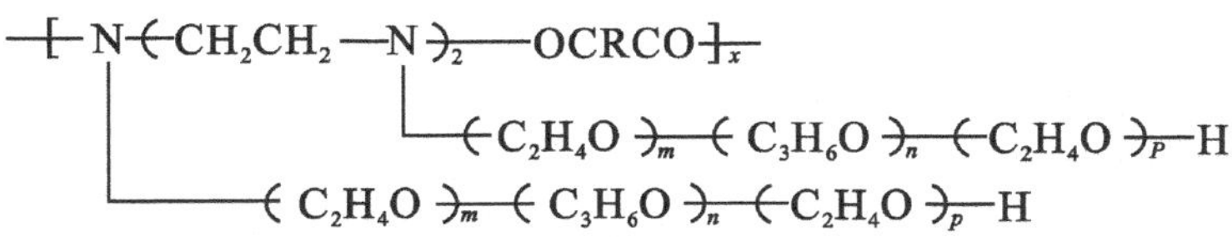

(Dicarboxylic chain-extending products of polyoxyethylene polyoxypropylene polyoxyethylene diethylenetriamine) (Barthold et al.1990)

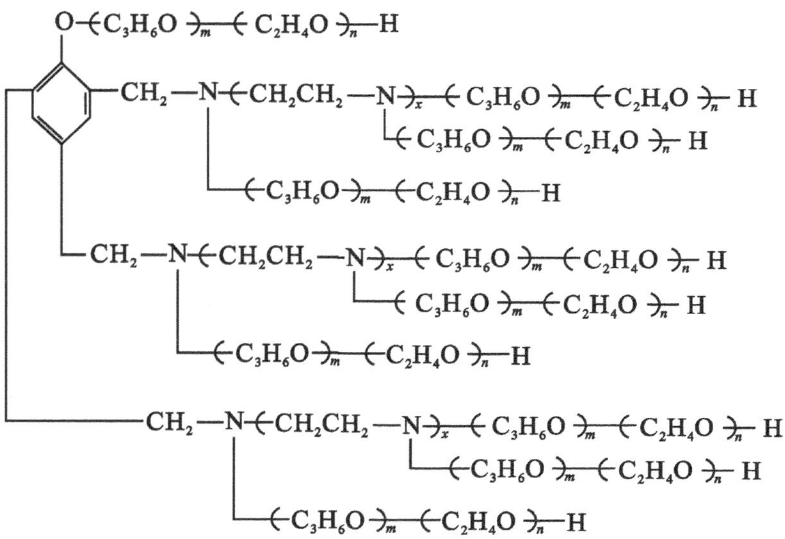

(Polyoxyethylene polyoxypropylene phenol amine resin, PFA type demulsifier)

12.1.2.3 Demulsification Mechanisms of the Demulsifiers for the W/O Emulsified Crude Oil

Different demulsifiers have different demulsification mechanisms (Xiao 2008).

Low molecular demulsifiers are water-soluble demulsifiers (HLB > 8). They are inverse emulsifiers relative to the emulsifiers of the W/O emulsified crude oil (HLB values in the range of 3–6), which can break W/O emulsified crude oil by counteracting effect.

Water-soluble demulsifiers of polymer demulsifiers also have this counteracting effect. The oil-soluble demulsifier is also a high-performance demulsifier, indicating that the demulsification mechanism is not mainly dependent on the counteracting effect. Polymer demulsifiers function by the following mechanisms.

Formation of Unstable Adsorption Film

The adsorption film formed by macromolecular demulsifier (especially the branched chain linear macromolecular demulsifier) replacing the primary emulsifier is not compact enough, thus has little protection effect.

Bridging of Water Droplets

Macromolecular demulsifier can adsorb on the surfaces of two or more water droplets simultaneously. These droplets connected by demulsifiers have more opportunity to collide and coalescence.

Solubilization of Emulsifiers
Under very low concentrations, macromolecular demulsifiers can form micelles which can increase the solubility of emulsifier molecules, resulting in the demulsification of emulsified crude oil.

12.1.2.4 The Developing Trend of Polymer Demulsifiers

The polymer demulsifiers have the following development tendencies.

Further Increasing of the Relative Molecular Weight
This trend is reflected in the usage of various chain extenders, such as aldehyde, binary carboxylic acid, multiple carboxylic acid (e.g., polyacrylic acid (Zhang et al. 2008)), epoxy derivatives, and polyisocyanates (Hart 1993).

From Water Soluble to Oil Soluble
As water content in the oilfield production is getting higher, the demulsification effects of the water-soluble demulsifiers which are mainly distributed in water phase, as a result, are getting worse. In contrast, oil-soluble demulsifiers are mostly distributed in oil phase, therefore, the active time of which can be prolonged and the effect can be increased.

From Linear to Branched Chain
The initiator from hydroxyl series is changed to phenolic resin, initiators from series to multiple ethylene amines.

Novel Demulsifiers Are Under Development
This is determined by the strong specificity of demulsifiers. Besides silicon-, nitrogen-, phosphorus-, and boron-containing demulsifiers, novel polymer demulsifiers using ethylene carbonate to replace oxyalkylene compounds are synthesized. Some examples are given below.

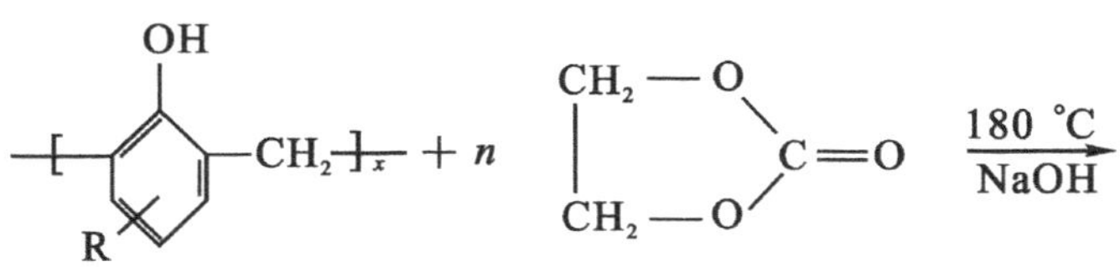

(Alkyl phenol formaldehyde resin) (Ethylene carbonate)

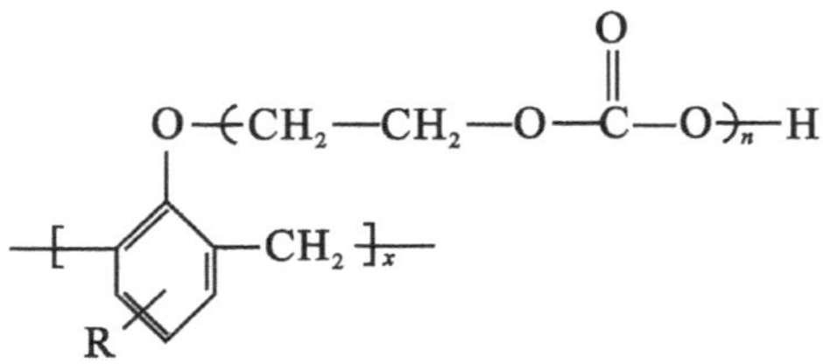

(Poly ethylene carbonate alkyl phenol formaldehyde resin)
(Stephenson et al.1993)

In addition, water-soluble polymer without oxyalkylene is also proposed as demulsifier. For example, the following copolymer can be used to break the W/O emulsified crude oil (Bhattacharyya 1992).

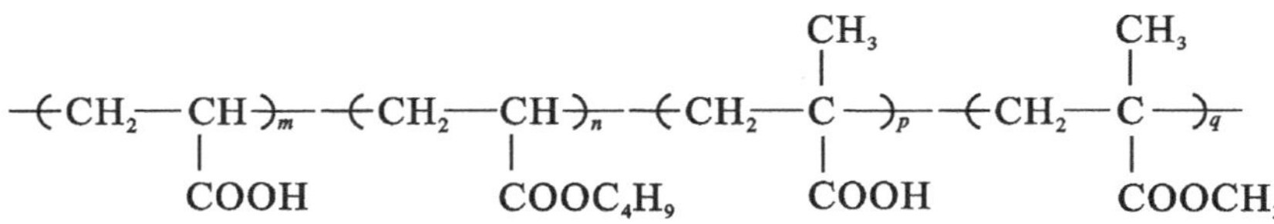

(Copolymer of acrylic acid, butyl acrylate, methyl acrylic acid and methyl methacrylate)

Use of Mixtures
This is the most effective way to overcome the specificity of polymer demulsifiers. For example, silicon-containing demulsifiers can be mixed with polyoxyethylene phenolic resin for application (Hoffmann et al. 1980); esterified polyoxyethylene phenolic resin can be mixed with non-esterified polyoxyethylene phenolic resin for application.

12.1.3 Demulsification of O/W Emulsified Crude Oil

12.1.3.1 Methods of the Demulsification of O/W Emulsified Crude Oil

Methods of the demulsification of O/W emulsified crude oil contain heating method, electronic method, and chemical method, which are usually used in combination.

Heating Method
The heating method for the demulsification of O/W emulsified crude oil functions the same as that of the O/W emulsified crude oil.

Electronic Method
The demulsification of O/W emulsified crude oil by electronic method is conducted in high-voltage alternating current field with medium frequency (1×10^3–2×10^4 Hz) or high frequency (>2×10^4 Hz) (in view of the conductivity of the water, one of the electrodes must be insulated). Under the action of electric field, the distribution ordering of the adsorbed layer of emulsifier is disturbed, which weakens the protection effect, resulting in the coalescence of the oil droplets and finally the demulsification.

Chemical Method
The chemical method for demulsification of O/W emulsified crude oil is achieved using demulsifiers.

12.1.3.2 Demulsifiers of the O/W Emulsified Crude Oil

Four kinds of demulsifiers can be used for the demulsification of O/W emulsified crude oil such as electrolytes, low molecular alcohols, surfactants, and polymers.

Available electrolytes include hydrochloric acid, sodium chloride, magnesium chloride, calcium chloride, aluminium nitrate, and zirconium oxychloride, etc.

Available low molecular alcohols can be categorized into water-soluble alcohols and oil-soluble alcohols; the former contains methanol, ethanol, and propanol, while the latter includes hexyl alcohol and heptanol alcohol, etc.

Available surfactants include cationic surfactants and anionic surfactants. Some examples are given below.

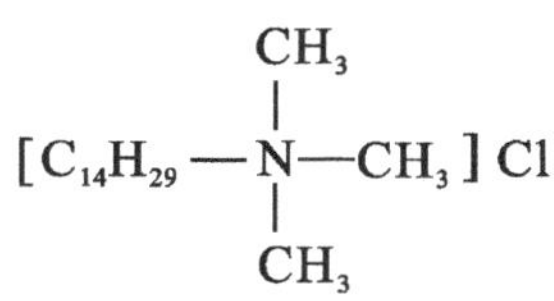

(Myristyltrimethyl ammonium chloride)

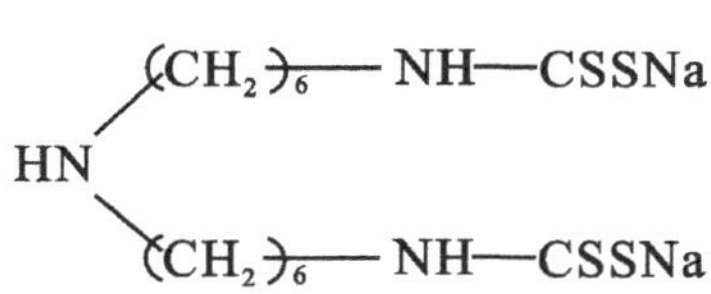

[Di-(hexamethylene)amine-di-(sodium dithiocarbamate)]
(Durham et al.1989)

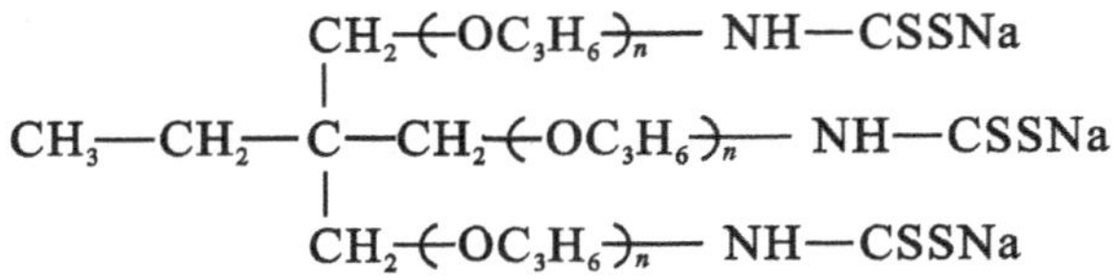

[Polypropylene oxide-2,2- bis(hydroxymethyl) *N*-butyl alcohol ether tri(sodium dithiocarbamate)] (Thompson et al.1989)

Available polymers include cationic polymers such as the following.

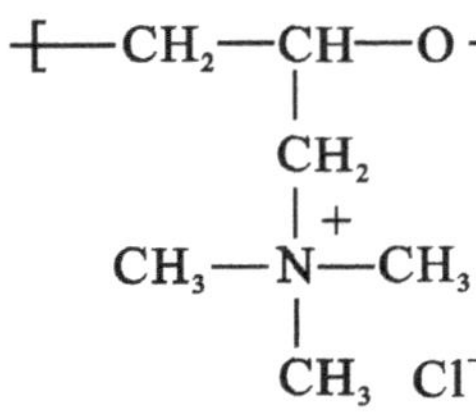

(Polyoxypropylene trimethyl ammonium chloride) (Mcdonald 1971)

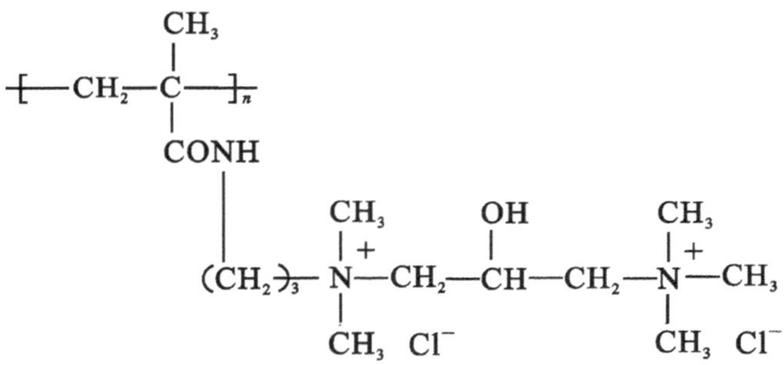

(Polymethyl acrylamide-1,3-propylidene-2'-hydroxy-1',3'-propylidene pentamethyl di-ammonia chloride) (Mcentier 1983)

Available polymers include nonionic polymers such as the following.

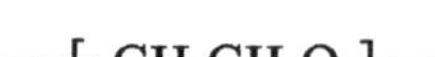

(Polyethylene glycol) (Bialek 1981)

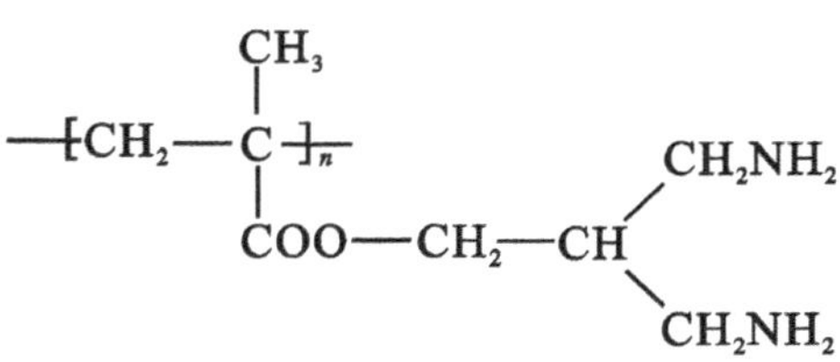

[Polymethylacrylic acid(1-amino-2-aminomethyl) propyl ester] (Braden et al.1991)

Available polymers also include nonionic–cationic polymers such as the following.

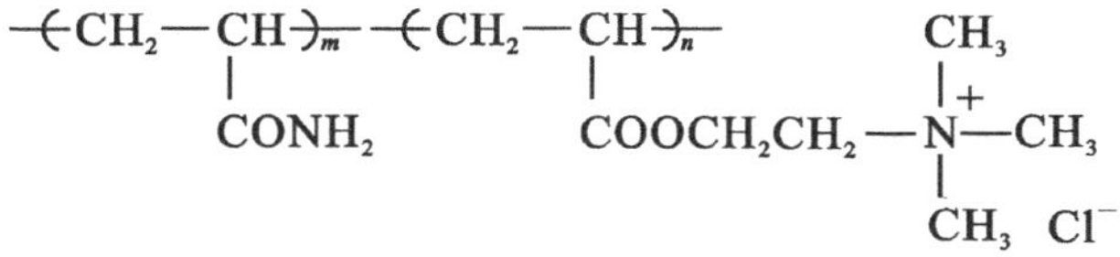

[Copolymer of acrylamide and (acrylic acid-1,2-ethylidene ester trimethyl ammonium chloride] (Hart et al.1998)

12.1.3.3 Demulsification Mechanisms of the Demulsifiers for the O/W Emulsified Crude Oil

Also, different demulsifiers have different demulsification mechanisms.

Electrolytes mainly function by reducing the electronegativity of the oil droplet surface and changing the hydrophilic–lipophilic balance of the emulsifier.

Low molecular alcohols function by changing the polarity of the oil and water phase (increasing polarity of the oil phase and decreasing the polarity of the water phase), so that the emulsifier can move toward the oil phase or the aqueous phase to break the emulsion.

Surfactants function by mechanisms of reacting with emulsifier (cationic surfactant), forming unstable adsorption film (anionic surfactant with branched structure), and counteracting effect (oil-soluble surfactant).

The nonionic polymers break emulsion by bridging mechanism. Cationic and nonionic–cationic polymers function not only through the bridging mechanism, but also by reducing the electronegativity of the oil surface.

12.1.3.4 Development Trend of Demulsifiers for O/W Emulsified Crude Oil

The demulsifiers for O/W emulsified crude oil have the following development trends.

- In four types of demulsifiers for O/W emulsified crude oil, the development of surfactant and polymer is dominant.
- In surfactant and polymer demulsifiers, the development of quaternary ammonium salt-type surfactant and polymer is dominant.
- The development of quaternary ammonium salt-type polymer shifts towards high degree of quaternary amination. The degree of quaternary amination here refers to the percentage of the polymer chains containing quaternary ammonium groups in the whole polymer chains.
- Due to the environment-friendliness, the development of quaternary ammonium salt-type natural polymer (such as quaternary ammonium salt type starch) has received special attention.

- Because different demulsifiers have different demulsification mechanisms, mixtures of different demulsifiers tend to be used for better effect (Newcombe 1980; Newcombe 1983).

12.2 Defoaming of the Foaming Crude Oil

12.2.1 Formation Mechanism of the Crude Oil Foam

The problems of foaming in crude oil are mainly encountered in the process of oil–gas separation and oil stabilization, during which natural gas (containing C_1–C_7 hydrocarbons, mainly alkanes) is released from crude oil by reducing pressure and (or) elevating temperature.

Oil and gas surface is generated when natural gas is released from the crude oil. The surfactants contained in crude oil, including the low molecular surfactants (such as fatty acids and naphthenic acids) and polymer surfactants (such as resin and asphaltene), can adsorb on the oil and gas surface and play different roles in the formation of crude oil foams.

The low molecular surfactants can spread to the surface of the oil and gas to reduce its surface tension, decrease the surface work required in foam generation, thus facilitating the formation of foam.

Due to the large molecule, polymer surfactant cannot diffuse to the surface of the oil and gas easily, thus contributing little to the generation of the crude oil foam. However when adsorbed on the surface of oil and gas, they can form high-strength surface films on the surface of the oil and gas due to their special structure (there are polar groups in the polycyclic aromatic hydrocarbon sheets and their branched chains), which play an important role to the stabilization of the crude oil foam.

Since the foam stability is closely related to the discharge rate of the liquid film between crude oil foams, the viscosity of crude oil has a direct impact on the stability of the crude oil foams.

Since the lipophilic moiety of the surfactant is attractive to the crude oil, the oil discharge is stopped at certain extent, maintaining certain film thickness of the liquid film. Therefore, the crude oil foam has certain stability.

In the foams, the bubble sizes are not uniform. Since the pressure inside the small bubble is greater than that inside the big bubbles, the gas in small bubbles can diffuse to the big bubbles through the liquid film. Therefore, small bubbles gradually become smaller until disappear, while large bubbles gradually become larger, resulting in the foam damage (Zhao 1991). This kind of foam failure mechanism is also present in crude oil foam. Since the presence of surfactant adsorption film can inhibit the gas permeability of the liquid films among large or small bubbles, the foam stability is improved.

Crude oil foam formation will seriously affect the stability of crude oil and gas separation effect, and hinder the normal measuring work.

12.2.2 Defoaming of the Foaming Crude Oil

Chemical agents that can eliminate the foams in crude oil are called crude oil defoamers. Crude oil defoamers can eliminate the foam of crude oil.

Crude oil defoamers can be divided into the following categories.

12.2.2.1 Solvent Defoamers

This kind of defoamer refers to the low molecular alcohols, ethers, alcohol ethers, and esters which are commonly used as solvents. When sprayed onto the oil foams, these defoamers can spread on the foams rapidly due to their low surface tension with air and low interfacial tension with oil, which will lead to the local thinning of the liquid film and the destruction of the foam.

The following solvents are available as defoamers.

$C_5H_{11}OH$

(Pentanol)

$C_8H_{17}OH$

(Octanol)

$C_3H_7—O—C_3H_7$

(Propyl ether)

$C_4H_9—O—C_4H_9$

(Butyl ether)

$HO\left[CH_2CH_2O\right]_2C_6H_{13}$

(Diethylene glycol hexyl ether)

$HO\left[CH_2CH_2O\right]_2C_7H_{15}$

(Diethylene glycol heptyl ether)

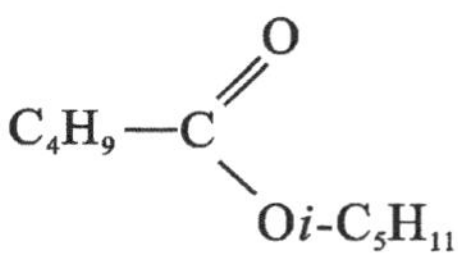

(Isoamyl valerate)

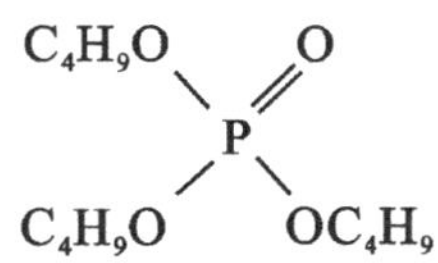

(Tributyl phosphate)

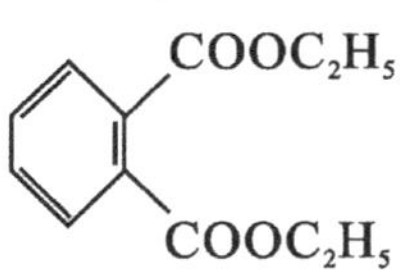

(Diethyl phthalate)

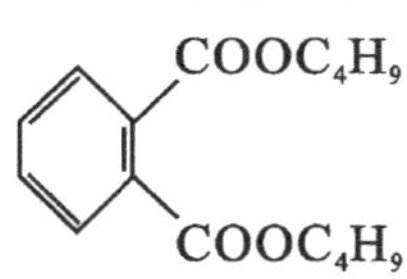

(Dibutyl phthalate)

12.2.2.2 Surfactant Defoamer

Surfactant defoamers refer to some surfactants with branched structure. When sprayed onto the oil foams, these defoamers can replace the original foam-stabilizing surfactant and form unstable protective film, resulting in foam destruction.

The following surfactants are available as defoamers.

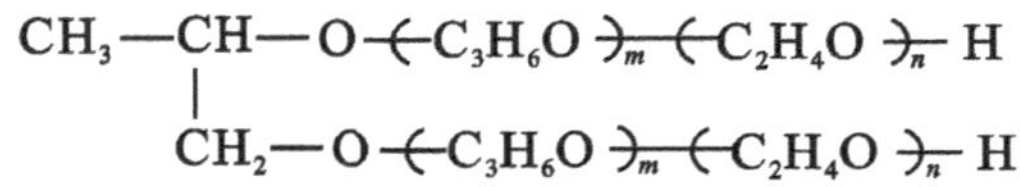

(Polyoxyethylene polyoxypropylene propylene glycol ether)

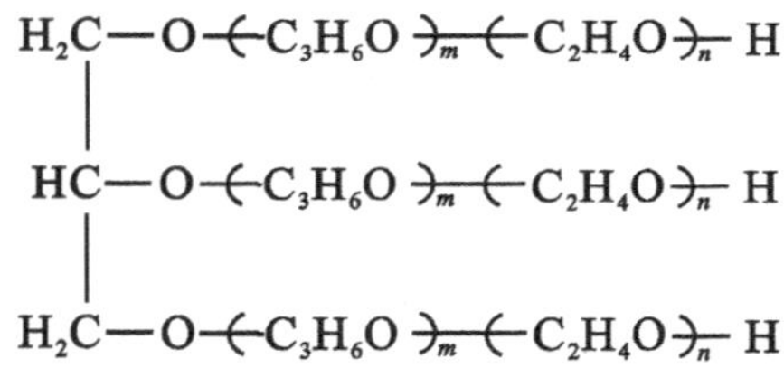

(Polyoxyethylene polyoxypropylene glyceryl ether)

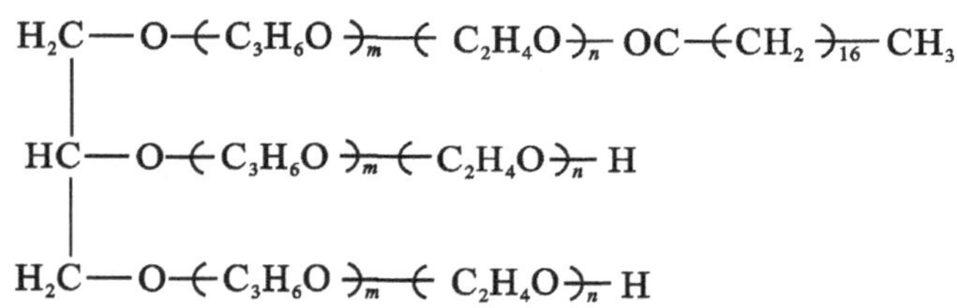

(Polyoxyethylene polyoxypropylene glyceryl ether stearate)

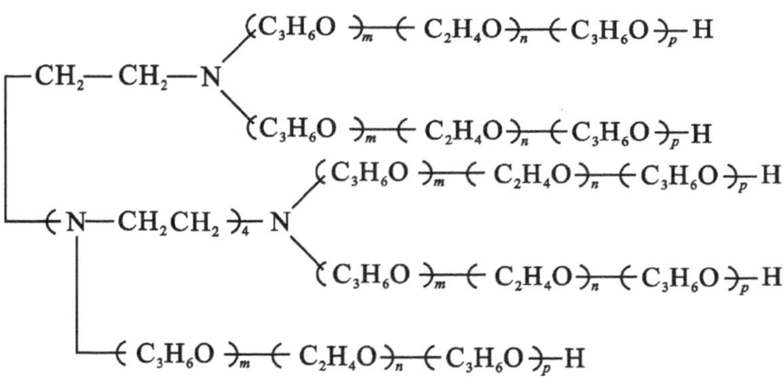

(Polyoxypropylene polyoxyethylene polyoxypropylene pentylenehexylamine)

12.2.2.3 Polymer Defoamers

This kind of defoamer refers to the polymer with low surface tension with air and low interfacial tension with oil, whose defoaming mechanism is the same as that of the solvent defoamers.

Polymer defoamers include mainly polysiloxanes, with some examples given below.

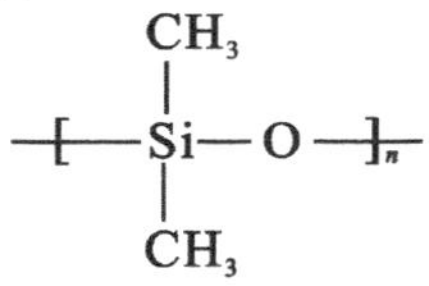

(Polydimethylsiloxane)

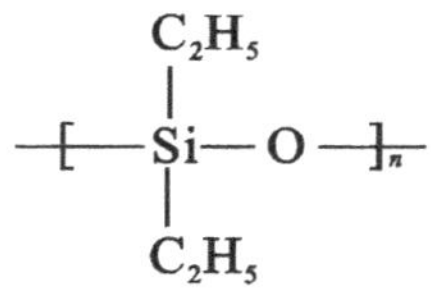

(Polydiethylsiloxane)

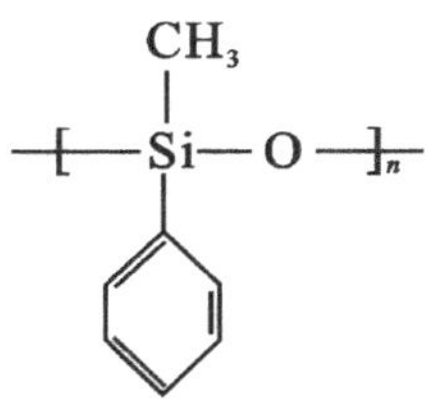

(Polymethylphenylsiloxane)

Fluorine-containing polysiloxane and polyether-modified polysiloxanes can also be used.

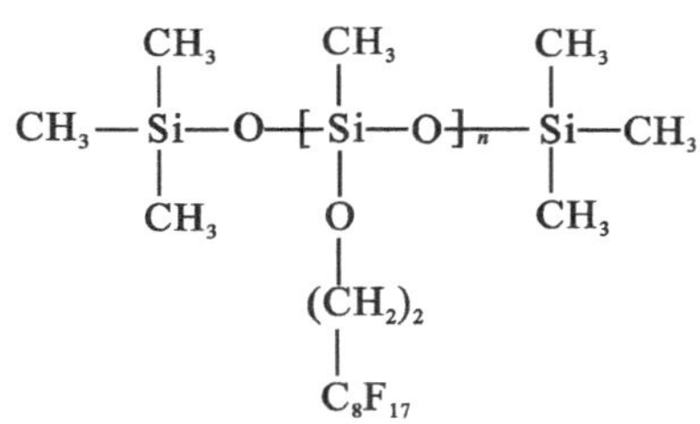

(Polymethylfluorodecylsiloxane) (Fink et al.1988)

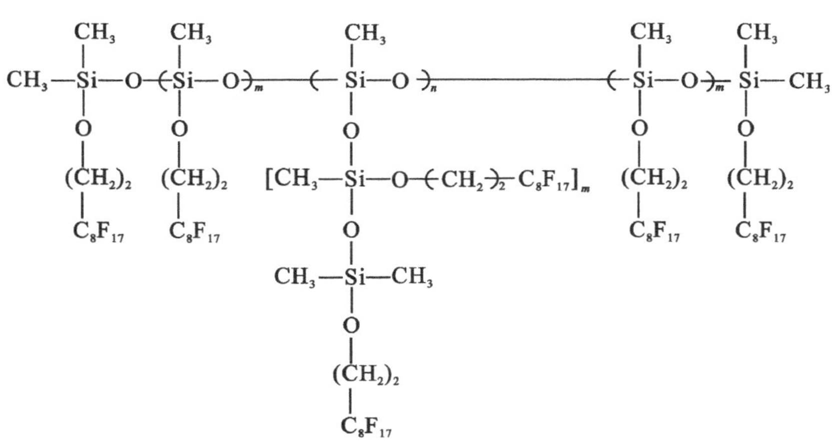

(Modified product of polymethylfluorodecylsiloxane) (Taylor 1991)

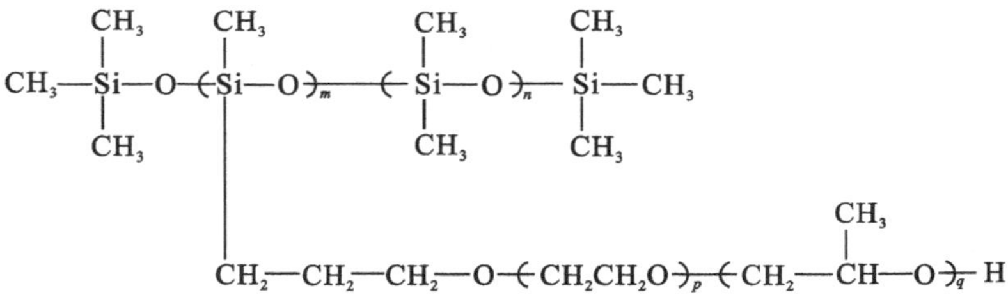

(Polyether-modified product of polysiloxane) (Ni et al.2009; Li et al.2005)

References

Barthold K, Baur R, Fikentscher R et al (1990) Alkoxylated ami-containing polyamines and their use for breaking oil-in-water and water-in-oil emulsion. US Patent 4,935,162, 19 June 1990

Bhattacharyya BR (1992) Water soluble polymer as water-in-oil demulsifier. US Patent 5,100,582, 31 Mar 1992

Bialek RF (1981) Tar sands emulsion-breaking process. US Patent 4,302,326, 24 Nov 1981

Braden ML, Allenson SJ (1991) Method for separating liquid from water using amine containing polymers. US Patent 5,021,167, 4 June 1991

Durham DK, Conkle UC, Downs HH (1989) Additive for clarifying aqueous systems. GB Patent 2,219,291, 6 Dec 1989

Fink HF, Koerner G, Berger R et al (1988) Perfluoroalkylethyloxy-substituted siloxanes. GB Patent 2,196,976, 11 May 1988

Hart PR (1993) Method for breaking emulsion in a crude oil desalting system. US Patent 5,256,305, 26 Oct 1993

Hart PR, Brown JM (1998) Method of resolving oil and water emulsions. US Patent 5,730,905, 24 Mar 1998

He F, Ma X (2005) An overview of crude oil demulsifier and selection evaluation method. Xinjiang Pet Sci & Technol 15(3):25–27, 30

Hoffmann H, Koerner G, Rossmy G et al (1980) Use of demulsifying mixtures for breaking petroleum emulsions. US Patent 4,183,820, 15 Jan 1980

Huang J (1993) Oil and gas and chemical water treatment process. China University of Petroleum, pp 85–90

Langdon WK, Camp RL (1980) Heteric/block polyoxyalkylene compound as crude oil demulsifiers. US Patent 4,183,821, 15 Jan 1980

Li C, Lu Y, Gong S et al (2005) The synthesis and compound of polyether modified polysiloxane defoaming agents. Hebei Chem Eng Ind (5):20–23

Liebold G, Oppenlaender K, Buettner E et al (1975) Demulsifying crude oil emulsions. US Patent 3,907,701, 23 Aug 1975

Mcdonald MT (1971) Quaternary adducts of polyepihalohydrin and use there of. US Patent 3,591,520, 6 July 1971

Mcentier EE, Mccoy DR (1983) Demulsification of bitumen emulsion using polymers of diquaternary ammonium monomers containing hydroxyl groups. US Patent 4,387,017, 7 June 1983

Meng L (2005) Application and development trend of crude oil demulsifier. Oil-Gas Field Surf Eng 24(4):18–19

Merchant P Jr, Lacy SM (1988) Water based demulsifiers formulation and process for its use in dewatering and desalting crude hydrocarbon oil. US Patent 4,737,265, 12 Apr 1988

Newcombe J (1980) Emulsion breaking with surfactant recovery. US Patent 4,216,079, 5 Sept 1980

Newcombe J (1983) Emulsion breaking of surfactant stabilized crude oil in water emulsions. US Patent 4,374,734, 5 Sept 1983

Ni X, Yuan G, Chen S et al (2009) Synthesis and compounding of polyether modified polysiloxane crude oil defoamer. Oil Chem 26(2):153–157

Oppenlaender K, Liebold G, Buettner E (1974) Crude oil demulsifiers. US Patent 3,786,081, 15 Jan 1974

Stephenson WK (1990) Alkoxylated vinyl polymer demulsifiers. US Patent 4,968,449, 6 Nov 1990

Stephenson WK, Deshazo JD (1993) Method of breaking crude oil emulsions using ethylene carbonate adducts of alkylphenolformaldehyde resins. US Patent 5,205,964, 27 Apr 1993

Taylor AS (1991) Fluorosilicone anti-foam additive. GB Patent 2,244,279, 27 Nov 1991

Thompson NES, Asperger RG (1989) Methods for treating hydrocarbon recovery operations and industrial waters. US Patent 4,826,625, 2 Jan 1989

Xiao Z (2008) Demulsification mechanism of emulsified crude oil and its influencial factors. J Oil Gas Technol 30(4):165–168

Zhao G (1991) Surfactants physical chemistry. Peking University Press, Beijing, pp 416–418

Zhao F (1994) Demulsifiers for cruid oil emulsion. J Univ Pet 18:104–113

Zhang Z, Lou S, Zhang F (2008) Synthesis of acrylic modified demulsifier. Chem Eng Oil & Gas 37(6):507–509, 542

Chapter 13
Pour Point Depression Transportation and Drag Reduction Transportation of Crude Oil

In order to improve the flow conditions of crude oil in long-distance pipeline transportation, the pour point depression and drag reduction are two important issues in crude oil gathering and transportation. In solving these problems, chemical methods are highly applicable methods (Luo 1997; Quan 1996; Zhao et al. 1996).

13.1 Pour Point Depression Transportation of Crude Oil

13.1.1 Classification of Crude Oil by Pour Point

Pour point refers to the maximum temperature when crude oil loses flowability under specified test conditions. Crude oil may lose its flowability for two reasons. One is the oil viscosity increases with decreasing temperature. When the viscosity increases to certain extent, oil loses flowability. The other reason is generated by the wax in crude oil. When the temperature is reduced to the wax precipitation temperature of crude oil, the wax crystals precipitate. With further lowering the temperature, the number of wax crystals increases. The wax crystals grow up and coalesce until a structure network throughout the crude oil is formed, resulting in the lost flowability of crude oil.

According to different pour points, crude oil can be divided into the following categories.

- Low Pour Point Crude Oil
  Low pour point crude oil refers to the crude oil with pour point lower than 0 °C, in which the mass fraction of wax is fewer than 2%.
- Medium Pour Point Crude Oil
  Medium pour point crude oil refers to the crude oil with pour point in the range of 0–30 °C, in which the mass fraction of wax lies in the range of 2–20%.

C. Dai and F. Zhao, *Oilfield Chemistry*,
https://doi.org/10.1007/978-981-13-2950-0_13

Fig. 13.1 Statistical relationship between pour point and wax content of crude oil

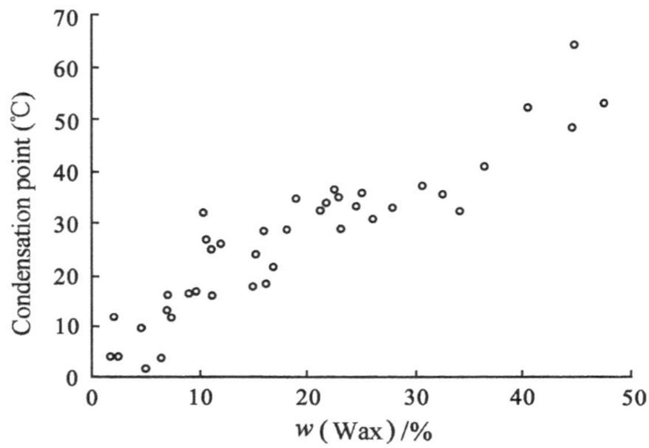

- High Pour Point Crude Oil
  High pour point crude oil refers to the crude oil with pour point higher than 30 °C, in which the mass fraction of wax is more than 20%.

As can be seen from the above classifications, the higher the pour point, the higher the wax content of crude oil is.

This rule can be seen from the statistical relationship diagram of pour point and wax content of crude oil (Fig. 13.1).

13.1.2 Viscosity–Temperature Curve of High-Wax Crude Oil

Medium pour point crude oil and high pour point crude oil are collectively called high-wax crude oil. By measuring the changes of oil viscosity with temperature at different shear rates, the viscosity–temperature curve shown in Fig. 13.2 can be obtained. As seen in Fig. 13.2, temperature has a significant effect on the viscosity of high-wax crude oil.

There are two feature points need to be further explained in Fig. 13.2. *A* is the wax precipitation point. When high-wax crude oil is cooled to temperature below that point, the wax crystals precipitate. The other point *B* is the abnormal point, from which a further reduction in temperature will lead to the change of viscosity of high-wax crude oil with shear rate, indicating that the crude oil has changed from Newtonian fluid into non-Newtonian fluid. Since the structures formed by wax crystals in high-wax crude oil are suffered from different degrees of destruction under different shear rates, when temperature is below the anomalous point temperature, the viscosity of high-wax crude oil changes with shear rate.

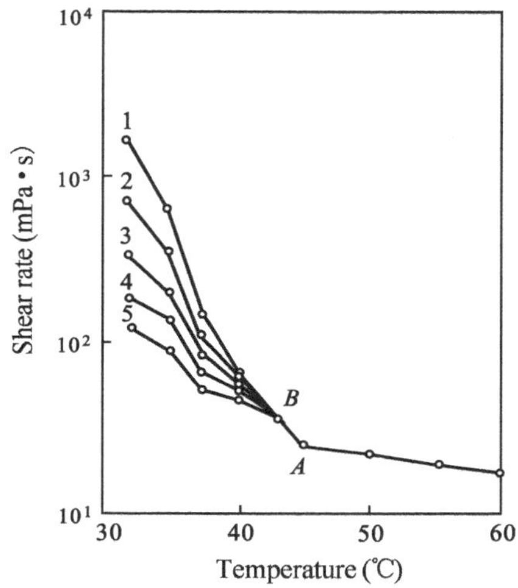

Fig. 13.2 Viscosity–temperature curve of high-wax crude oil. Shear rate: 1—8.1 s^{-1}; 2—24.3 s^{-1}; 3—72.9 s^{-1}; 4—218.7 s^{-1}; 5—656.0 s^{-1}

13.1.3 Pour Point Depression Transportation of Crude Oil

Pour point depression transportation of crude oil refers to the transportation of crude oil treated by pour point depression methods in long-distance pipeline.

There are several ways for the depression of pour points.

13.1.3.1 Physical Methods

This is a kind of heat treatment method. The crude oil is heated to the optimum heat treatment temperature first before it is cooled at a certain rate. In this way, the purpose of reducing the pour point of crude oil is achieved (Luo 1983; Bao 1985; Ge et al. 2009).

- Effect of heat treatment on the viscosity–temperature curve of crude oil.

Figure 13.3 is the viscosity–temperature curve of crude oil before and after heat treatment (Liu 1986).

As seen in Fig. 13.3, after the heat treatment, the viscosity–temperature curve of crude oil has undergone the following changes.

- After the wax precipitation point, the viscosity of crude oil decreases.
- The temperature range where the crude oil has Newton rheological behavior has widened, indicating the reduction of abnormal point.
- After the abnormal point, the change of oil viscosity with shear rate gets smaller.

Table 13.1 shows that pour point of crude oil has dropped significantly after heat treatment.

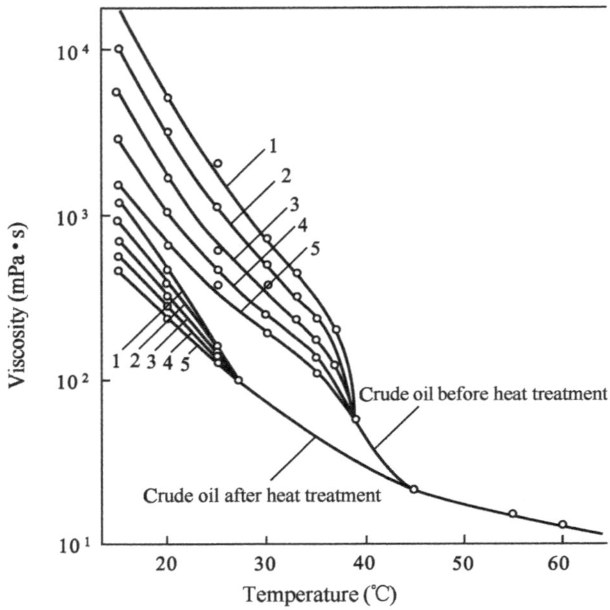

Fig. 13.3 Viscosity–temperature curves of crude oil before and after heat treatment

After the heat treatment, the changes in the viscosity–temperature curve of crude oil are generated by the effect of temperature on the presence status of each component in crude oil.

- Effect of heat treatment on the presence status of each component in crude oil.

  Heating can produce the following effects on the components of crude oil.

- Wax crystals completely dissolve in crude oil; wax disperses in oil as molecular state.
- The dispersity of asphaltene stacks is increased due to the influence of weakened hydrogen bonds and intensified heat movement, which means the stack size of asphaltenes is reduced; however, the number is increased.
- The adsorption amount of colloid on the surface of asphaltene stack is reduced due to the intensified heat movement, while the colloid content in crude oil increases correspondingly.

The changes of the presence status of each component in crude oil caused by heating cannot be restored immediately after the cooling. This means when oil is cooled to temperature below the wax precipitation point, the wax is precipitated in a condition where the content of asphaltene stack and colloid is higher than that before heating. Asphaltene stack can function by acting as crystal nucleus, while colloid can function by cocrystallization with wax and adsorption mechanism; thus, the wax crystals precipitated by treated crude oil are more disperse and loose and have reduced ability in forming structure. Therefore, the pour point of crude oil after the heat treatment is reduced.

Table 13.1 Effect of heat treatment on the pour point of crude oil

Crude oil production	*w* (wax) (%)	*w* (colloid + asphaltene) (%)	Condensation point before heat treatment (°C)	Heat treatment temperature (°C)	Condensation point after heat treatment (°C)
Daqing field	34.5	8.43	32.5	70	17.0
Zhongyuan oilfield	10.4	21.2	32.0	85	21.0
Jianghan oilfield	10.7	24.2	26.0	80	14.0
Huoshaoshan oilfield	20.5	20.9	20.5	70	7.0

13.1.3.2 Chemical Methods

Chemical method refers to the method of adding pour point depressant into crude oil. Crude oil pour point depressant is a type of chemical which can reduce the pour point of crude oil (Fielder et al. 1986; Slater et al. 1986; Zhang 1990; Misra et al. 1995).

Two main types of crude oil pour point depressants are used in the chemical method. One is the surfactant-type pour point depressant, such as petroleum sulfonate, polyoxyethylene alkyl amines, and hexamer triethanolamine oleate (Hafiz 2007), which functions by adsorbing on the surface of the wax crystals, inhibiting wax from forming network structure throughout the system. The other is polymer-type pour point depressant, which contains both the nonpolar part which can cocrystallize with wax molecules and polar part which can distort the wax crystal structures in the main chain and/or branched chain, thereby playing the role of pour point depression (Li 1996).

Polyacrylate is a typical polymer-type pour point depressant, whose state in oil is shown in Fig. 13.4. It can be seen that this kind of pour point depressant has many structures same to the wax molecule, including nonpolar part which can cocrystallize with wax molecules and polar part (alkyl) which can distort the wax crystal structures during the wax precipitation. Therefore, polyacrylate has significant pour point depression effect.

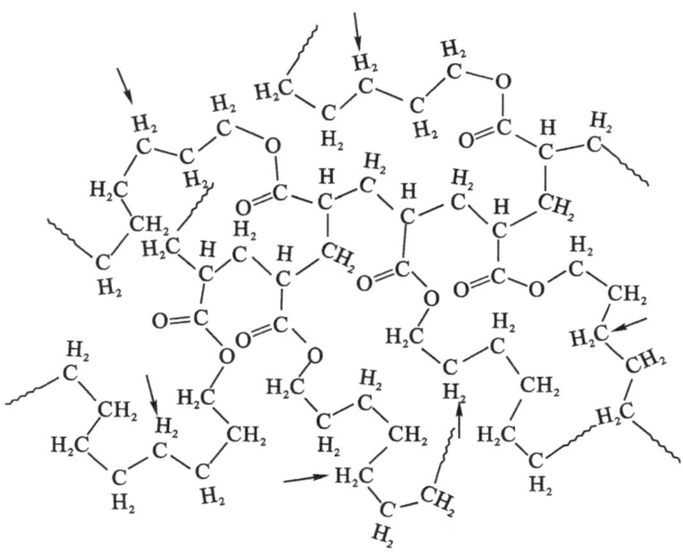

Fig. 13.4 Polyacrylate dissolved in crude oil

The following are some important polymeric crude oil pour point depressants.

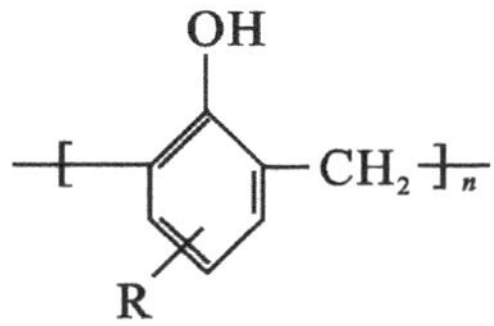

R: C_{12}–C_{36}

(Alkylphenol formaldehyde resin)
(Martella et al. 1992; Hiebert et al. 1998)

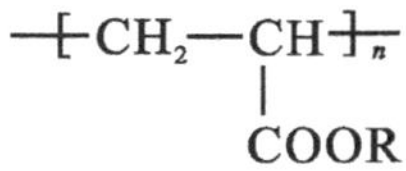

R: C_{14}–C_{40}

(Polyacrylate) (Choourfoer et al. 1973; Ding et al. 1999;Yang et al. 2001; Yang et al. 2009)

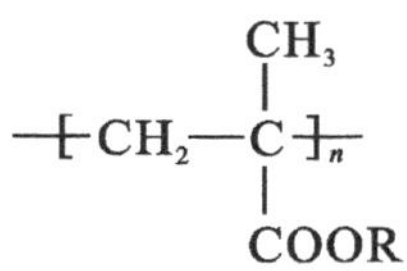

R: C_{12}–C_{38}

(Polymethacrylate) (Wilburn et al. 1990)

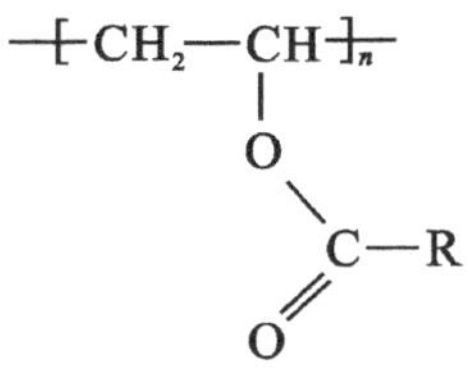

R: C_{15}–C_{35}

(Polyvinyl carboxylate)

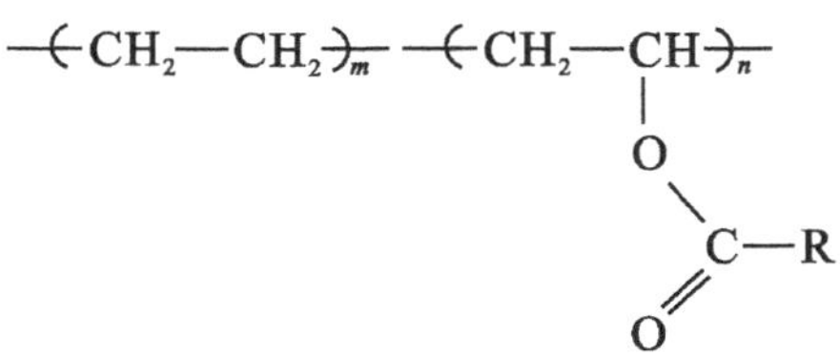

R: C_1–C_{25}

(Copolymer of ethylene and vinyl carboxylate)
(Day et al. 1973; Machado et al. 2001; Marie et al. 2005; Yang et al. 2009)

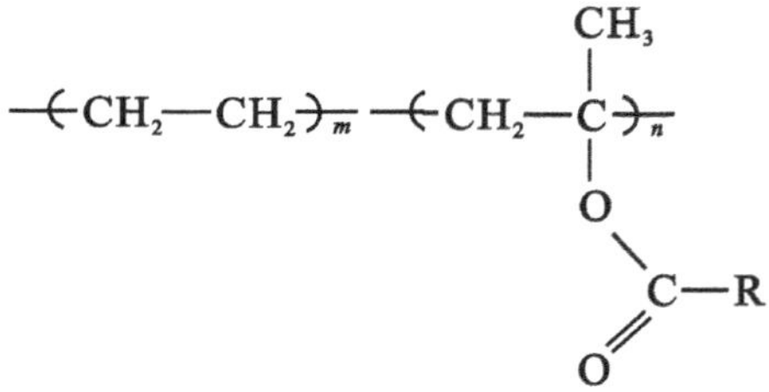

R: C_1–C_{25}

(Copolymer of ethylene and allyl carboxylate)

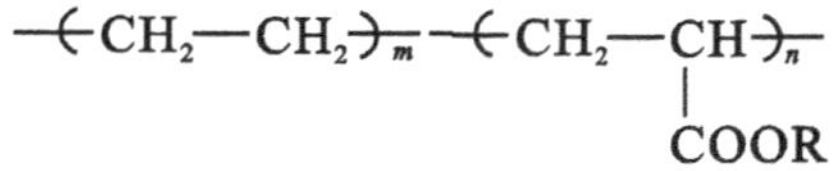

R: C_1–C_{26}

(Copolymer of ethylene and acrylate)

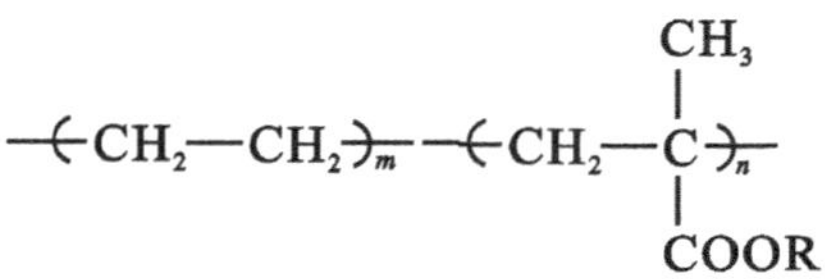

R: C_1–C_{26}

(Copolymer of ethylene and methyl acrylate)

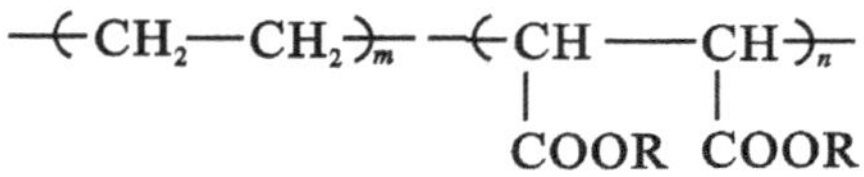

R: C_1–C_{26}

(Copolymer of ethylene and maleate) (Newberry 1985)

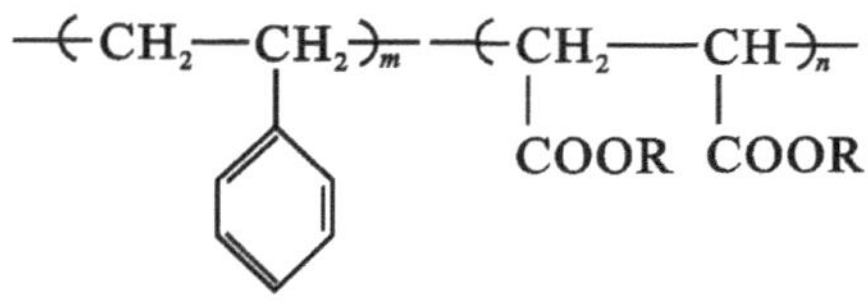

R: C_{14}–C_{40}

(Copolymer of styrene and maleate) (Al-Sabagh et al. 2009)

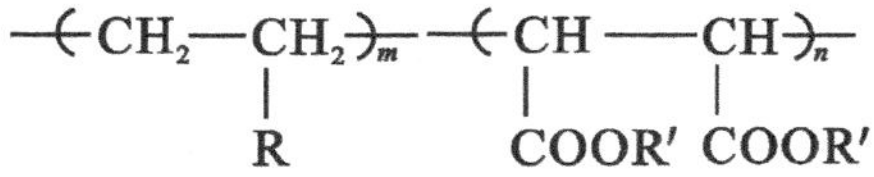

R: C_{20}–C_{28};
R′: C_{14}–C_{40}

(Copolymer of α-olefin and maleate) (Liao 1998)

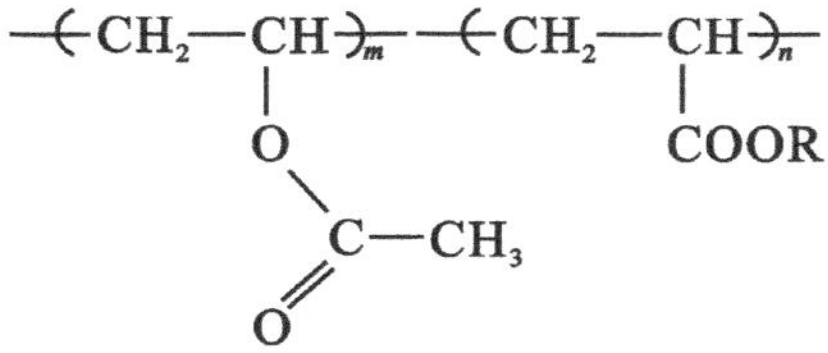

R: C_{14}–C_{40}

(Copolymer of vinyl acetate and acrylate)

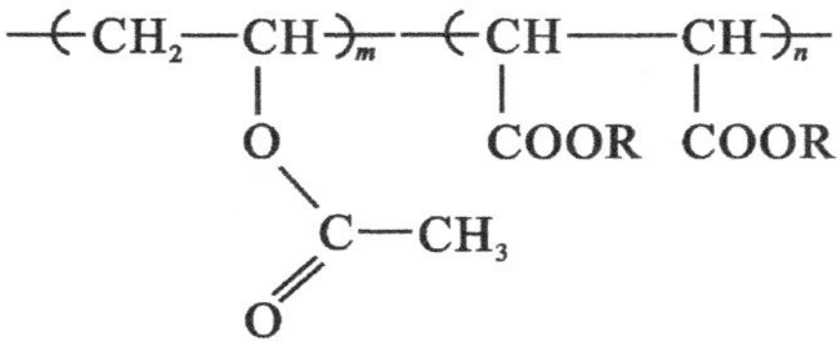

R: C_{14}–C_{40}

(Copolymer of vinyl acetate and maleate)

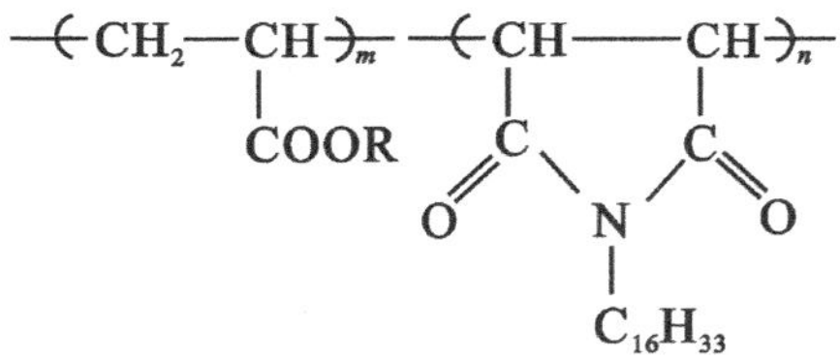

R: C_{14}–C_{40}

(Copolymer of acrylate and *N*-cetyl maleimide) (Deshmukh et al. 2008)

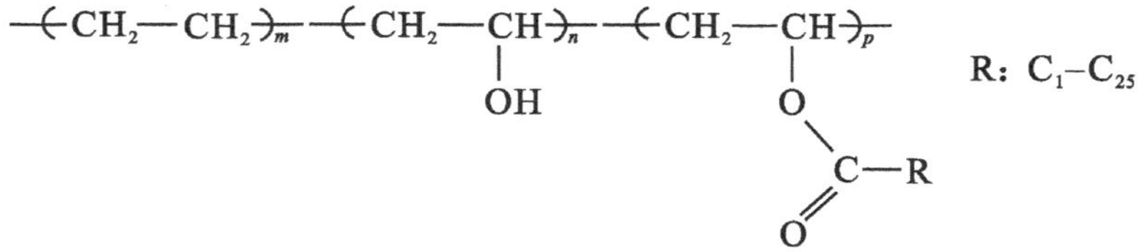

(Copolymer of ethylene, vinyl alcohol and vinyl carboxylate)

(Copolymer of ethylene, acrylic acid and acrylate)

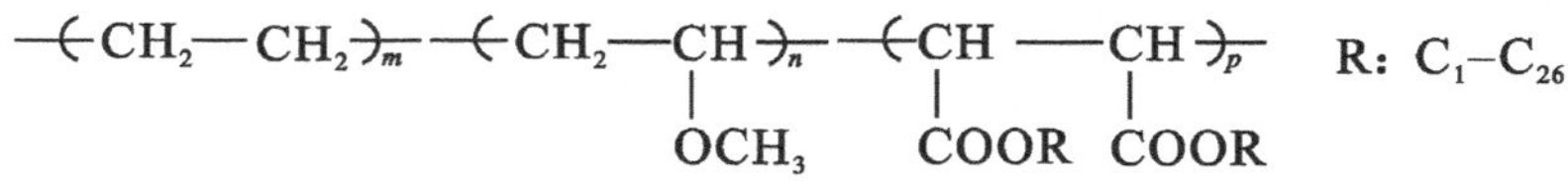

(Copolymer of ethylene, vinyl methyl ether and maleate)

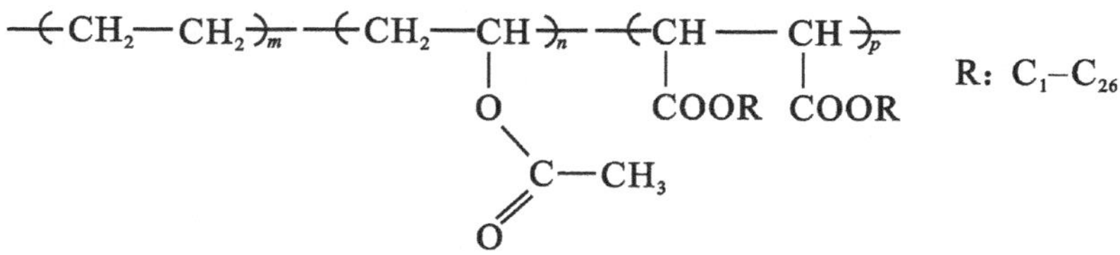

(Copolymer of ethylene, vinyl acetate and maleate)

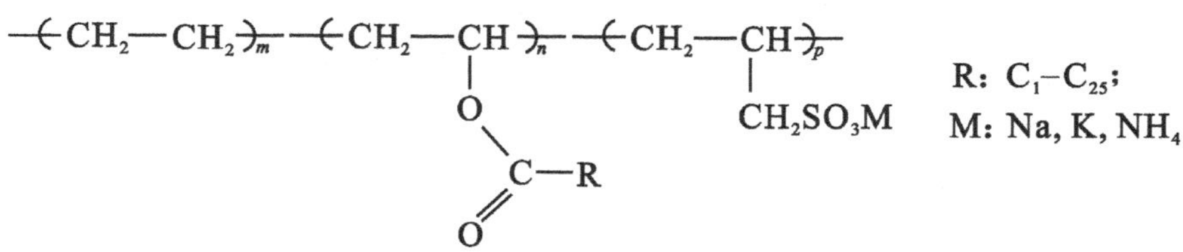

(Copolymer of ethylene, vinyl carboxylate and propylene sulfonate)

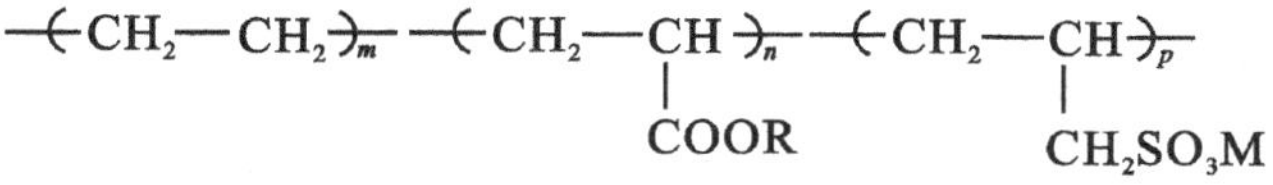

R: C_1–C_{26};
M: Na, K, NH_4

(Copolymer of ethylene, acrylate and propylene sulfonate)

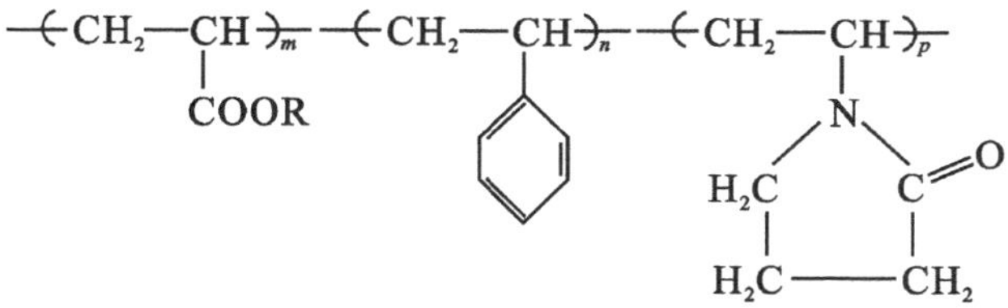

R: C_{14}–C_{40}

(Copolymer of acrylate, styrene and *N*-vinylpyrrolidone) (Kuzmic et al. 2008)

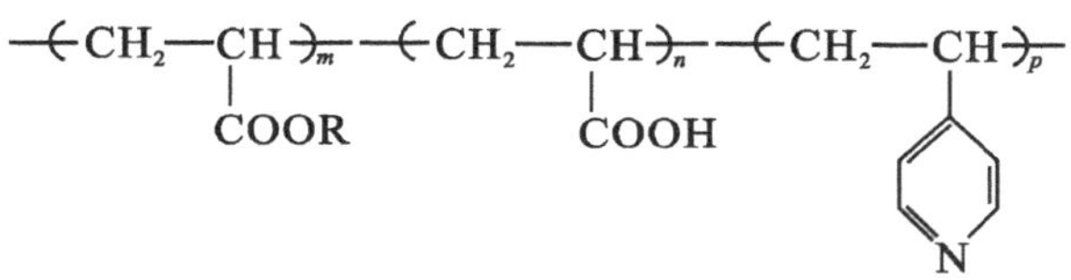

R: C_{14}–C_{40}

(Copolymer of acrylate, acrylic acid and vinylpyridine)

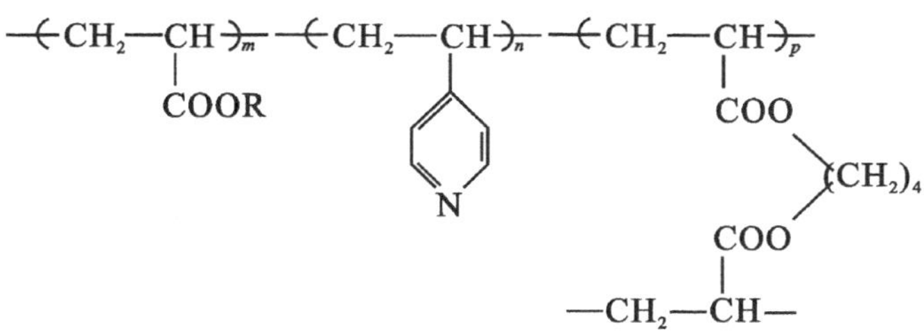

R: C_{14}–C_{40}

(Copolymer of acrylate, vinylpyridine and butanediol diacrylate)

Table 13.2 shows the effects of pour point depressants on the pour point of some high-wax crude oil.

Polymer crude oil pour point depressant has best effect when the average carbon number of its nonpolar portion is close to that of wax. Therefore in high-wax crude oil pour point depressant, besides the selection of different polymers, the average carbon number of the nonpolar portion of the polymer should be optimized (Manka et al. 2001; Li et al. 2002). Figure 13.5 shows the diagram of alkane carbon number distribution of wax in crude oil. It can be seen from the figure that the alkane carbon number of wax in this kind of crude oil lies in the range of 18–38, with a summit at 24 (i.e., the average carbon number of the wax). For this kind of crude oil, the polymer with an average carbon number of 24 in the nonpolar portion should have the best pour point depression effect (Barthell et al. 1982).

Table 13.2 Effects of pour point depressants on the pour points of crude oil

Crude oil production	Pour point before adding agents (°C)	Pour point depressant	Mass concentration of pour point depressant (mg/L)	Pour point after adding agents (°C)
Zhongyuan oilfield	33	Ethylene and vinyl acetate copolymer	50	13
Jianghan oilfield	25	Ethylene and vinyl acetate copolymer	100	15
Maling oilfield	16	Polyacrylate	50	2
Qinghai oilfield	32	Ethylene, acrylate, and propylene sulfonate copolymer	150	12

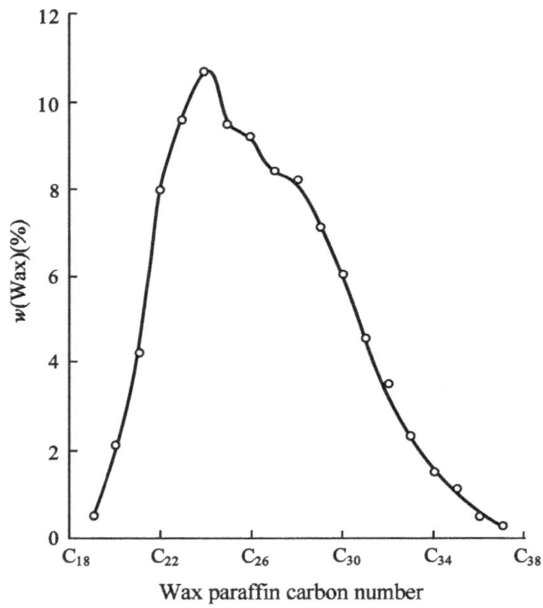

Fig. 13.5 Alkane carbon number distribution of wax in a kind of crude oil

13.1.3.3 Chemical–Physical Method

This is a kind of combination method, requiring the addition of pour point depressant in crude oil followed by heat treatment.

In order to compare the effects of single heat treatment and combination treatment, viscosity–temperature curves of crude oil under the circumstances of untreated, heat treatment, and combination treatment are measured.

Figure 13.6 shows the viscosity–temperature curves of a kind of high-wax crude oil in the above three cases (Chen et al. 1984). During the heat treatment, the crude oil is heated to 85 °C before cooling. During the combination treatment, 100 mg/L pour point depressant (copolymer of ethylene and vinyl acetate) is added into the crude oil at 60 °C, and then the crude oil is heated to 85 °C before cooling.

As seen in Fig. 13.6, compared with the crude oil after heat treatment, crude oil after combination treatment has better low-temperature fluidity, expressed by the fact that it has lower crude oil viscosity after the wax precipitation point, and it bears Newtonian fluid characteristics in a wider temperature range (i.e., the temperature where the abnormal point appears is lower).

It is shown in Table 13.3 that the crude oil after combination treatment bears lower pour point than the crude oil after heat treatment does.

The reason why crude oil after combination treatment has better low-temperature fluidity than that after single heat treatment is because both the natural pour point depressant (colloid and asphaltene) and additional polymer pour point depressant take effect in the crude oil after combination treatment. In other words, combination treatment is the extension and strengthening of heat treatment in pour point depression effect. In some circumstances where the crude oil properties cannot satisfy the requirements in pipeline transportation, combination treatment can play important role.

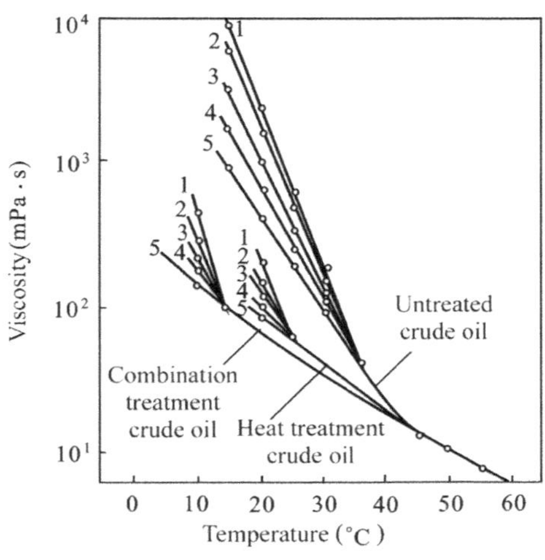

Fig. 13.6 Viscosity–temperature curves of a kind of high-wax crude oil under the circumstances of untreated, heat treatment, and combination treatment. Shear rate: 1—4.5 s^{-1}; 2—8.1 s^{-1}; 3—13.5 s^{-1}; 4—24.3 s^{-1}; 5—40.5 s^{-1}

Table 13.3 Effects of heat treatment and combination treatment on the pour point of crude oil

Oilfield	Pour point before treatment (°C)	Pour point after heat treatment (°C)	Pour point after combination treatment (°C)
Daqing	32.5	17.0	12.3
Jianghan	26.0	14.0	6.0
Renqiu	34.0	17.0	13.5
Hongjingzi	17.0	8.0	1.5

13.2 Drag Reduction Transportation of Crude Oil

13.2.1 Flow Pattern and Its Resistance

Reynolds number is a dimensionless number that is used to characterize the flow state of fluid in the tube. It is generally defined as follows:

$$Re = vd/\upsilon \tag{13.1}$$

where v is average velocity of the fluid (m/s); d is inner diameter of the pipe (m); υ is the kinematic viscosity of the fluid (m^2/s).

According to the Reynolds number, fluid flow in the pipe can be divided into two types: laminar flow and turbulent flow. In laminar flow, the flow resistance of the fluid is produced by the fluid momentum exchange between adjacent flow layers. In turbulent flow, the flow resistance of the fluid is caused by the vortices with random scale and motion.

Although the vortices generated by turbulent flow are random, the vortices tend to gradually decompose and generate smaller vortices.

As the smaller the vortex scale, the larger the energy viscous loss. Therefore, the energy of the small vortices would finally be exhausted by the viscous force of fluid, turning into thermal energy.

13.2.2 Drag Reduction Transportation of Crude Oil

Drag reduction transportation of crude oil refers to the transportation of crude oil in turbulent flow state added with drag reducing agent in long-distance pipeline (Lester 1985; Berge et al. 1996).

There are many vortices which gradually get smaller in the crude oil in turbulent flow state. The pipeline energy transmits from bigger vortices to smaller vortices and finally is exhausted in the form of thermal energy. Hence, crude oil in turbulent flow state needs to consume a large amount of pipeline transportation energy. In order to reduce the energy consumption, the drag reducing agent can be added to the crude oil in transportation process.

Crude oil drag reducing agent refers to the chemical agent that can reduce the transportation resistance of crude oil in the turbulent flow state. Crude oil drag reducing agents are all oil-soluble polymers which exist in oil mainly in coiled state. The polymer molecules that are present in this state are elastic.

When drag reducing agents are present in the turbulent state of the crude oil, vortices at all levels can transfer energy to the drag reducing agent molecules, of which elastic deformations occur to store the energy. The energy can be released and returned to vortices when stress of drag reducing agent relaxes to maintain the turbulent flow state of the fluid, which reduces the external energy, required to maintain this state and thus reduces resistance (Yin et al. 2002).

It is only when the crude oil is in a turbulent flow state that the drag reducing agent takes effect (Savins 1964). The following polymers can be used as crude oil drag reducing agents.

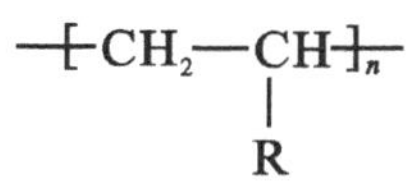

R: C_6–C_{26}

(Poly α-olefin) (Culter et al. 1972; Pei et al. 2005; Shi et al. 2006; Liu et al. 2007; Fu et al. 2009)

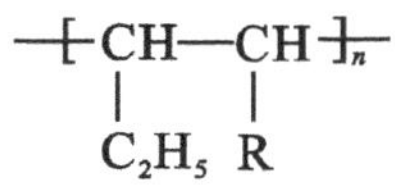

R: C_6–C_{26}

(Polyethylalkylethylene) (Mack 1985)

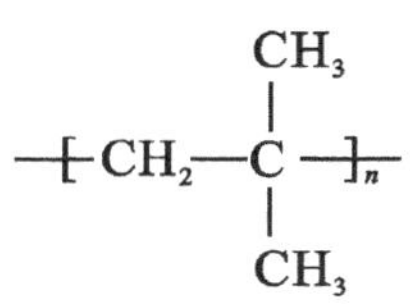

(Polyisobutene)

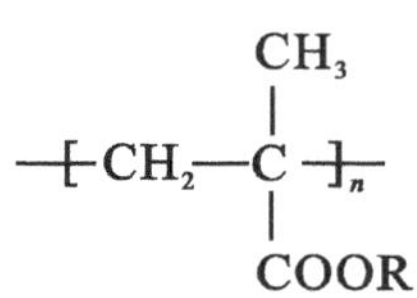

R: C_6–C_{26}

(Polymethacrylate)

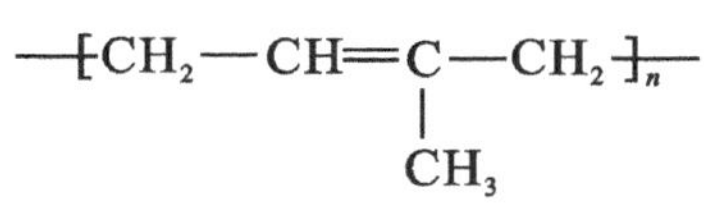

(Polyisoprene)

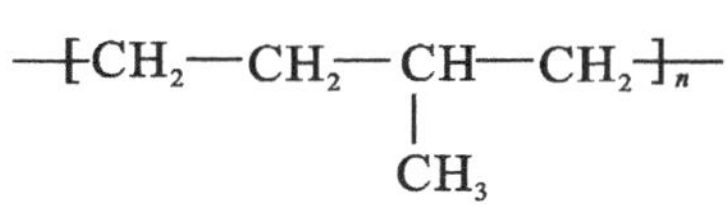

(Hydrogenated polyisoprene) (Meier et al.1974)

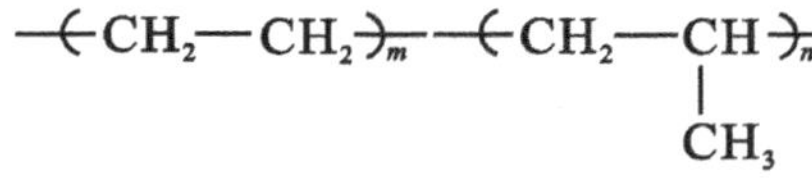

(Copolymer of ethylene and propylene) (Seymour et al. 1971; Johnston et al. 1994)

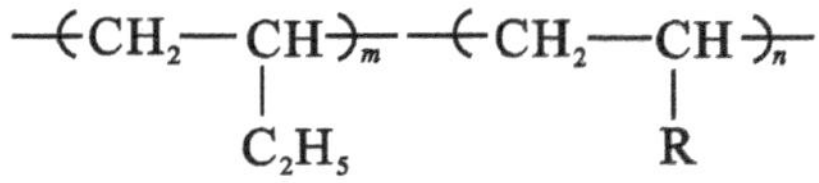

R: C_6–C_{26}

(Copolymer of 1- butane and α-olefin) (Motier 1985)

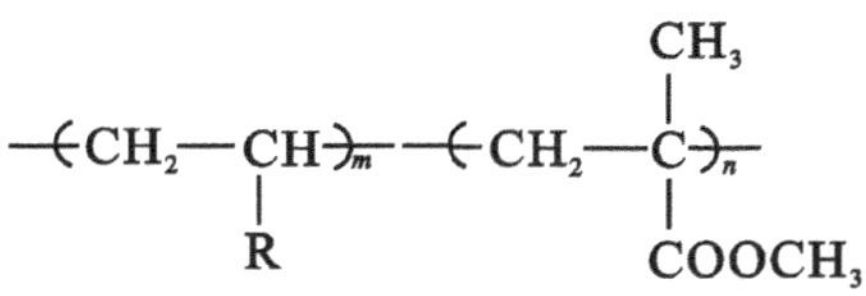

R: C_6–C_{26}

(Copolymer of α-olefin and methacrylate)

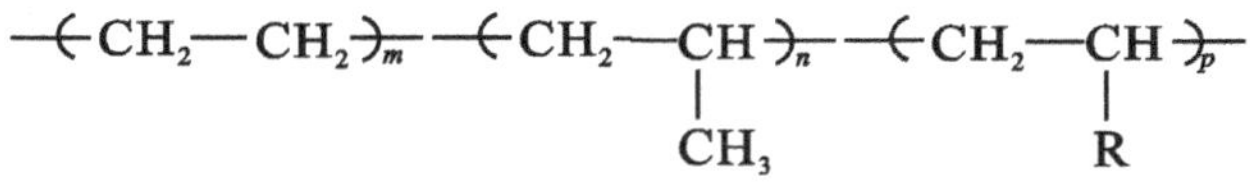

R: C_6–C_{26}

(Copolymer of ethylene, propylene and α-olefin)

The drag reduction effect can be evaluated by the drag reducing rate and the flow increasing rate. In the case of constant pipe flow, the drag reducing rate is defined as follows:

$$DR = \frac{\Delta p_2 - \Delta p_1}{\Delta p_1} \times 100\% \tag{13.2}$$

where DR is drag reducing rate; Δp_1 is the pipeline transportation resistance without drag reducing agent, MPa; Δp_2 is the pipeline transportation resistance with drag reducing agent, MPa.

In the case of constant transportation resistance, the flow increasing rate is defined as follows:

$$FI = \frac{Q_2 - Q_1}{Q_1} \times 100\% \tag{13.3}$$

where *FI* is flow increasing rate; Q_1 is the pipeline flow without drag reducing agent, m^3/h; Q_2 is the pipeline flow with drag reducing agent, m^3/h.

In general conditions of crude oil pipeline transportation, the relationship between the pipeline transportation resistance and the pipeline flow is as follows:

$$\Delta p = 0.0246 \cdot \frac{Q^{1.75} \cdot \nu^{0.25}}{d^{4.75}} \cdot L\rho g \tag{13.4}$$

where Δp is the pipeline transportation resistance, Pa; Q is the pipeline flow, m^3/s; ν is the kinematic viscosity of crude oil, m^2/s; d is the pipe diameter, m; L is the pipe range, m; ρ is the crude oil density, kg/m^3; g is the acceleration of gravity, m/s^2.

The relationship between the drag reducing rate and the flow increasing rate can be deduced from Eq. 13.4.

$$FI = \left[\left(\frac{1}{1 - DR}\right)^{0.55} - 1\right] \times 100\% \tag{13.5}$$

A drag reducing agent is added to crude oil to conduct drag reducing test, from which the pipeline transportation resistances with and without drag reducing agent are acquired. The drag reducing rate and the flow increasing rate of this drag reducing agent can be calculated according to Eqs. 13.2 and 13.5, respectively, as shown in Table 13.4.

As shown in Table 13.4, the pipeline transportation resistance reduces dramatically by adding a small amount of drag reducing agent.

The following factors have an important influence on the drag reduction effect of drag reducing agent.

The Property of Crude Oil

The lower the viscosity and density of crude oil, the easier it is to reach the turbulence condition, which is favorable to the function of drag reduction agent. High water content of crude oil will affect the dissolution of drag reducing agent, thus affect its drag reducing efficiency.

Table 13.4 Drag reduction of pipeline oil[b] by drag reducing agent[a]

Mass concentration of drag reducing agent (mg/L)	Pipeline transportation resistance[c] (MPa)	Drag reducing rate (%)	Flow increasing rate (%)
0.0	3.41	0.0	0.0
20.9	3.02	11.4	6.9
28.3	2.87	15.8	10.0
57.4	2.62	23.1	15.5

Note [a]The drag reducing agent is Poly-α-olefin

[b]The viscosity of oil is 22.8 mPa·s; the density is 0.832 g/cm³; the average oil temperature is 46.7 °C

[c]The pipe diameter is 0.72 m; the pipeline oil flow is 2,847 m³/hour

The Structure of Drag Reducing Agent

The relative molecular weight of drag reducing agent should not be too low or too high (preferably in the range of 10^5–10^6). Too low relative molecular weight will affect the efficiency of drag reduction, while drag reducing agents with too high relative molecular weight have poor oil solubility and are subject to sheared degradation.

The main chain of drag reducing agent should have certain amount and certain length of branched chains (such as copolymer of ethylene, propylene, and α- olefin), which can provide appropriate flexibility to the drag reducing agent molecules and also protective effects to the main chains, improving the shear stability and drag reduction efficiency of the drag reducing agent.

The Conditions of Pipeline Transportation

The higher the transportation temperature, the lower the viscosity of the crude oil, the more beneficial it is to the function of drag reducing agent. Faster flow rate and smaller pipe diameter will contribute to a greater Reynolds number and higher turbulence degree, which is beneficial for the function of drag reducing agent. However, when the flow rate is too fast, the drag reducing agent will be degraded and the efficiency of drag reducing agent will be decreased (Fig. 13.7).

The increase of concentration of drag reducing agent will lead to the enhancement of drag reduction efficiency. However when the concentration is over a certain value, the increase of drag reduction efficiency will not be significant (Fig. 13.8). Therefore, crude oil drag reducing agents have their best application concentration.

Drag reducing agent has played an important role in pipeline oil transportation (Meng et al. 1987; Yu 1990; Hu 1997; Zhen et al. 2006; Li et al. 2009).

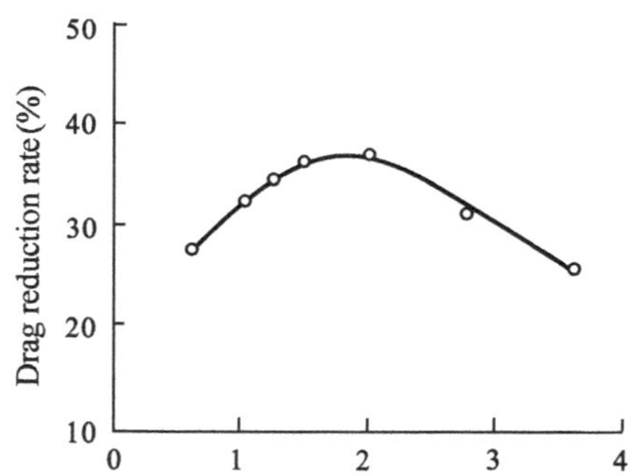

Fig. 13.7 Effect of pipeline flow rate on drag reduction rate of drag reducing agent

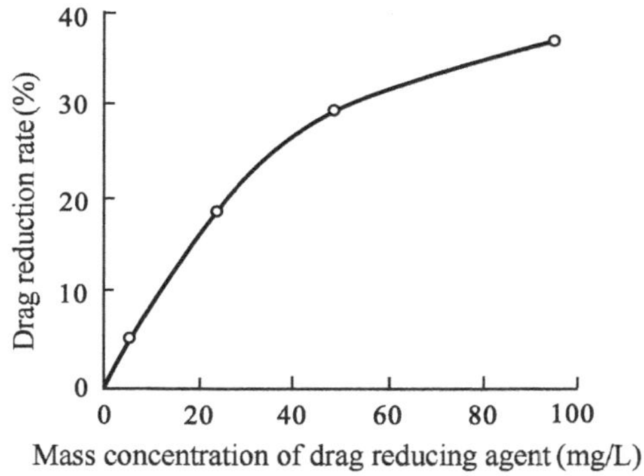

Fig. 13.8 Effect of mass concentration of drag reducing agent on drag reduction rate

References

Al-Sabagh AM, Noor El Din MR, Morsi RE et al (2009) Styrene maleic anhydride copolymer ester as flow improvers of waxy crude oil. J Petrol Sci Eng 65(3–4):139–146

Bao C (1985) Research on the heat treatment technology of wax crude. Acta PetEi Sin 6(3):85–95

Barthell E, Capelle A, Chmellr M et al (1982) Copolymer of *n*-alkyl acrylates and maleic acid and derivatives thereof and their use as crystallisation inhibitors for paraffin bearing crude oils. GB Patent 2,082,604, 10 Mar 1982

Berge BK, Solsvik O (1996) Increased pipeline throughput using drag reducer additives (DRA): field experiences. SPE 36835

Chen J, Liao D, Chen Y (1984) Synthestical and optimum heal treatment of crude oil. Oil & Gas Storage Transp 3(2):1–9

Choufoer JH, Van Kerkvoort WJ, Vander Meij PH et al (1973) Process for the preparation of a crude oil composition with depressed pour point. US Patent 3,776,247, 4 Dec 1973

Culter JD, Mcclaflin GG (1972) Method of friction loss reduction in oleaginous fluids flowing through conduit. US Patent 3,692,676, 19 Sept 1972

Day JJ, Zajac J (1973) Method for increasing the mobility of waxy crude oils. US Patent 3,735,770, 29 May 1973

Deshmukh S, Bharambe DP (2008) Synthesis of polymeric pour point depressants for Nada crude oil (Gujarat, India) and its impact on oil rheology. Fuel Process Technol 89(3):227–233

Ding X, Qi G, Yang S (1999) Thermodynamic analysis for the interaction of polyacrylate with wax in heptane. Polymer 40(14):4139–4142

Fielder M, Johnson RW (1986) The use of pour point depressant additive in the Beatrice field. SPE 15888

Fu W, Guan M, Li H, Hu Z (2009) Study on the performance of oil-soluble polymer drag reducing agent. Pet Process Petrochem 40(5):66–70

Ge T, Zhao J (2009) Research on the pattern of heat treatment on crude. Pet Plan & Eng 20(4):24–28

Hafiz AA, Khidr TT (2007) Hexa-triethanol amine oleate esters as pour point depressant for waxy crude oils. J Petrol Sci Eng 56(4):296–302

Hiebert GL, Detar MB (1998) Pour point depressants and their use. US Patent 5,707,946, 13 Jan 1998

Hu T (1997) Applied experiments on drag reducer in home oil transportation pipelines. Oil & Gas Storage Transp 16(6):11–14

Johnston RL, Fry LG (1994) Drag reducers for flowing hydrocarbons. US Patent 5,376,697, 27 Dec 1994

Kuzmic AE, Radosevic M, Bogdanic G et al (2008) Studies on the influence of long chain acrylic esters polymers with polar monomers as crude oil flow improver additives. Fuel 87(13–14):2943–2950
Lester CB (1985) Drag reducing agent, the basics of drag reduction. Oil & Gas J 83(5):51–56
Li J (1996) Polymer depressants and their applications. Oil & Gas Storage Transp 15(10):7–11
Li C, Zhang C, Sun D (2002) The influence of structure of pour point depressant on rheological properties of crude oil and its rheological mechanism. Chemistry 65(11):762–766
Li J, Dai Y, Hu J et al (2009) Result analysis on field test of HG DRA in Zhongyuan-Luoyang oil pipeline. Oil & Gas Storage Transp 28(3):36–37
Liao K (1998) Study on MAOC pour point depressant. Pet Process Petrochem 29(1):28–30
Liu J (1986) Experimental research heat treatment of Da Qing's output oil. Oil & Gas Storage Transp 5(3):25–31
Liu X, Li H, Bo W et al (2007) Preparation of oil soluble drag reducer by solution polymerization. Fine Chem 24(5):512–516
Luo T (1983) Heat treatment of waxy crude in China. Acta PetEi Sin 4(2):75–84
Luo T, Han J (1997) China develops the pipeline transportation technology of drugged crude oil. Oil & Gas Storage Transp 16(12):5–6
Machado ALC, Lucas EF, Gonzalez G (2001) Poly(ethylene covinyl) (EVA) as wax inhibitor of a Brazilian crude oil: oil viscosity, pour point and phase behavior of organic solution. J Petrol Sci Eng 32(24):159–165
Mack MP (1985) Polymerization process for drag reducing substances. US Patent 4,493,903, 15 Jan 1985
Manka JS, Ziegler KL (2001) Factors affecting the performance of crude oil wax control additives. SPE 67326
Marie E, Chevalier Y, Eydoux F et al (2005) Control of n-alkanes crystallization by ethylene-vinyl acetate copolymer. J Colloid Interface Sci 290(2):406–418
Martella DJ, Jaruzelski JJ, Chen JJ (1992) Method of preparing alkyl phenolformaldehyde condensates. US Patent 5,118,875, 2 Jun 1992
Meier DJ, Kruka VR (1974) Methods and compositions for reducing the frictional drag of flowing fluids. US Patent 3,801,508, 2 Apr 1974
Meng Q, Zhang Z (1987) The test and application of friction reducing agent in Tie-Da Pipeline. Oil & Gas Storage Transp 6(1):43–50
Misra S, Baruah S, Singh K (1995) Paraffin problems in crude oil production and transportation: a review. SPE Prod & Facil 10(1):50–54
Motier J (1985) Compositions for and method of reducing hydrocarbon fluid friction loss in conduits. US Patent 4,527,581, 9 Jul 1985
Newberry ME (1985) Particulate composition. US Patent 4,518,509, 21 May 1985
Pei X, Chen W, Zhang F (2005) Study of synthesis of drag reduction agent with α-alkene. Henan Pet 19(3):81–83
Quan Z (1996) On modified crude oil pipeline transportation Process. Oil & Gas Storage Transp 15(1):1–6
Savins JG (1964) Drag reduction characteristics of solutions of macromolecules in turbulent pipe flow. SPEJ 4(3):203–214
Seymour EV, Winkler DE (1971) Process of reducing friction loss in flowing hydrocarbon liquids. US Patent 3,559,664, 2 Feb 1971
Shi Y, Li H, Zhou Q et al (2006) Synthesis of drag reducer for crude oil froma-olefin over $TiCl_4/Al(i\text{-}Bu)_3$ Catalyst. Petrochem Technol 35(4):371–375
Slater G, Davis A (1986) Pipeline transportation of high pour point New Zealand crude using pour point depressants. SPE 15656
Wilburn BE, Heilman WJ (1990) Methacrylate pour point depressants and compositions. US Patent 4,956,111, 11 Sept 1990
Yang F, Li C et al (2009) Depressive effects evaluation of ethylene-vinyl acetate copolymer on waxy crude oils. J China Univ Pet 33(5):108–113

Yang F, Li C et al (2009) Depressive effect of polyacry late (PA) pour point depressant on waxy crude oils. J Petrochem Univ 22(2):20–25
Yang Y, Qi G et al (2001) Synthesis of poly (*n*-behenyl acrylate) and it's application as flow improver. Acta PetEi Sin (Pet Process Sect) 17(5):60–65
Yin G, Gao H et al (2002) The study on the mechanism of drag reduction by polymer additives. Oil & Gas Storage Transp 21(7):1–2, 12
Yu S (1990) Resistance-reducing technic application in petroleum pipeline of our country. Oil & Gas Storage Transp 9(2):16–22
Zhang F (1990) Pipeline of coagulant applied technology. Oil & Gas Storage Transp 9(5):11–20
Zhao F, Mao W et al (1996) Applying depressant and drag reducer in fully loaded pipeline. Oil & Gas Storage Transp 15(8):10–13
Zhen Z, Cao D (2006) Field test on pipeline transportation with DRA in Huangdao-Qilu Sinopec oil pipeline. Oil & Gas Storage Transp 25(2):33–37

Chapter 14
Natural Gas Treatment

Natural gases are combustible gases produced from the stratum,which mainly contain saturated hydrocarbon gases (such as methane, ethane, propane, and butane) and a small amount of non-hydrocarbon gases (such as carbon dioxide, hydrogen sulfide, and mercaptan). The natural gas produced from gas field is called gas-field gas, while that produced from oilfield is called oilfield gas (or oilfield associated gas). In order to protect environment and develop and utilize natural gas, a series of treatments need to be conducted (Turner et al. 1997).

14.1 Dehydration of Natural Gas

14.1.1 The Necessity of the Dehydration of Natural Gas

Under the formation condition, natural gas is in contact with water. Therefore, natural gas always contains a certain amount of water vapor, which can cause the following hazards.

- In the presence of acid gas such as hydrogen sulfide and carbon dioxide, the condensation of water vapor may cause severe corrosion of pipeline.
- In certain conditions, natural gas and water can form hydrate which would cause pipe plugging.
- When using low-temperature separation method to recycle the light hydrocarbon in the natural gas, water vapor will condense to water and freeze, which would also cause pipe blockage.

Therefore, natural gas should be dehydrated to certain degree, based on different requirements and using different methods (Li et al. 2009).

C. Dai and F. Zhao, *Oilfield Chemistry*,
https://doi.org/10.1007/978-981-13-2950-0_14

14.1.2 Representations of Water Content in Natural Gas

The water content of natural gas can be expressed by absolute humidity and relative humidity.

Absolute humidity of natural gas refers to the quality of the water vapor in unit volume of natural gas, and the unit is g/m^3.

If natural gas and water reach phase equilibrium under certain conditions, the water content in natural gas corresponding to the absolute humidity is the saturated water vapor content in natural gas.

Therefore, we can define the relative humidity of natural gas.

Relative humidity of natural gas refers to the ratio of absolute humidity of natural gas to saturated water vapor content of natural gas.

The water content of natural gas can also be expressed by its water dew point.

The water dew point of natural gas refers to the temperature when saturated water vapor content equals to the absolute humidity under certain pressure. Therefore, the lower the water dew point of natural gas, the lower the water content in natural gas. In order to ensure the safe transportation of the natural gas, the water dew point must be 5–15 °C lower than the ambient temperature along the pipeline.

Figure 14.1 shows the relationship of the saturated water vapor content in the natural gas with temperature and pressure, where the relative density of natural gas is 0.6 and the water contains no salt. For other relative densities of natural gas and various water salt contents, we can look up the corresponding correction coefficient from Figs. 14.2 and 14.3, and then calculate correct results by the following formula.

$$w(H_2O) = w_0(H_2O) \cdot \alpha \cdot \beta \tag{14.1}$$

where $w(H_2O)$ is saturated water vapor content in natural gas in real conditions; $w_0(H_2O)$ is saturated water vapor content in natural gas in the diagram; α is correction coefficient of natural gas relative density; β is correction coefficient of salt content in water.

Figure 14.1 is also called water dew point diagram of natural gas, through which the water dew point of natural gas under different conditions can be found.

With known relative density of natural gas, temperature, pressure, water vapor content and the salt content in water, α and β can be looked up from Figs. 14.2 and 14.3. Assuming that the water vapor content under above condition is $w(H_2O)$, $w_0(H_2O)$ can be calculated through Eq. 14.1. Then according to $w_0(H_2O)$ and the pressure, water dew point can be looked up in Fig. 14.1.

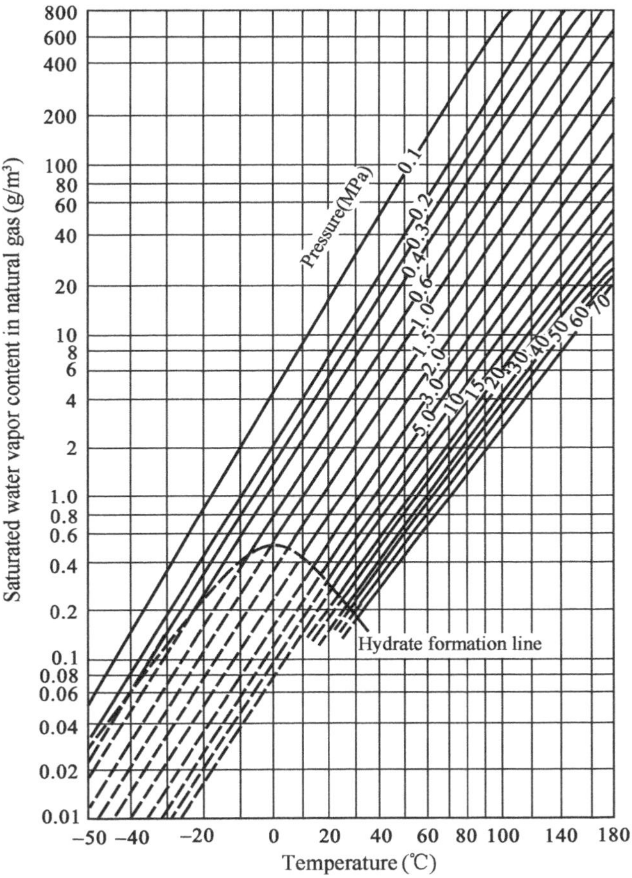

Fig. 14.1 Changes of saturated water vapor content in natural gas with temperature and pressure

Fig. 14.2 Correction coefficient (α) of the relative density of natural gas

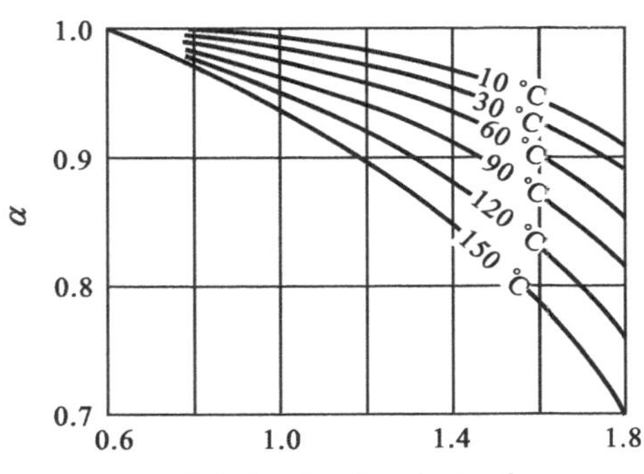

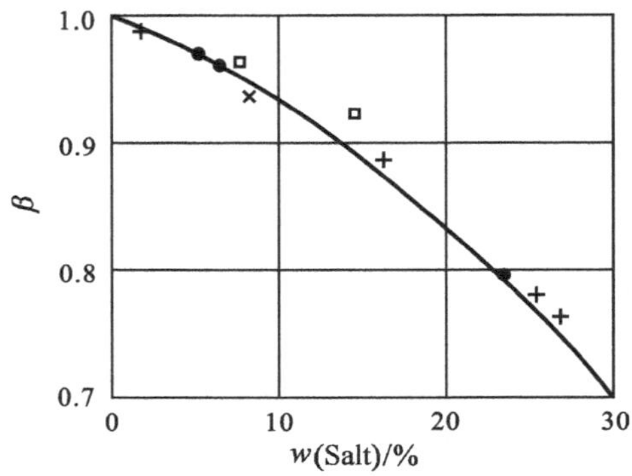

Fig. 14.3 Correction coefficient (β) of the salt content in water

14.1.3 Dehydration Methods of Natural Gas

The following methods can be used to remove the water vapor in natural gas.

Cooling Method

As can be seen from Fig. 14.1, with the decrease of temperature and the increase of pressure, the saturated water vapor content in natural gas is reduced; therefore, the cooling method or pressure boost cooling method can be applied to remove the water vapor in natural gas. The dehydration temperature must be 5–7 °C lower than the water dew point of natural gas in pipeline.

Adsorption Method

This is a method using adsorbent to remove water vapor in natural gas.

In this method, the solids which are applied as adsorbents should have large specific surface area (surface area per unit mass of adsorbent), high porosity, water selectivity, thermal and chemical stability, certain mechanical strength, recyclability, wide sources, and low cost.

Available adsorbents include activated alumina, silica gel and molecular sieve (Table 14.1).

Molecular sieve has uniform micropore structure of porous crystal. The micropore diameter of molecular sieve is close to the diameter of water molecules (0.28 nm), and the polarity of the molecular sieve is close to that of water, therefore molecular sieve bears excellent water absorption capability. The water content of natural gas after molecular sieve dehydration can be as low as 0.1–10 g/m^3.

While removing water vapor in natural gas by adsorption method, wet natural gas can be pumped into adsorption tower filled with adsorbent. The natural gas which comes out of the absorption tower is dry gas. After the adsorbent in adsorption tower reaching its moisture capacity, hot gas can be infused to remove the water on the adsorbent surface (regeneration). The adsorbent after regeneration can still be used for natural gas dehydration.

Table 14.1 Adsorbents used for the dehydration of natural gas

Adsorbent		Molecular formula	Specific surface (m^2/g)	Porosity (%)	Regeneration temperature (°C)	Introductions
Activated alumina		Al_2O_3	210–350	50–65	180–450	Amorphous porous body, $w(Al_2O_3)$ = 90–94%
Silica gel		SiO_2	550–830	50–65	120–230	Amorphous porous body, $w(SiO_2)_{max}$ = 90%
Molecular sieve	4A	$Na_2O \cdot Al_2O_3 \cdot mSiO_2^a$	700–900	55–60	150–310	Crystal porous body, uniform micropore structure, the diameter of micropores is 0.45 nm
	5A	$CaO \cdot Al_2O_3 \cdot mSiO_2^a$	700–900	55–60	150–310	Crystal porous body, uniform micropore structure, the diameter of micropores is 0.55 nm

Note [a]m = 2.0–2.5

Absorption Method
This is a method using absorbent to remove water vapor in natural gas (Gavlin et al. 1998).

In the absorption method, the solvents which are applied as absorbents should have high water solubility, low natural gas solubility, thermal and chemical stability, low viscosity, low steam pressure, low foaming and emulsifying tendency, non-corrosiveness, recyclability, wide sources, and low cost.

The commonly used absorbent is glycol, mainly diethylene glycol, triethylene glycol, and tetraethylene glycol (Table 14.2). These absorbents can combine with water through hydrogen bond to remove the water in natural gas.

Due to the high boiling point of glycol, it can be regenerated and reused by distillation method and gas stripping method.

14.2 Deacidification of Natural Gas

The acid gases in natural gas mainly include hydrogen sulfide and carbon dioxide. The presence of hydrogen sulfide would aggravate the corrosion of pipeline and equipment, the environmental pollution and also cause catalyst poisoning in the

Table 14.2 Absorbents used for the dehydration of natural gas

Absorbent	Molecular formula	Density (g/cm^3)	Freezing point (°C)	Boiling point (°C)	Water solubility	Thermal decomposition temperature (°C)	Regeneration temperature (°C)
Diethylene glycol	$HO-[CH_2CH_2O]_2-H$	1.118	−8.3	245.0	Easily soluble	164.4	149–163
Triethylene glycol	$HO-[CH_2CH_2O]_3-H$	1.125	−7.2	287.4	Easily soluble	206.7	117–196
Tetraethylene glycol	$HO-[CH_2CH_2O]_4-H$	1.128	−5.6	327.3	Easily soluble	237.8	204–224

post-processing process. The existence of carbon dioxide can not only increase the corrosion of pipeline and equipment, but also reduce the calorific value of natural gas (calorific value produced by unit mass of natural gas). Therefore, the acid gas in natural gas should be removed (Maddox 1982; Mimura et al. 1998).

When removing hydrogen sulfide and carbon dioxide in natural gas, other sulfides such as carbon disulfide (CS_2), carbonyl sulfur (COS), and mercaptan (RSH) will also be removed.

The following methods can be applied to remove acid gas from the natural gas.

14.2.1 Adsorption Method

This is a method using adsorbent to remove acid gas.

The following adsorbents can be used.

Chemical Adsorbent

This is a kind of adsorbent which can react with acid gas during the adsorption.

In the field use, the chemical adsorbent is mainly sponge iron.

The chemical composition of sponge iron is iron oxide (Fe_2O_3), which has special selectivity to the hydrogen sulfide. When hydrogen sulfide adsorbs on the face of sponge iron, it can be removed by the following reaction.

$$2Fe_2O_3 + 6H_2S \rightarrow 2Fe_2S_3 + 6H_2O$$

The used sponge iron can be regenerated through the reaction with the oxygen in the air.

$$2Fe_2S_3 + 3O_2 \rightarrow 2Fe_2O_3 + 6S$$

Although sponge iron can react with carbon dioxide, the reaction is not complete. Therefore, the sponge iron cannot be used for removal of carbon dioxide.

In addition to sponge iron, zinc oxide and calcium oxide can also be used as chemical adsorbent. However, these chemical adsorbents cannot be recycled as sponge iron.

Physical Adsorbent

This is a kind of adsorbent which removes the acid gas by means of physical adsorption.

In the field use, the physical adsorbent is mainly molecular sieve. As the gas is acid, acid-resistant molecular sieve should be used.

The acid-resistant ability of molecular sieve is closely related to the silica–alumina ratio in its composition. Here the silica–alumina ratio refers to the mole ratio of silicon dioxide to ferric oxide in the molecular sieve, usually in the range of 2.0–2.5 (Table 14.1); however, the silica–alumina ratio of acid-resistant molecular sieve is in the range of 4–10.

The acid gases adsorbed in molecular sieve can be desorbed by aerating hot natural gas so that the molecular sieve can be regenerated and reused.

In addition to the molecular sieve, activated carbon and silica gel can also be used as physical adsorbent.

14.2.2 Absorption Method

This is a method using absorbent to remove acid gas.

Two kinds of absorbents can be used.

Chemical Absorbent

This is a kind of absorbent which can react with acid gas during the absorption.

Monoethanolamine is a kind of chemical absorbent which can remove the acid gas under low temperature (25–40 °C).

$$\underset{\text{(Monoethanolamine, MEA)}}{2HOCH_2CH_2NH_2} + H_2S \underset{\text{High temperature}}{\overset{\text{Low temperature}}{\rightleftharpoons}} (HOCH_2CH_2NH_3)_2S$$

$$2HOCH_2CH_2NH_2 + CO_2 + H_2O \underset{\text{High temperature}}{\overset{\text{Low temperature}}{\rightleftharpoons}} (HOCH_2CH_2NH_3)_2CO_3$$

$$HOCH_2CH_2NH_2 + RSH \underset{\text{High temperature}}{\overset{\text{Low temperature}}{\rightleftharpoons}} HOCH_2CH_2NH_3SR$$

By heating method (above 150 °C) and gas stripping method, the acid gas absorbed in the absorbent can be removed to regenerate the absorbent.

Chemical absorbents similar to monoethanolamine include the following alcohol amines.

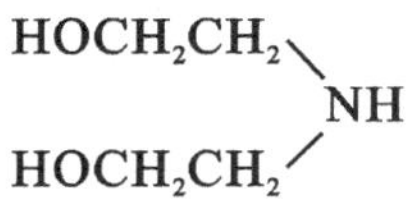

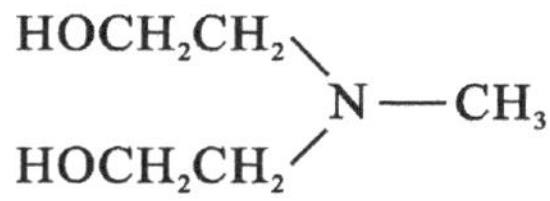

(Diethanolamine, DEA) (Methyldiethanolamine, MDEA)

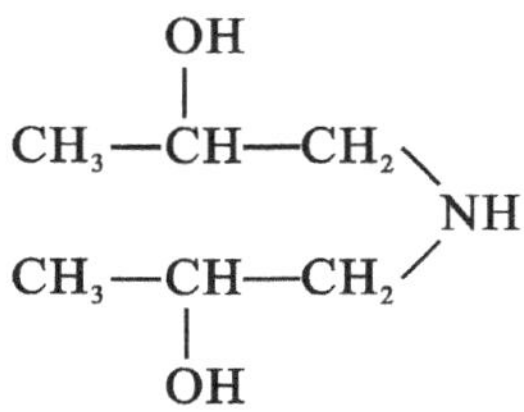

(Triisopropanolamine, TIPA)

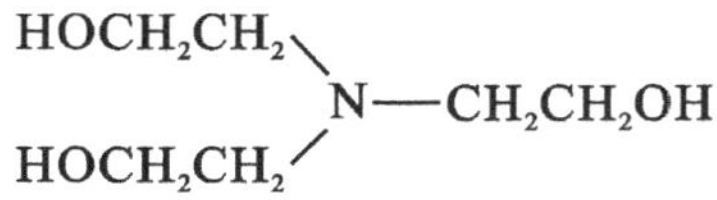

(Triethanolamine, TEA)

$$H{-}[OCH_2CH_2]_2{-}NH_2$$

(Diethyleneglycolamine, DEGA)

Alcohol amine method refers to the method to remove acid gas using alcohol amine as absorbent, in which alcohol amine is generally prepared into aqueous solution with a mass fraction of 10–30%.

Compound use of different alcohol amines can improve the effect of removing acid gas (Schubert 1997).

Sodium hydroxide is also a kind of chemical absorbent. When acid gases pass through sodium hydroxide solution, they can be removed through the following reaction.

$$H_2S + 2NaOH \rightarrow Na_2S + 2H_2O$$
$$CO_2 + 2NaOH \rightarrow Na_2CO_3 + H_2O$$
$$H_2S + Na_2S \rightarrow 2\ NaHS$$
$$CO_2 + 2\ Na_2S + H_2O \rightarrow Na_2CO_3 + 2\ NaHS$$
$$H_2S + Na_2CO_3 \rightarrow NaHS + NaHCO_3$$
$$CO_2 + Na_2CO_3 + H_2O \rightarrow 2NaHCO_3$$

Sodium hydroxide solution cannot be regenerated after using.

The method of using sodium hydroxide solution to remove acid gas is called alkali washing method.

Physical Absorbent

This is a kind of absorbent which removes acid gas by means of physical absorption.

The available absorbents are shown as follows.

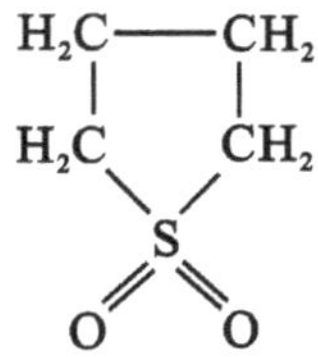

(Sulfolane)

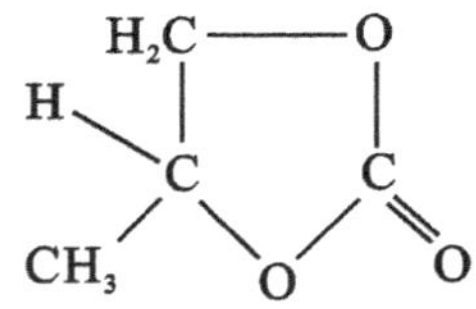

(Propylene carbonate)

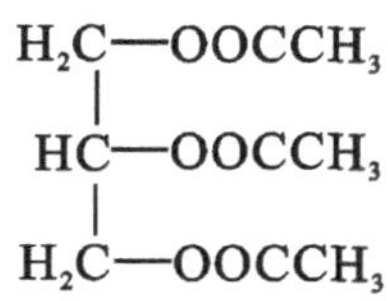

(Triacetin)

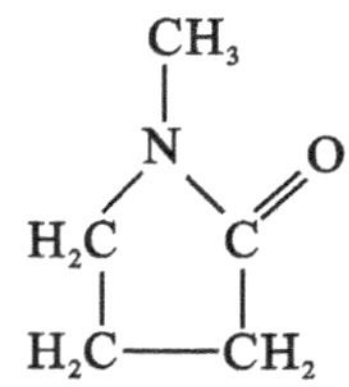

(*N*-methyl pyrrolidone)

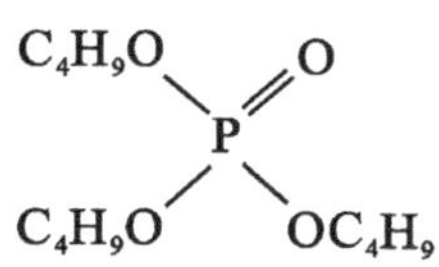

(Tributyl phosphate)

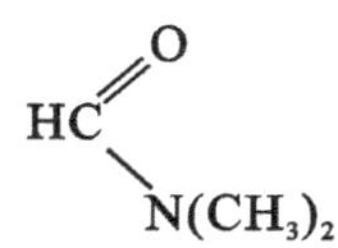

(Dimethylformamide)

$$CH_3 {+} CH_2CH_2O {+}_n CH_3 \qquad n\text{: } 2\text{–}7$$

(Polyethylene glycol dimethylether)

$$CH_3 {+} CH_2CH_2O {+}_n C_3H_7 \qquad n\text{: } 2\text{–}7$$

(Polyethylene glycol methylpropylether)

These physical absorbents can remove acid gases by dissolving. After absorption of acid gas, these absorbents can be regenerated by heating and gas stripping method.

Chemical absorbent and physical absorbent can be mixed for use. One way to remove acid gas is called sulfone amine method, in which the absorbent is prepared by the mixture of sulfone and alcohol amine. The sulfone commonly used in sulfone amine method is sulfolane, while the commonly used alcohol amine is diisopropyl alcohol amine. The compound absorbent has the following features.

- Absorption variety. Many kinds of acid gases can be removed such as CO_2, COS, H_2S, RSH, RSR.
- High degree of purification. The content of hydrogen sulfide in purified natural gas can be as low as 6 mg/m^3.

- Large absorption capacity. 1 m^3 absorbent can absorb up to 120 m^3 acid gas.
- Low consumption. Both the circulation quantity and evaporation consumption of absorbent are few.

Now the compound absorbents are commonly used.

14.3 Inhibition of the Generation of Natural Gas Hydrate

Under certain conditions, natural gas can react with water to generate hydrate which may block the natural gas pipeline. In order to solve this problem, the generation condition, structure, and inhibition methods of the natural gas hydrate should be understood.

14.3.1 Formation Conditions of Natural Gas Hydrate

The formation of natural gas hydrate has two conditions.

- Water exists in natural gas.
- There are sufficiently low temperature and sufficiently high pressure.

Figure 14.4 is the phase diagram of propane hydrate, from which the two formation conditions of natural gas hydrate can be understood. It can also be observed from Fig. 14.4 that propane will no longer produce propane hydrate when the temperature is higher than 5.5 °C. This critical temperature is usually referred as the

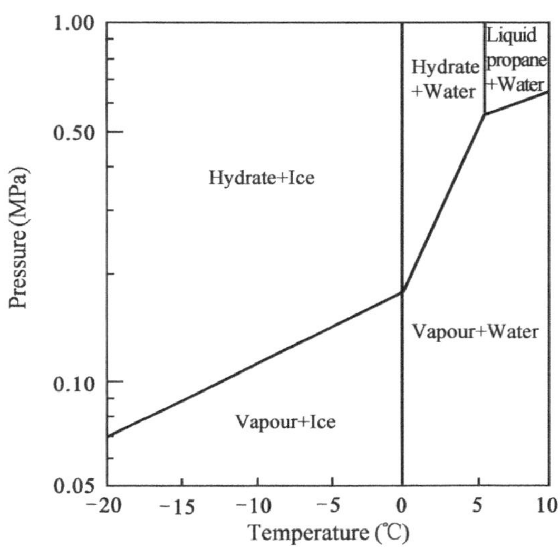

Fig. 14.4 Phase diagram of propane hydrate (Cox 1983)

Table 14.3 Critical hydrate formation temperature of each component in natural gas

Composition	CH_4	C_2H_6	C_3H_8	i-C_4H_{10}	n-C_4H_{10}	CO_2	H_2S
Hydrate generated critical temperature (°C)	47.0	14.5	5.5	2.5	1.0	10.0	29.0

critical hydrate formation temperature. The critical hydrate formation temperature of each component in natural gas can be seen in Table 14.3.

14.3.2 The Structure of Natural Gas Hydrate

Natural gas hydrate has three kinds of basic cavity composed of water molecules (Hight 1994). One is small cavity (Fig. 14.5a), while the other two are big cavities (Fig. 14.5b and 14.5c). Small cavity a is a dodecahedron which is composed by 12 pentagons, denoted as 5^{12}. The big cavity b is tetrakaidecahedron which is composed by 12 pentagons and 2 hexagons, denoted as $5^{12}6^2$. The big cavity c is hexadecahedron which is composed by 12 pentagons and 4 hexagons, denoted as $5^{12}6^4$. In these three basic kinds of cavities, the vertices are water molecules and the attachment lines are hydrogen bonds. The three basic cavities in natural gas hydrate form two kinds of unit cells, the cell type I and type II. Type I is composed of small cavity a and big cavity b; while type II is composed of small cavity a and big cavity c. Table 14.4 shows the characteristics of the two kinds of unit cells in natural gas hydrate.

According to Table 14.4, we can calculate the number of water molecules combined with each component of natural gas.

$$CH_4 \cdot 6H_2O \quad C_2H_6 \cdot 8H_2O \quad C_3H_8 \cdot 17H_2O$$
$$i - C_4H_{10} \cdot 17H_2O \quad n - C_4H_{10} \cdot 17H_2O \quad H_2S \cdot 6H_2O \quad CO_2 \cdot 6H_2O$$

The natural gas hydrate crystal composed of unit cells mentioned above will first separate out on the interface between natural gas and water and disperse in water, then these crystals grow, aggregate and deposit, causing blockage to the natural gas pipeline.

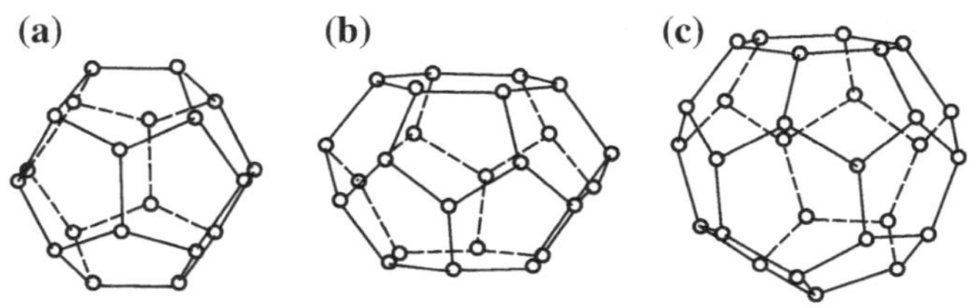

Fig. 14.5 Basic cavity of natural gas hydrate. **a** Dodecahedron small cavity; **b** Tetrakaidecahedron big cavity; **c** Hexadecahedron big cavity

Table 14.4 The characteristics of the unit cell of nature gas hydrate

Cell	I		II
The number of water molecules per cell	46		136
The number of cavities per cell	Small	2	16
	Big	6	8
Cavity diameter (nm)	Small	0.397	0.391
	Big	0.430	0.473
Gas molecule that can enter and stabilize the cavity	Small	CH_4, H_2S, CO_2	CH_4, H_2S, CO_2
	Big	C_2H_6 except for the above gas	C_3H_8, n-C_4H_{10}, i-C_4H_{10} except for the above gas

14.3.3 Inhibition Methods of the Generation of Natural Gas Hydrate

There are several methods to inhibition the generation of natural gas hydrate.

Pressure Reduction
For natural gas at certain temperature and with certain composition, the pressure can be dropped to hydrate formation pressure (can be understood from Fig. 14.4).

Maintaining Certain Temperature
For natural gas at certain pressure and with certain composition, the temperature can be maintained above the hydrate formation temperature (can also be understood from Fig. 14.4).

Reducing the Water Content in Natural Gas
The water content of natural gas can be decreased by dehydration to reduce the water dew point of natural gas. When temperature is higher than the water dew point of natural gas, natural gas will not produce natural gas hydrate (can be understood from Fig. 14.1).

Employment of Natural Gas Hydrate Inhibitor
There are three kinds of natural gas hydrate inhibitors (Kelland et al. 1995; Kelland et al. 1994). One kind is alcohol such as methanol, ethanol, or ethylene glycol. Because alcohol is completely miscible with water, the tendency of natural gas forming hydrate on the surface of alcohol solution is significantly lower than that on the surface of water. Meanwhile, the greater the alcohol content, the stronger tendency it can inhibit the crystallization precipitation of hydrate. According to the composition of natural gas, when w (alcohol) is in the range of 10–60%, hydrate crystallization precipitation can be effectively controlled, thus inhibiting the formation of natural gas hydrate.

Another kind is the surfactant (some examples are given below). Because surfactants can disturb (produce distortion) the crystalline surface of the precipitated or grown natural gas hydrate by adsorption, which is adverse to the further grow up and aggregation of the crystals, thus inhibiting the deposition of natural gas hydrate on the gas pipeline surface.

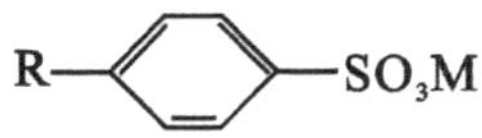

R: C_8–C_{22};
M: Na, K, NH_4

(Alkylbenzene sulfonate)

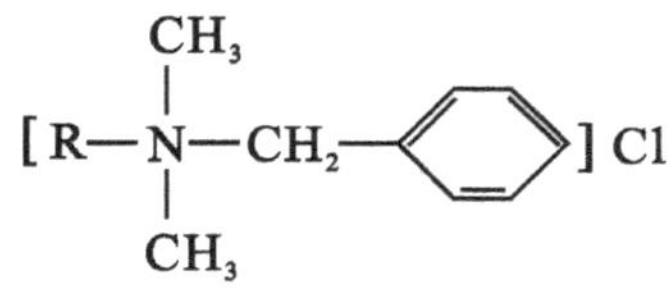

R: C_8–C_{18}

(Alkylbenzyl dimethyl ammonium chloride)

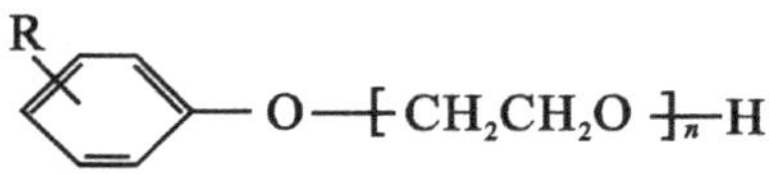

R: C_8–C_9;
n: 3–5

(Polyoxyethylene alkyl phenol ether)

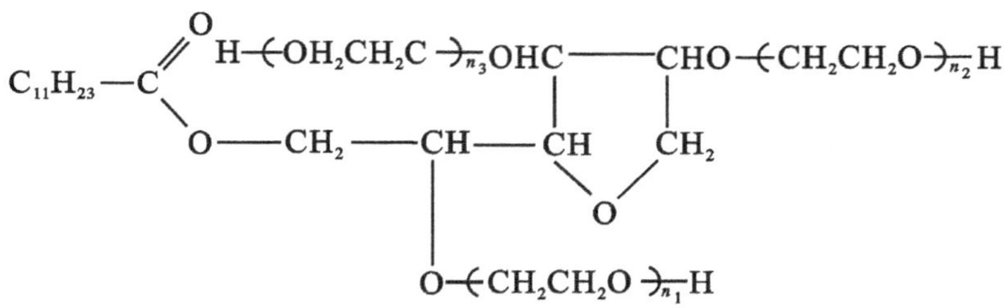

(Sorbidemonolaurate polyoxyethylene ether) (Lee et al. 2005)

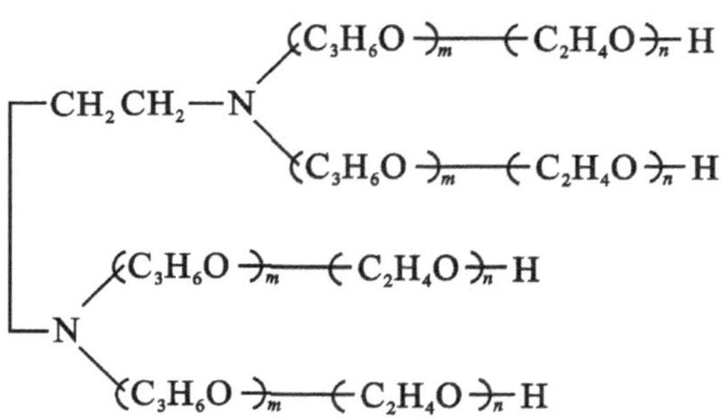

(Polyoxyethylene polyoxypropylene glycylamine) (Kelland et al. 2009)

The last kind is polymer (some examples are given below).

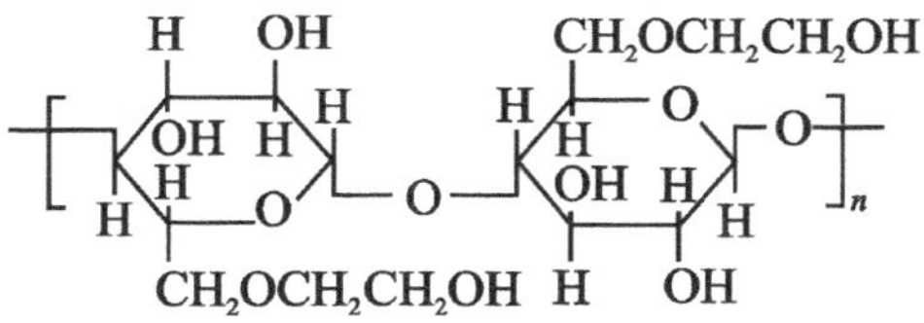

(Hydroxyethyl cellulose)

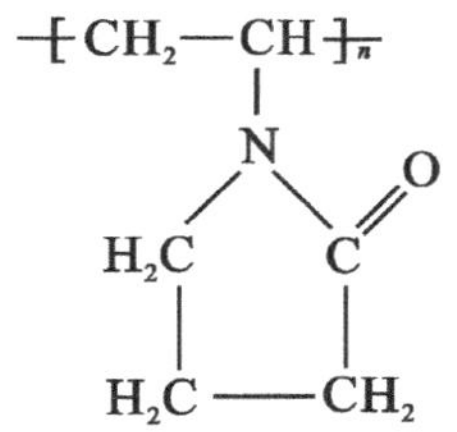

(Poly-*N*-vinylpyrrolidone)
(Notz et al. 1996; Tang et al. 2006)

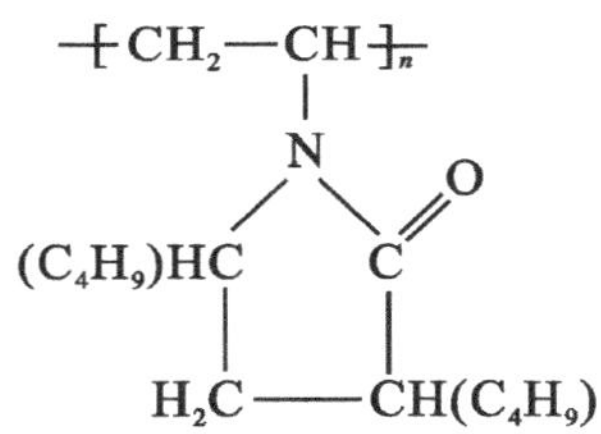

(Butyl derivative of Poly-*N*-vinylpyrrolidone)

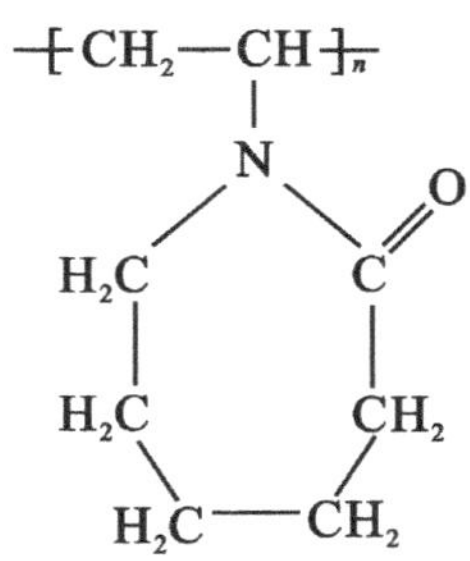

(Poly-*N*-vinylcaprolactam)
(Zhang et al. 2007)

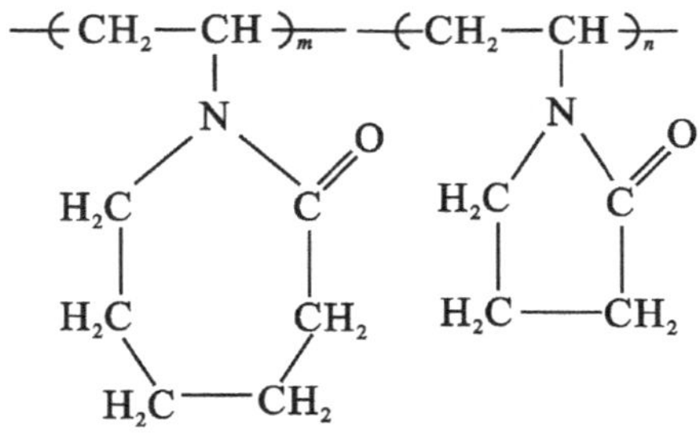

(Copolymer of *N*-vinylcaprolactam and *N*-vinylpyrrolidone)

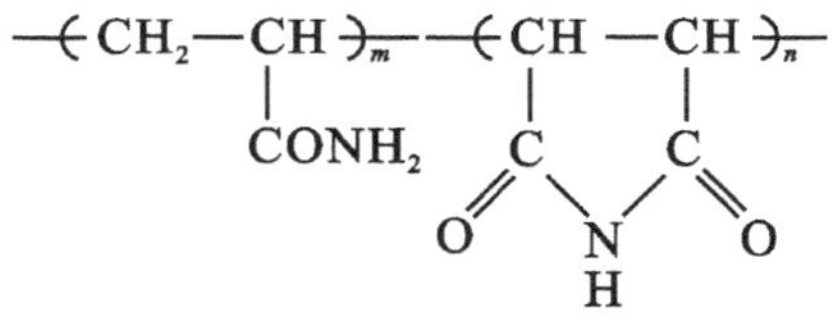

(Copolymer of acrylamide and maleimide) (Costello et al.1998)

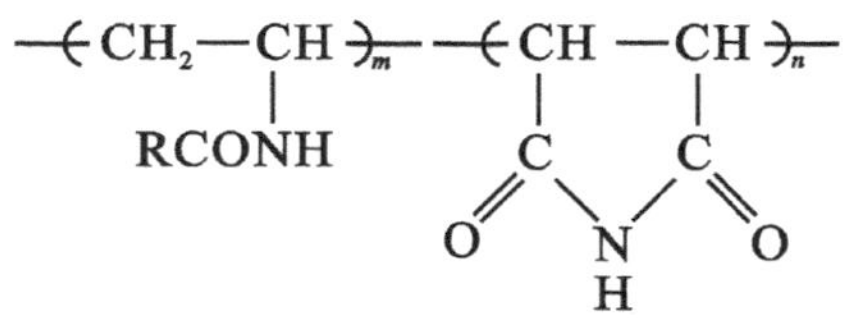

(Copolymer of *N*-Vinylamide and maleimide)

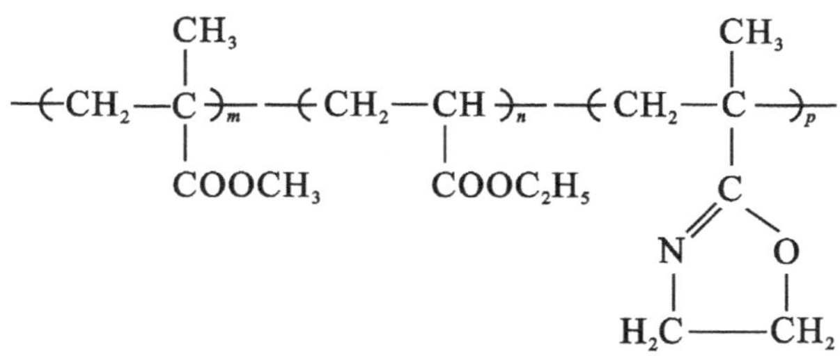

(Copolymer of methyl methacrylate, ethyl acrylate and isopropenyl oxazoline) (Collek et al. 1996)

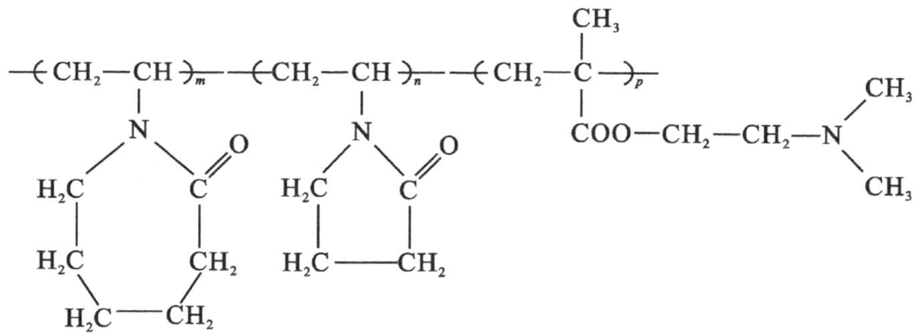

(Copolymer of *N*-vinylcaprolactam, *N*-vinylpyrrolidone and dimethylamino methacrylic acid-1,2-ethylidene ether) (Wu et al. 2000)

The above polymers are all nonionic polymers. In addition, anionic polymers and cationic polymers can also be used (Klug et al. 1998). These polymers are water-soluble polymers. The precipitated or grown crystals of natural gas hydrate can combine with polymer links so that they are separated from each other. In this way even if these crystals can continue to grow up, they cannot aggregate or deposit on the surface of natural gas pipeline. Thus, the blockage of pipeline would be prevented. By comparing the function mechanisms of the above-mentioned three kinds of natural gas hydrate inhibitors, it can be seen that the alcohol inhibitor controls the generation of natural gas hydrates in the crystallization stage, the surfactant inhibitor controls the generation of natural gas hydrates in the crystal growth stage, and the polymer inhibitor controls the generation of natural gas hydrates in the coalescence and deposition stage. Therefore, they are all called natural gas hydrate inhibitors.

Natural gas hydrate inhibitors are usually used in mixtures. For example, the compound inhibitor containing 2.5×10^3 mg/L copolymer of *N*-vinyl caprolactam and *N*-vinyl pyrrolidone, 1.0×10^3 mg/L alkyl benzyl dimethyl ammonium chloride and 2.5×10^3 mg/L *n*-tetrapentyl ammonium chloride has good inhibitory effect on the generation of natural gas hydrate (Duncum et al. 1997).

References

Collek S, Oelfke RH (1996) Method for inhibiting hydrate formation. US Patent 5,583,273, 10 Dec 1996

Costello CA, Berluche E, Oelfke RH et al (1998) Maleimide copolymers and method for inhibiting hydrate formation. US Patent 5,744,665, 28 Apr 1998

Cox JL (1983) Natural gas hydrates, properties, occurrence and recovery. Butterworth Publishers, Massachusetts, p 57

Duncum SN, Edwards AR, James K et al (1997) Hydrate inhibition. GB Patent 2,313,841, A, 10 Dec 1997

Gavlin G, Goltsin B (1998) Gas dehydration process. WO Patent 98/1,854,2, 7 May 1998
Hight MA (1994) State-of-the-art survey on hydrate formation. SPE 28507
Kelland MA, Svartaas TM, Andererersen LD (2009) Gas hydrate antiagglomerant properties of polypropoxylates and some other demulsifiers. J Pet Eng 64(1–4):1–10
Kelland MA, Svartaas TM, Dybvik LA (1994) Control of hydrate formation by surfactants and polymers. SPE 28506
Kelland MA, Svartaas TM, Dybvik LA (1995) Studies on new gas hydrate inhibitors. SPE 30420
Klug P, Feustel M, Frenz V (1998) Additives to inhibit the formation of gas hydrate. WO Patent 98/2,255,7, 28 May 1998
Lee JD, Englezos P (2005) Enhancement of the performance of gas hydrate kinetic inhibitors with polyethylene oxide. Chem Eng Sci 60:5323–5330
Li J, Niu L (2009) Application of natural gas dehydration process in Fushan oilfield. Pet Plan & Eng 20(2):33–35
Maddox RN (1982) Gas conditioning and processing, volume 4, Gas and liquid sweetening. Gampbell Petroleum Series, Norman, Oklahoma, U.S.A
Mimura T, Shimojo S, Iijima M et al (1998) Method for the removal of carbon dioxide and hydrogen sulfide from a gas containing these gases. EP Patent 827,772, 11 Mar 1998
Notz PK, Bumgardner SB, Schaneman BD et al (1996) Application of kinetic inhibitors to gas hydrate problems. SPE Prod & Facil 11(4):256–260
Schubert CN (1997) Method and composition for sweetening liquified petroleum gas. CA Patent 2,186,806, 6 Apr 1997
Tang C, Fan S (2006) Research works about PVP inhibition performance on natural gas hydrate. Nat Gas Ind 26(3):125–128
Turner HM, Bigger JM, Meyer JP (1997) Gas treatment method. US Patent 5,693,297, 2 Dec 1997
Wu D, Hu Y, Yang J (2000) Progress in study of new inhibitor for natural gas hydrate. Nat Gas Ind 20(6):95–97
Zhang W, Yin Z, Liu X et al (2007) Experimental study of polyvinylcaprolactan as kinetic methane hydrate inhibitor. Nat Gas Ind 27:105–107

Chapter 15
Oilfield Wastewater Treatment

The water separated from the liquid produced from oil well is called oilfield wastewater. Oilfield wastewater contains suspended solids and crude oil droplets; meanwhile, problems of corrosion, scale formation, and bacterial reproduction also exist. Therefore, corresponding chemicals are needed for the treatment of oilfield wastewater, which are referred as wastewater treatment agents.

There are six main purposes in the wastewater treatment, namely oil removal, oxygen removal, flocculation of suspended solids, scale inhibition, corrosion inhibition, and sterilization. Correspondingly, the wastewater treatment agents including oil remover, deoxidizer, flocculant, scale inhibitor, corrosion inhibitor, and bactericide are applied.

15.1 Oil Removal of the Wastewater

Crude oil exists in the wastewater in the form of oil droplets. Diffused electric double layers (as shown in Fig. 15.1) form on the surfaces of oil droplets due to the adsorption of anionic surfactants (e.g., carboxylate surfactants), rendering the surfaces of oil droplets negatively charged. Through the function of oil removers, the oil droplets in wastewater are easier to aggregate and float so that they can be removed in the settling tanks.

Chemical additives that can reduce the oil content in wastewater are called oil removers. There are two categories of oil removers.

15.1.1 Cationic Polymers

Cationic polymers with the following structures can be employed.

C. Dai and F. Zhao, *Oilfield Chemistry*,
https://doi.org/10.1007/978-981-13-2950-0_15

Fig. 15.1 Negatively charged oil droplet surface in oilfield wastewater

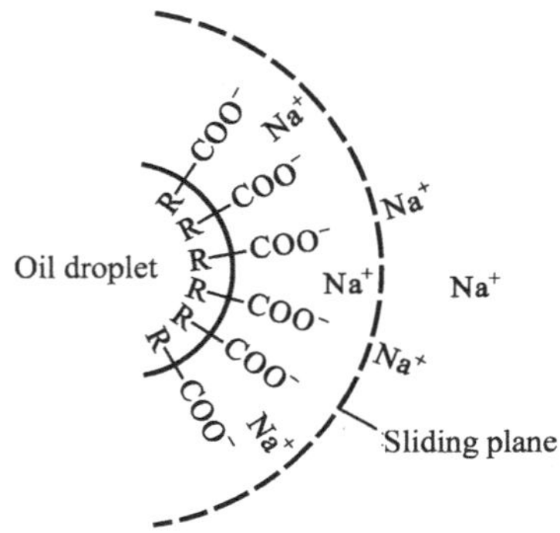

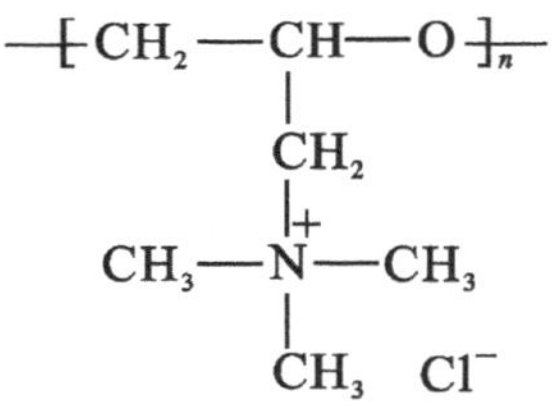

(Polyoxyproylene trimethyl ammonium chloride)

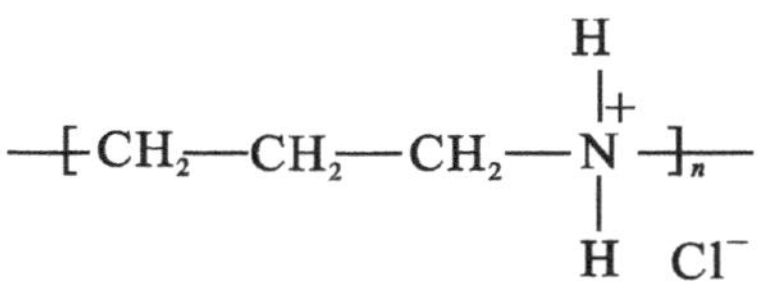

(Poly-1,3-propylene ammonium chloride)

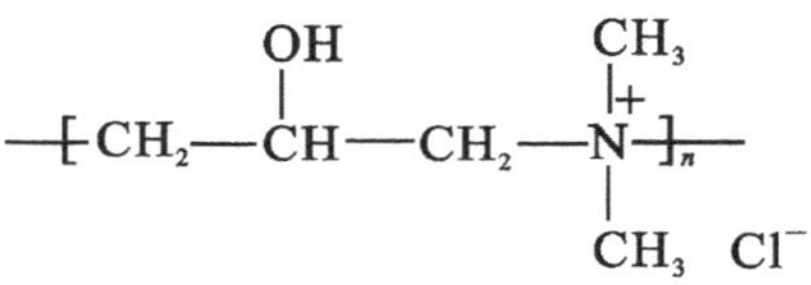

(Poly-2-hydroxy-1,3-propylene dimethyl ammonium chloride)

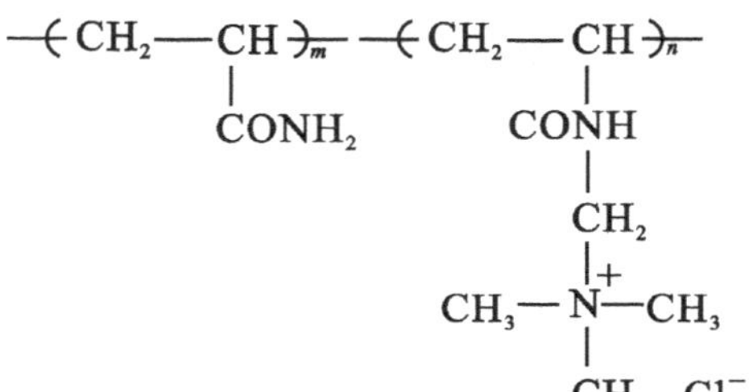

(Copolymer of acrylamide and acrylamide methylene trimethyl ammonium chloride)

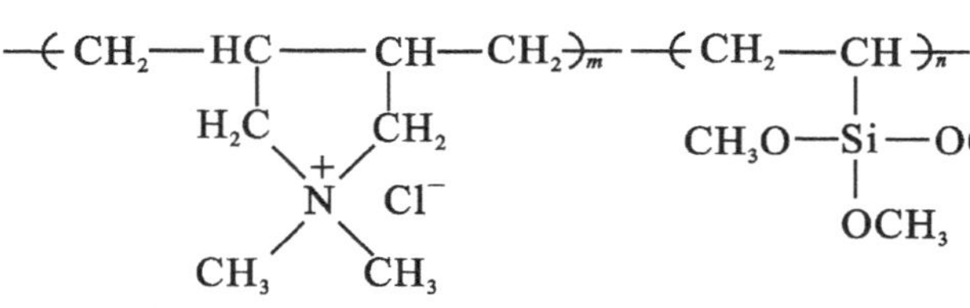

(Copolymers of diallyl dimethyl ammonium chloride and vinyl trimethoxysilane) (Sivakumar et al. 1996)

These cationic polymers can neutralize the negative charge on the surface of oil droplets and bridge the oil droplets and thus have good oil removal effect.

15.1.2 Branched Surfactants

Branched surfactants with the following structures can be employed.

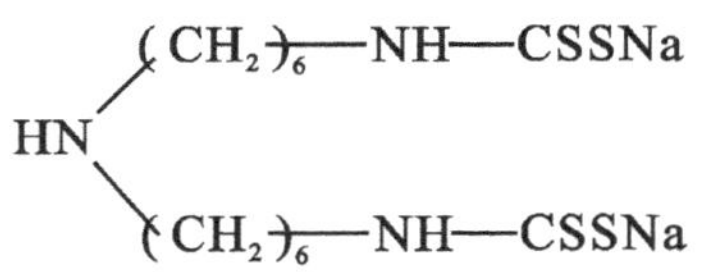

(Bis-hexamethylene amine bis-sodium dithiocarbamate) (Durham et al. 1989)

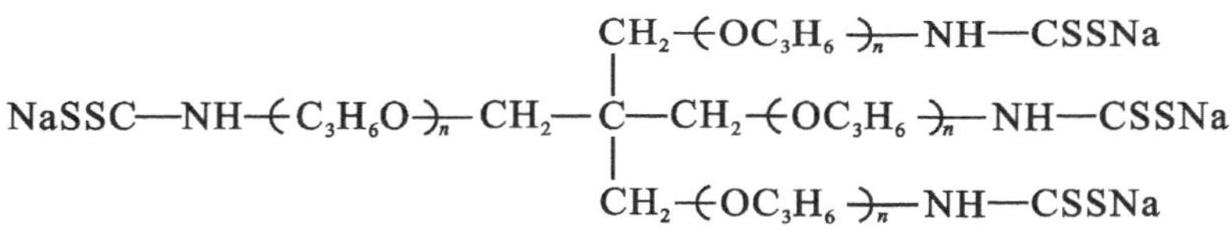

[Polyoxypropylene glycol ether-4-methylbenzal-*N*, *N*'-diethyl di (sodium dithiocarbamate)]

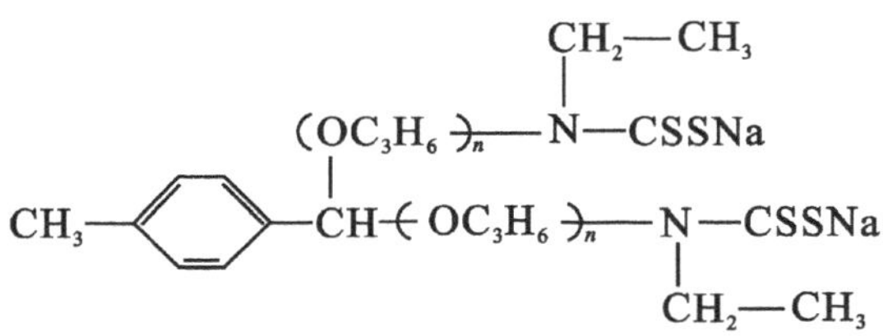

[(Polyoxypropylene-2,2-bis(hydroxymethyl)-n-butyl ether-*N*-sulfomethyl-tris (sodium dithiocarbamate)] (Thompson et al. 1990)

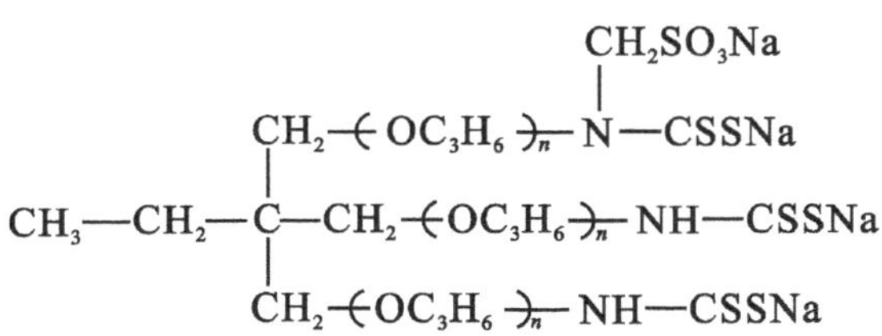

[(Polyoxypropylene pentaerythrilol ether-tetra(sodium dithiocarbamate)] (Rivers 1992)

These branched surfactants can replace the original adsorption film on the surface of oil droplets and weaken the protection effect of it, making oil droplets easier to aggregate and float and finally be separated from the wastewater.

15.2 Oxygen Removal of the Wastewater

The dissolved oxygen in wastewater can exacerbate metal corrosion and thus should be removed. Methods including heating, gas stripping, and vacuumizing can be applied to remove the dissolved oxygen in water; however, the most common method is the application of oxygen removers.

The chemicals that can remove dissolved oxygen in wastewater are generally called oxygen removers, which are also reducing agents.

The following oxygen removers can be applied.

Sulfite

Sulfites such as sodium sulfite, sodium hydrogen sulfite, sodium thiosulfate, and sodium hyposulfite can be applied, which can remove dissolved oxygen through the following reactions.

$$2Na_2SO_3 + O_2 \rightarrow 2Na_2SO_4$$
$$2NaHSO_3 + O_2 \rightarrow 2NaHSO_4$$
$$Na_2S_2O_3 + O_2 + H_2O \rightarrow 2NaHSO_3$$
$$Na_2S_2O_4 + O_2 + H_2O \rightarrow NaHSO_4 + NaHSO_3$$

As the reaction rate between oxygen and sulfites at room temperature is quite slow, catalysts need to be added. Divalent metal salts of Cu, Mn, Ni, and Co can be used as catalysts, among which the divalent metal salt of Co works the best.

Formaldehyde

Formaldehyde can remove oxygen through the following reaction (Knight 1973).

$$2CH_2O + O_2 \rightarrow 2HCOOH$$

Hydrazine

Hydrazine can remove oxygen through the following reaction (Schremp et al. 1961).

$$NH_2 - NH_2 + O_2 \rightarrow N_2 \uparrow + 2H_2O$$

Thiourea

Thiourea can remove oxygen through the following reaction.

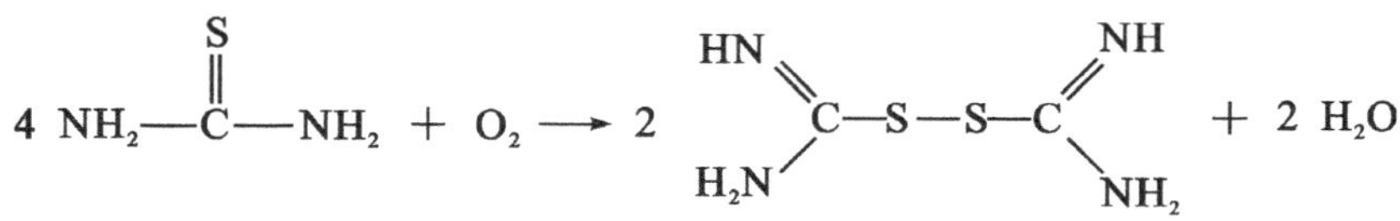

Isoascorbic Acid

Isoascorbic acid can remove oxygen through the following reaction (Son et al. 1985).

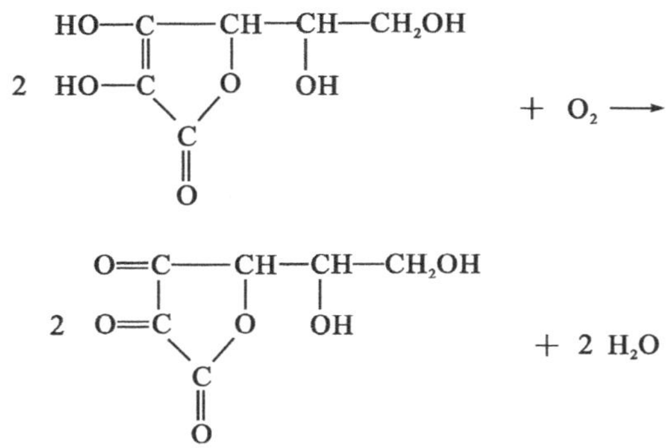

15.3 Flocculation of Suspended Solids in Wastewater

The suspended solids in the wastewater are mainly clay particles, which are mutually exclusive due to the negative charges on their surfaces, making it hard for them to aggregate, settle, and be removed from wastewater.

Flocculation is a common method used to remove suspended solids in wastewater. Chemicals that can form flocs with suspended solids and make them settle quickly are called flocculants.

Flocculants are usually composed of the following two chemicals.

15.3.1 Coagulants

Coagulants refer to the chemical agents which can neutralize the negative charges on the suspended solid surfaces such as hydroxyl aluminum, hydroxyl iron, and hydroxyl zirconium, by providing high-valent multinuclear olation complex ions.

Besides, chemicals such as aluminum trichloride, aluminum sulfate, sodium aluminate, potassium alum, ammonium alum, ferric chloride, zirconium oxychloride can have the same function as hydroxyl aluminum, hydroxyl iron, and hydroxyl zirconium, by forming multinuclear olation complex ions through hydrolysis, complexation, and hydroxyl bridge effects.

15.3.2 Coagulant Aids

Coagulant aids are chemicals which can bridge on the surfaces of suspended solids and make them settle quickly. Water-soluble polymers such as polyacrylamide, partially hydrolyzed polyacrylamide (HPAM), polyethylene glycol, polyvinyl alcohol, carboxymethyl starch, hydroxyethyl starch, carboxymethyl cellulose, hydroxyethyl cellulose, guar gum, carboxymethyl guar gum, hydroxypropyl guar gum, sodium alginate, and xanthan gum all can be used as coagulant aids.

As cationic polymers have functions of both the coagulant and the coagulant aid, they are ideal flocculants (Qu et al. 2006; Lu et al. 2009).

15.4 Scale Inhibition of Wastewater

The inorganic substances precipitated from water with very low solubility are referred as scale. The most common scales in oilfields are calcium carbonate scale, calcium sulfate scale, strontium sulfate scale, and barium sulfate scale. The deposit of scale on the surfaces of pipelines, equipments, and strata are called scale formation. Scale inhibitors are used to alleviate or prevent the scale-forming ions from forming scale. The following are the main scale inhibitors.

15.4.1 Condensed Phosphates

Condensed phosphates are classified as chain-like condensed phosphates (e.g., tripolyphosphate) and annular condensed phosphates (e.g., hexametaphosphate). These scale inhibitors can only be used under 50 °C, otherwise they may partially hydrolyze to produce orthophosphates which may react with divalent metal ions (e.g., Ca^{2+}) to form scale.

15.4.2 Phosphonates

Phosphonates can be obtained by the reaction of multi-amino compounds with formaldehyde and phosphate acid followed by neutralization of alkali.

Phosphonates with the following structures can be applied as scale inhibitors.

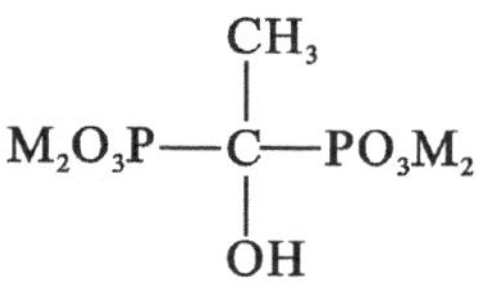

(Hydroxy ethylidene bisphosphonates, HEDP)

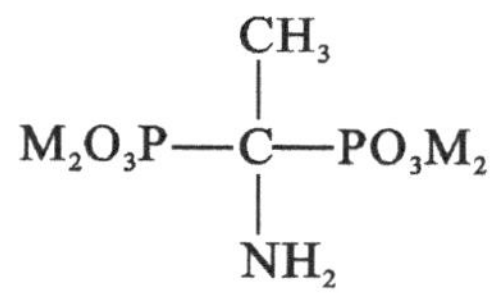

(Amino ehtylidene diphosphonates, AEDP)

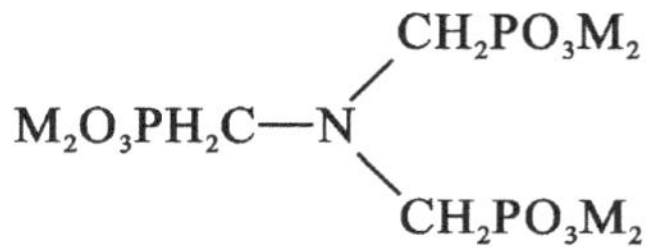

(Amino trimethylene phosphonates, ATMP) (Doyle et al. 1992)

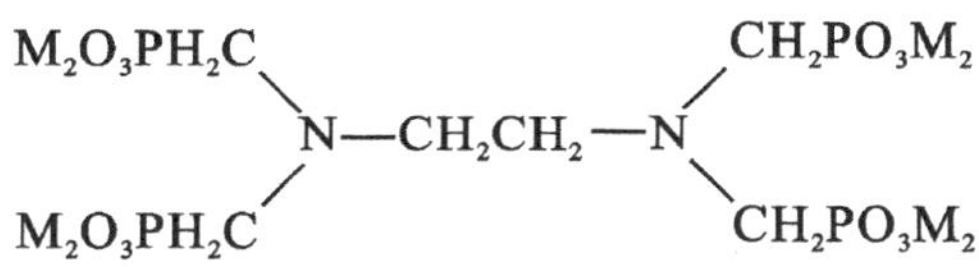

(Ehtylenediamine tetramethylene phosphonates, EDTMP) (Ralston 1973)

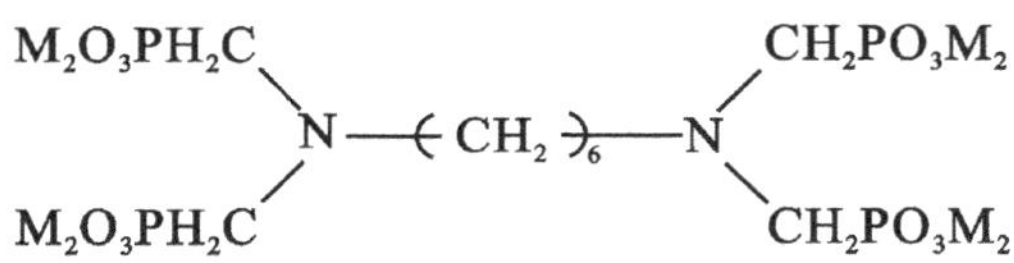

(Hexamethylene diamine tetramethylene phosphonates, HMDTMP) (Gill et al. 1990)

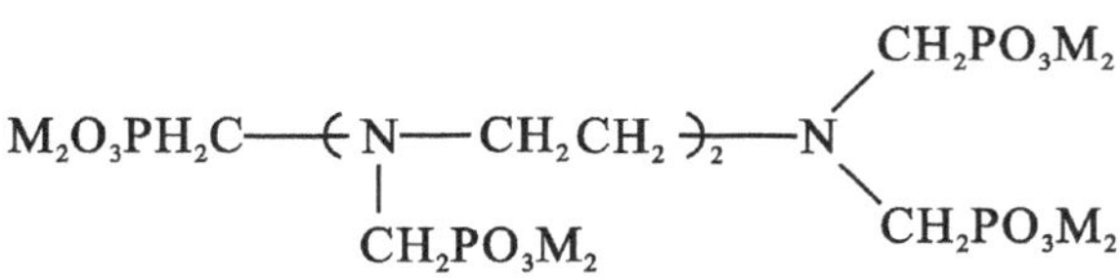

(Diethylene triamine pentamethylene phosphonates) (Mazzolini et al. 1992)

Phosphonates have better thermal stability and lower dosage compared with condensed phosphates, which make them ideal scale inhibitors.

15.4.3 Amino Polycarboxylates

Amino polycarboxylates can be obtained by the reaction of amino compounds and chloroacetate under alkaline condition.

Amino polycarboxylates with the following structures can be applied as scale inhibitors.

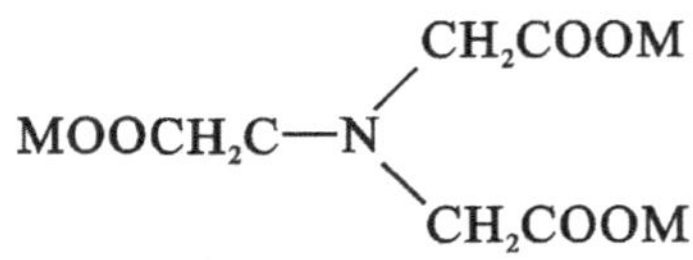

(Nitrilo triacetates, NTA)

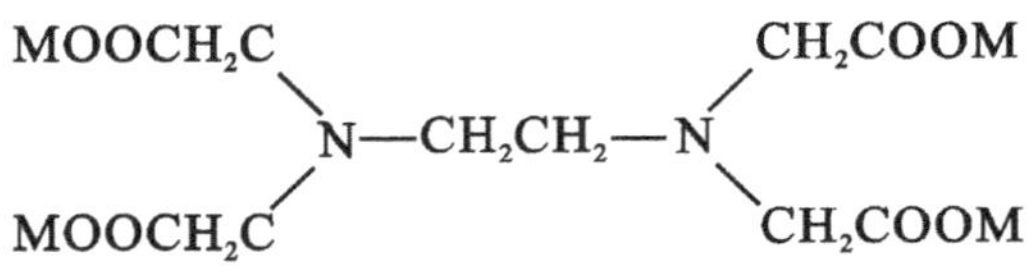

(Ethylendiamine tetraacetates, EDTA)

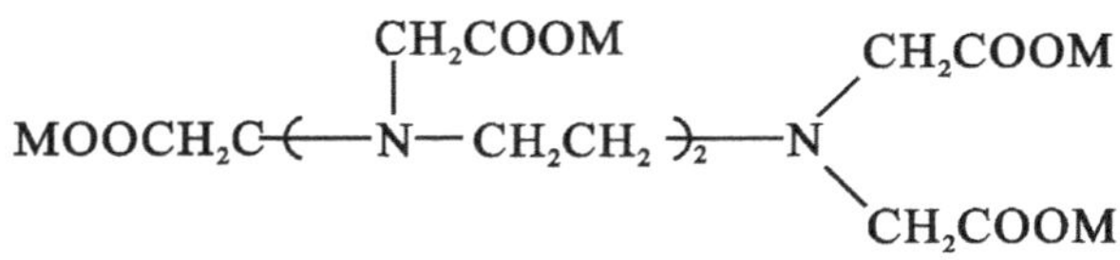

(Diethylenetriamine pentaacetates, DTPA)

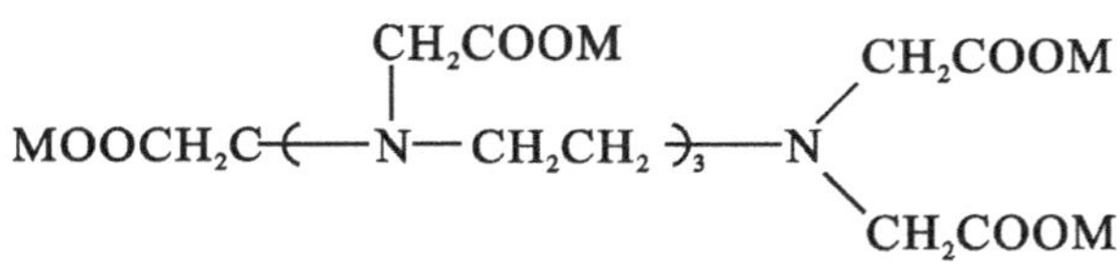

(Trietnylenetetramine hexaacetates, TTHA)

Amino polycarboxylates are thermally stable scale inhibitors.

15.4.4 Surfactants

The following surfactants can be applied as scale inhibitors (Tate et al. 1976; Wang et al. 1997).

$R{-}O{+}CH_2CH_2O{+}_n SO_3M$

(Polyoxyethylene alkyl ether sulfates)

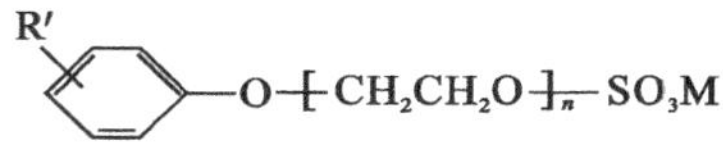

(Polyoxyethylene alkyl phenol ether sulfates)

$R{-}O{+}CH_2CH_2O{+}_n CH_2CH_2SO_3M$

(Polyoxyethylene alkyl ether sulfonates)

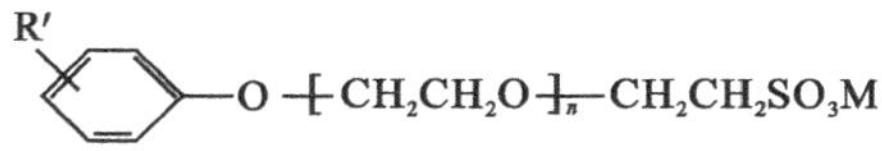

(Polyoxyethylene alkyl phenol ether sulfonates)

$R{-}O{+}CH_2CH_2O{+}_n CH_2COOM$

(Polyoxyethylene alkyl ether carboxylates)

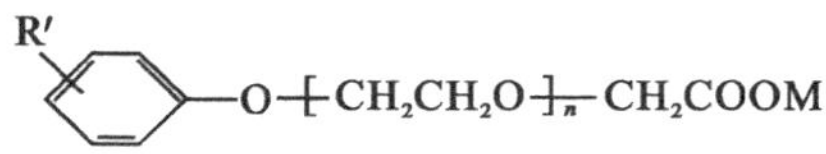

(Polyoxyethylene alkyl phenol ether carboxylates)

In the above chemical structures, R represents C_{12}–C_{18}; R′ represents C_8–C_{14}; $n = 2$–10; M represents Na^+, K^+ or NH_4^+.

Among these surfactants, sulfonate surfactants and carboxylate surfactants have obviously superior thermal stability than sulfate surfactants because they will not hydrolyze like the latter.

15.4.5 Polymers

Low molecular weight (<50,000) homopolymers and copolymers with carboxyl chains can be used as scale inhibitors. Some examples are given below (Xia et al. 2008).

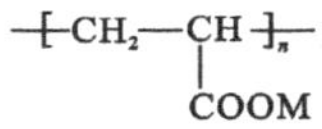

(Polyacrylates)
(Greenfield 1975)

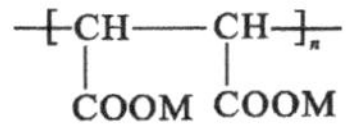

(Polymaleates)

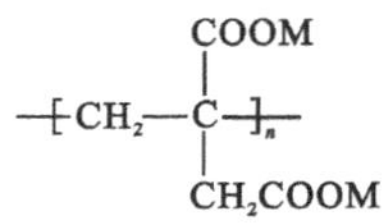

(Polycarbomethyl acrylates)
(Miller et al. 1991)

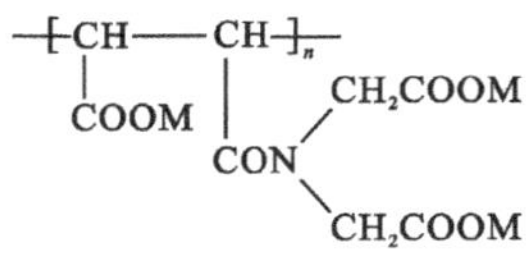

(Polymaleate dicarboxylic dimethylamide)
(Boffardi et al. 1976)

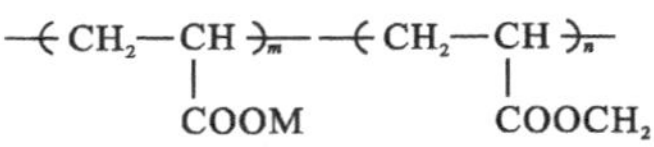

(Copolymers of acrylate and methacrylate)

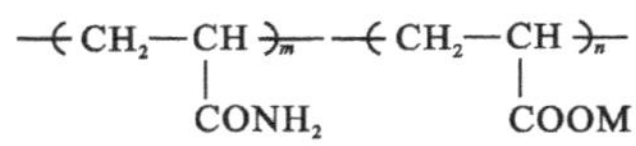

(Copolymers of acrylamide and acrylates) (Song et al. 1977)

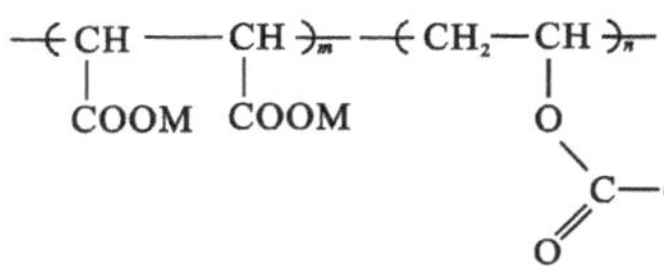

(Copolymers of Maleate and vinyl acetate) (Johnson et al. 1973)

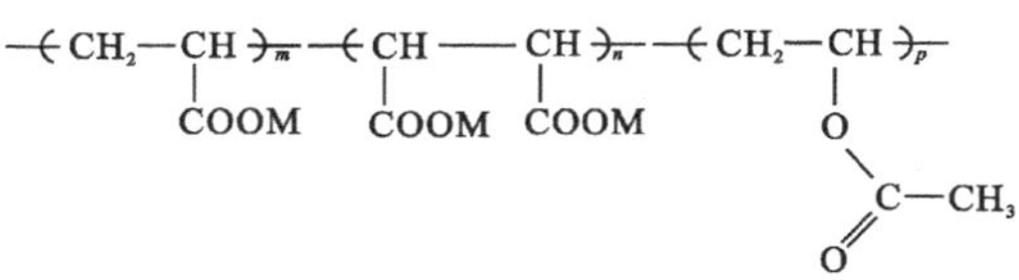

(copolymers of acrylates, maleates and vinyl acetate, AA-MA-VAC)
(Wu et al. 2009)

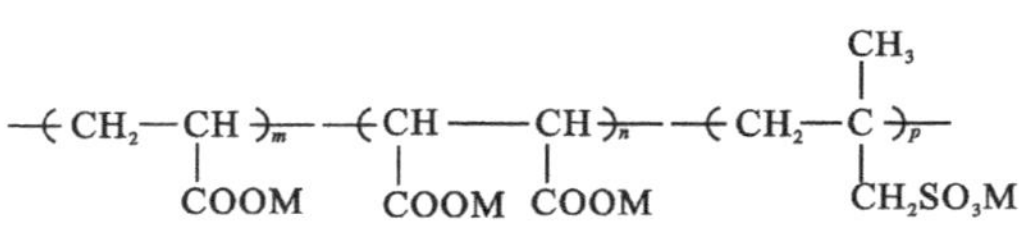

(Copolymers of acrylate, maleate and methallyl sulfonate, AA-MA-MAS)
(Yu et al. 2005)

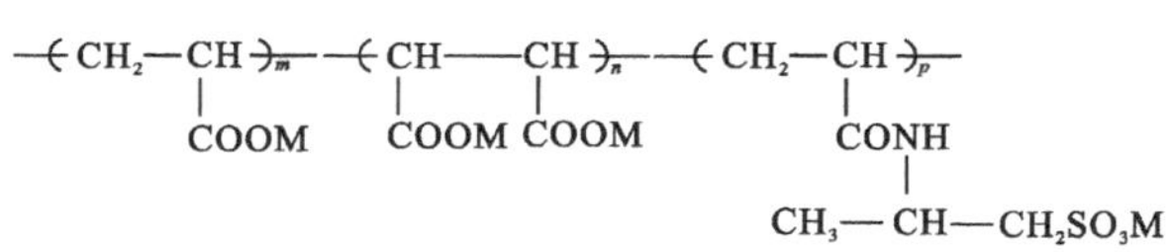

(Copolymers of acrylate, maleate and 2-acrylamido-propyl sulfonate,
AA-MA-APS) (Hann et al. 1998)

The following polymer scale inhibitors receive special attention due to their non-toxic and biodegradable properties.

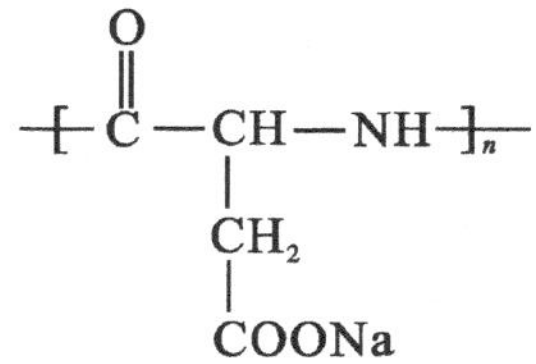

(Sodium polyaspartate) (Shao et al. 2001)

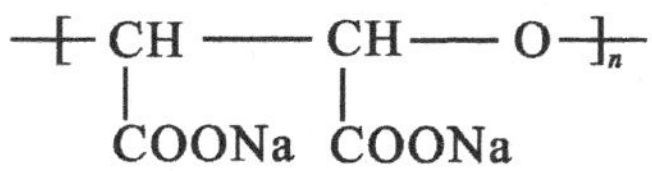

(Polyethylene oxide sodium succinate)
(Liu et al. 2008; Han et al. 2009)

Some polymers without carboxyl links can also be used as scale inhibitors.

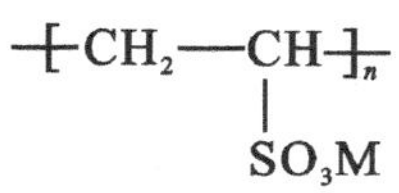

(Polyvinyl sulfonate, PVS)
(Falk et al. 1992)

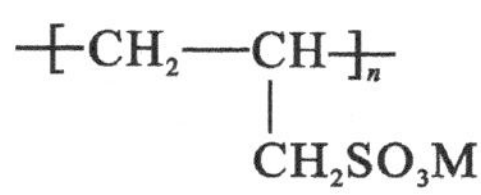

(Polypropylene sulfonate, PPS)

Polymers have advantages of good thermal stability and low dosage, which make them ideal scale inhibitors.

The scale inhibitors mentioned above can inhibit scale formation by two mechanisms related to adsorption. One is the mechanism of lattice distortion. The normal state scale surfaces are distorted by the adsorption of scale inhibitors, which inhibit or partially inhibit the further growth of scale particles, keeping the scale-forming ions in saturated state and making the formed loose scale easy to be swept away by water currents. The other is the mechanism of electrostatic repulsion. Scale inhibitors can adsorb on the scale surfaces to form diffused electric double layers (Fig. 15.2), which make the scale surfaces negatively charged, inhibiting the aggregation of scale particles.

Moreover, the anions generated by the dissociation of some scale inhibitors in water can form stable water-soluble cyclic structures with scale-forming cations by reaction and complexation to inhibit scale formation. The following are some structures formed by inhibitor anions with Ca^{2+} through reaction and complexation.

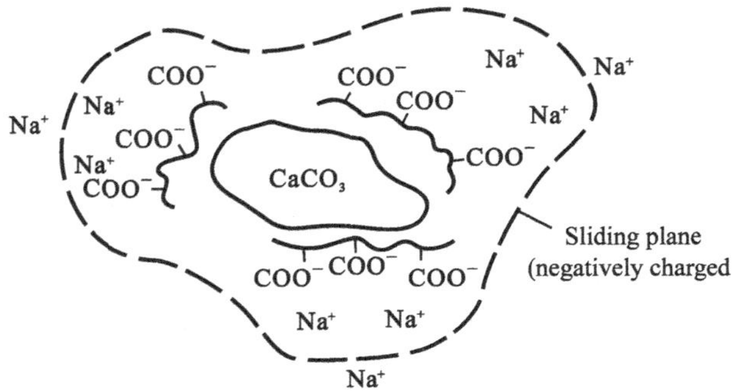

Fig. 15.2 Diffused electric double layers formed by polymers containing carboxylic chains on $CaCO_3$ surface

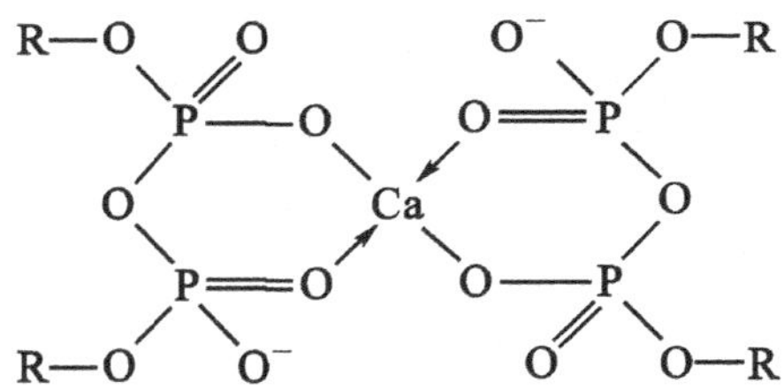

(Pyrophosphates and Ca^{2+})

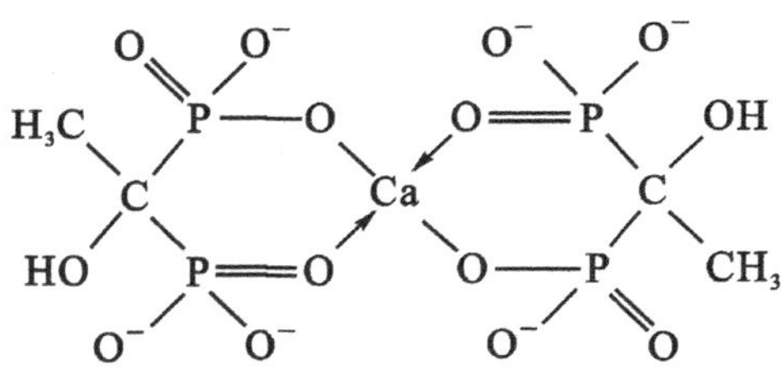

(HEDP and Ca^{2+})

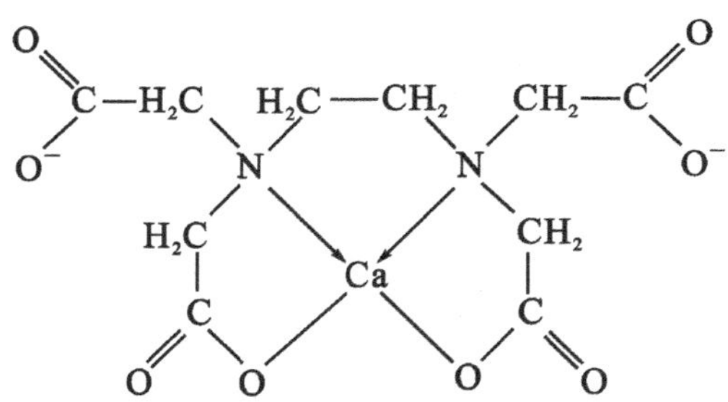

(EDTA and Ca^{2+})

15.5 Corrosion Inhibition of Wastewater

Wastewater is a neutral media with a pH value of 6–8, so neutral media corrosion inhibitors can be applied. According to their function mechanisms, corrosion inhibitors can be divided into the following three types.

15.5.1 Oxidation Film Inhibitor

This type of corrosion inhibitor plays the role of corrosion inhibition by generating dense protective films through oxidation. Because the generated protective films will promote the anodic passivation of metal, this kind of corrosion inhibitor is also called passivation film corrosion inhibitor.

Dichromate (such as $Na_2Cr_2O_7$ and $K_2Cr_2O_7$) is a kind of oxidation film corrosion inhibitor, which produces Cr_2O_3 and Fe_2O_3 through the following oxidizing reactions.

$$Cr_2O_7^{2-} + H_2O \rightarrow 2H^+ + 2CrO_4^{2-}$$
$$CrO_4^{2-} + 3Fe(OH)_2 + 4H_2O \rightarrow Cr(OH)_3 + 3Fe(OH)_3 + 2OH^-$$
$$2Cr(OH)_3 \rightarrow Cr_2O_3 + 3H_2O$$
$$2Fe(OH)_3 \rightarrow Fe_2O_3 + 3H_2O$$

These oxides form a passivation film of iron–ferric oxide–chrome oxide on the steel surface to control the steel corrosion.

Molybdate (e.g., Na_2MoO_4) also belongs to this kind of inhibitor. Under the existence of dissolved oxygen, it can form a passivation film of iron–ferric oxide–molybdenum oxide on the steel surface to control the steel corrosion.

The other chemicals belonging to this kind of inhibitors include nitrite (such as $NaNO_2$, NH_4NO_2), tungstate (e.g., Na_2WO_4), vanadate (e.g., NH_4VO_3), selenate (e.g., Na_2SeO_4), antimonate (e.g., Na_3SbO_4), acetate (e.g., CH_3COONa), benzoate (e.g., 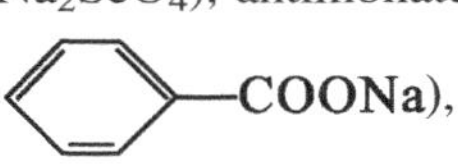

), methyl benzoate (e.g., 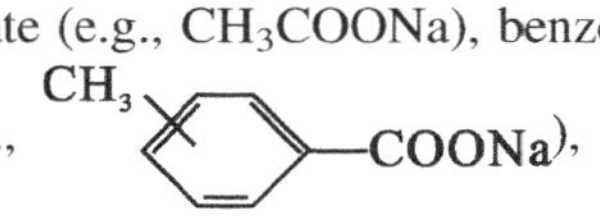

), and salicylate (e.g., 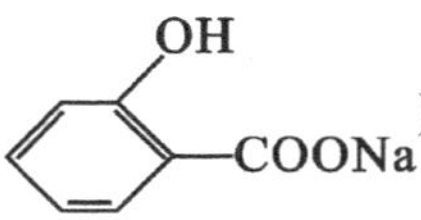

), etc.

15.5.2 Precipitation Film Corrosion Inhibitor

This kind of corrosion inhibitor prevents corrosion by forming a precipitation film on the anode or cathode surface of corrosion cell.

Sodium silicate can react with corrosion product Fe^{2+} on the anode surface, producing iron silicate precipitation film to inhibit corrosion.

$$Fe^{2+} + Na_2O \cdot mSiO_2 \rightarrow FeO \cdot mSiO_2 + 2Na^+$$

Zinc sulfate can react with OH^- produced by cell reaction on the cathode surface, forming a zinc hydroxide precipitation film to inhibit corrosion.

$$Zn^{2+} + 2OH^- \rightarrow Zn(OH)_2 \downarrow$$

In addition, sodium hydroxide, sodium carbonate, sodium dihydrogen phosphate, disodium hydrogen phosphate, trisodium phosphate, sodium hexametaphosphate, sodium tripolyphosphate, sodium gluconate, sodium amino trimethylene phosphonate (ATMP), sodium ethylenediamine tetramethylene phosphonate (EDTMP), sodium 1-Hydroxyethylidene-1,1-diphosphonate (HEDP) belong to this kind of inhibitor.

Figure 15.3 is a relational graph of steel corrosion rate versus concentration of some oxide film inhibitors and precipitation film inhibitors. It shows that when concentrations of these corrosion inhibitors surpass certain value, they all have corrosion inhibition function; however, when the concentrations are insufficient, some inhibitors even promote corrosion, which is due to the localized corrosion on the cathode surface. Considering the important role of a complete film (including oxidization film and precipitation film) in corrosion inhibition, it is suggested that corrosion inhibitor with 10–30 times of normal application concentration is used to pre-treat the metal surface, and then the concentration of corrosion inhibitor is restored to the normal level.

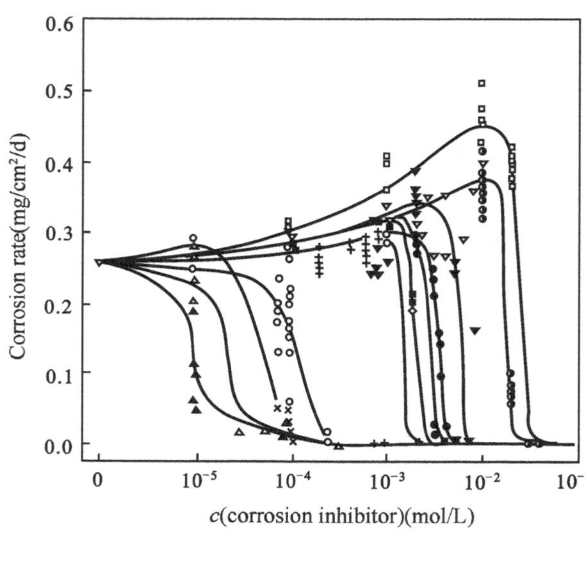

Fig. 15.3 Relationship of steel corrosion rate versus concentration of some neutral medium corrosion inhibitor (Cohen 1979). □—$1/2Na_2Co_3$; △—$NaNO_2$; ▽—NaOH; ×—$1/2Na_2WO_4$; ◑—$1/3Na_2HPO_4$; ▲—$1/2Na_2MoO_4$; ▼—$1/3Na_3PO_4$; ■—C_6H_5COONa; +—CH_3COONa; ●—$1/2Na_2SiO_3$; ○—$1/2Na_2CrO_4$

15.5.3 Adsorption Film Corrosion Inhibitor

This kind of corrosion inhibitor prevents corrosion by forming an adsorption film on the anode and cathode surface of corrosion cell. The following are some adsorption film corrosion inhibitors.

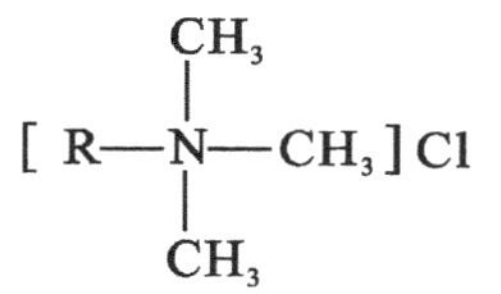

(Alkyltrimethyl ammonium chloride, R: C_{12}–C_{18})

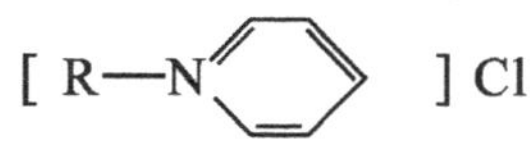

(Alkyl pyridinium chloride, R: C_{12}–C_{18})

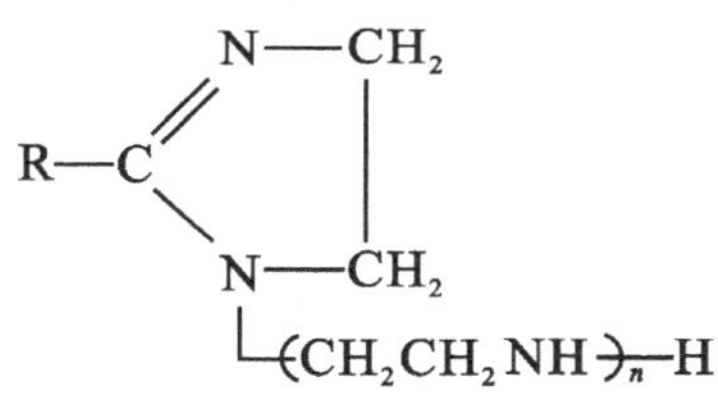

(1-polyaminoethyl-2-alkylimidazoline, R: C_{11}–C_{17})

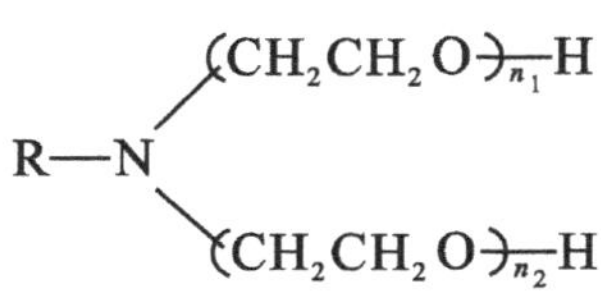

(Polyoxyethylene alkylamine, R: C_{12}–C_{18}; n_1+n_2: 5–50)

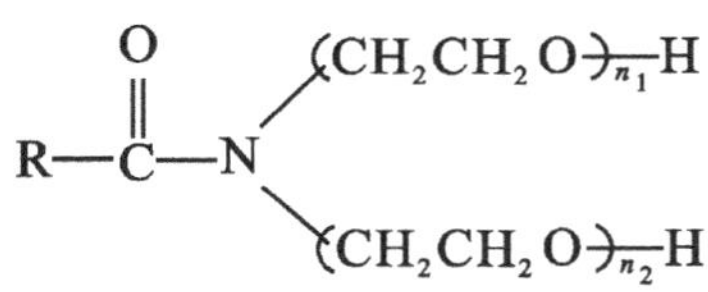

(Polyoxyethylene amide, R: C_{11}–C_{17}; n_1+n_2: 5–50)

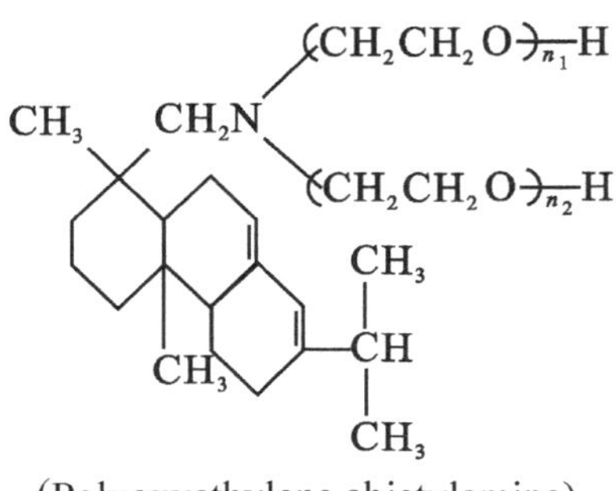

(Polyoxyethylene abietylamine)

The oxygen remover can inhibit corrosion by removing the dissolved oxygen from water, and bactericide can inhibit corrosion by suppressing the reproduction of sulfate-reducing bacteria; hence in some sense, these chemicals can also be taken as corrosion inhibitors.

Normally, the wastewater corrosion inhibitors are used in mixtures, such as the mixture of phosphate and zinc salt, the mixture of dichromate and zinc salt, the mixture of molybdate and zinc salt, the mixture of molybdate and organic phosphonate, the mixture of dichromate and polyoxyethylene rosin amine, the mixture of thiocarbamide and polyoxyethylene rosin amine, the mixture of ammonium thiocyanate and cinnamic aldehyde, and the mixture of molybdate (Henson et al. 1990), zinc salt, and gluconic salt.

15.6 Sterilization of Wastewater

The bacteria in wastewater are mostly sulfate-reducing bacteria (SRB), iron bacteria, and saprophytic bacteria, which can cause metal corrosion, stratum jamming, and chemical deterioration; therefore, sterilization is needed (Ning 1998; Jin et al. 2008; Zhao et al. 2008). Bactericides are chemicals that can sterilize bacteria with small dosage.

There are two kinds of bactericides.

15.6.1 Oxidizing Bactericide

This kind of bactericide kills bacteria through oxidation.

Chlorine is a typical oxidizing bactericide, which produces hypochlorous acid in water.

$$Cl_2 + H_2O \rightarrow HClO + HCl$$

Hypochlorous acid is not stable and can decompose.

$$HClO \rightarrow [O] + HCl$$

The atomic oxygen produced by such reactions can kill bacteria by oxidation.

In addition, the bactericides listed below can also kill bacteria by oxidation.

O_3	ClO_2	$NaClO$
(Ozone)	(Chlorine dioxide)	(Sodium hypochlorite) (He et al. 2008)

$Ca(ClO)_2$	K_2FeO_4	$KMnO_4$
(Calcium hypochlorite)	(Potassium ferrate) (Han et al. 2008)	(Potassium permanganate)

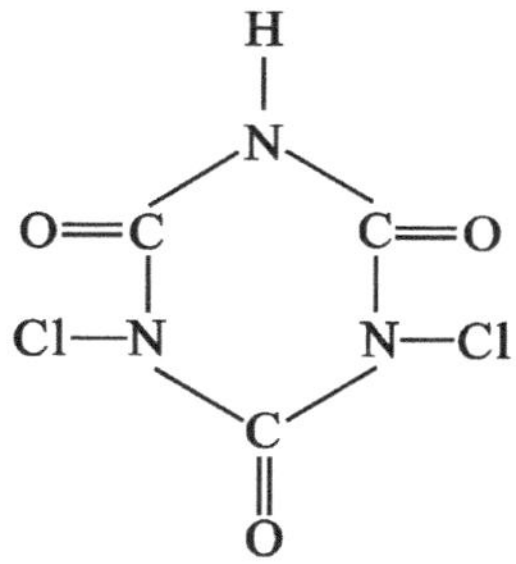

(Dichloroisocyanic acid)

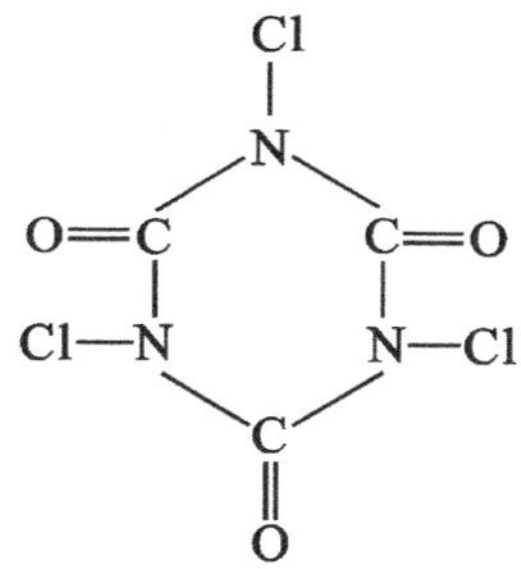

(Trichloroisocyanuric acid)

They can produce atomic oxygens in water for sterilization.

$$O_3 \rightarrow O_2 + [O]$$
$$2ClO_2 + H_2O \rightarrow 2HCl + 5[O]$$
$$NaClO \rightarrow NaCl + [O]$$
$$Ca(ClO)_2 \rightarrow CaCl_2 + 2[O]$$
$$K_2FeO_4 + 4H^+ \rightarrow 2H_2O + 2K^+ + Fe^{2+} + 2[O]$$
$$2KMnO_4 + 6H^+ \rightarrow 3H_2O + 2K^+ + 2Mn^{2+} + 5[O]$$

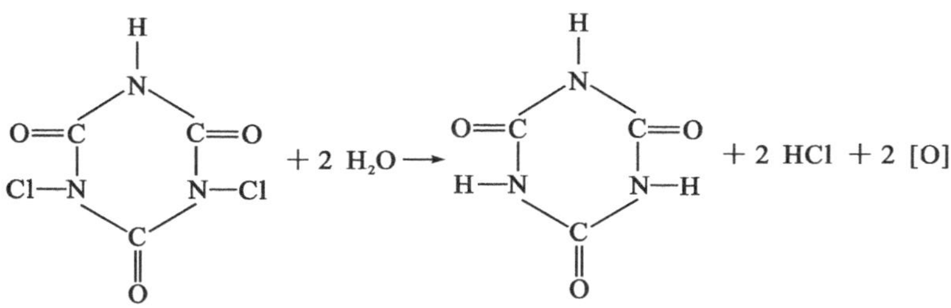

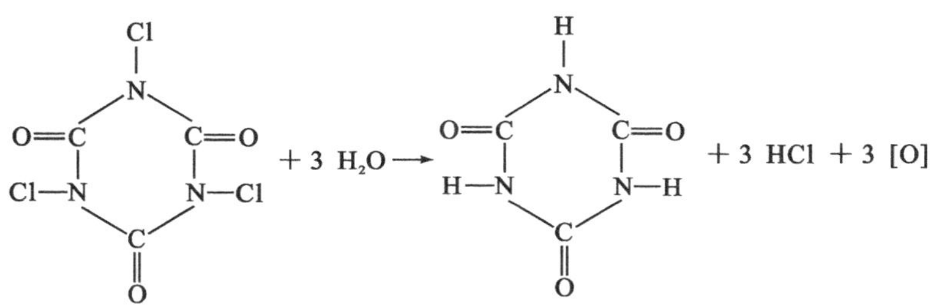

15.6.2 Non-oxidizing Bactericides

This type of bactericide can be separated into adsorption-type bactericide and infiltration-type bactericide by their mechanisms.

Adsorption-type bactericide works by adsorbing on the surfaces of bacteria to inhibit their normal metabolism. Because the bacteria surfaces normally carry negative charges, quaternary ammonium compounds are very effective adsorption-type bactericides. The following are some examples.

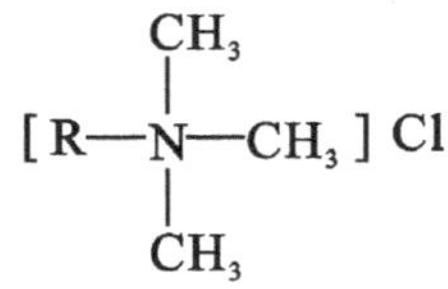

(Alkyltrimethyl ammonium chloride, R: C_8-C_{18}; preferably C_{12}-C_{16})

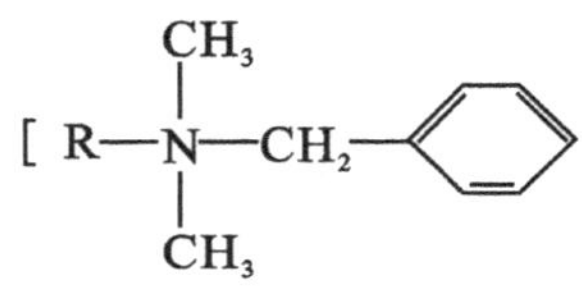

(Alkylbenzyl dimethyl ammonium chloride, R: C_8-C_{18})

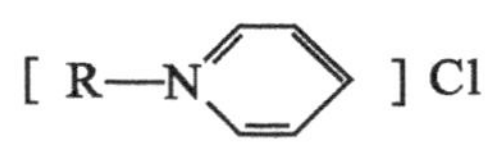

(Alkyl chloride pyridine, R: C_8-C_{18})

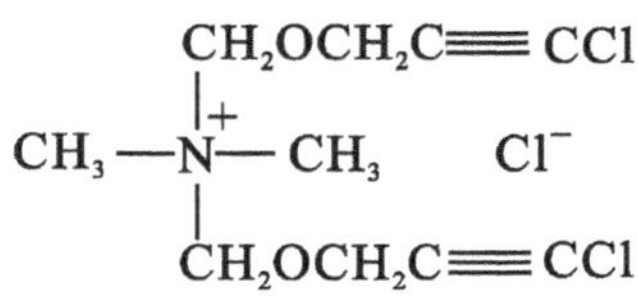

(Dimethyl dimethylene chlorinated acetylene propyl ether ammonium chloride) (Quinlan et al. 1980)

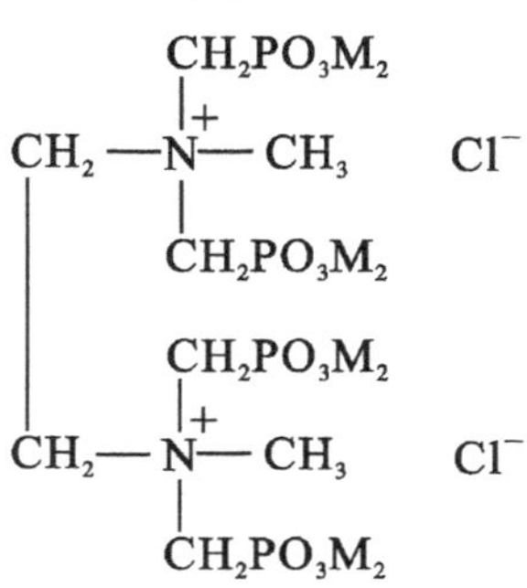

(Reaction product of ethylenediaminse tetramethyl phosphate and chloromethane) (Quinlan 1978)

Among these bactericides, some have multiple quaternary ammonium nitrogens and some are surfactants, therefore strengthening their using effects.

In addition, some bactericides contain structures which can inhibit corrosion, prevent scaling, and stabilize clay, making them bear simultaneously properties of corrosion inhibition, anti-scaling, and clay stabilizing.

Infiltration-type bactericides can infiltrate into the cytoplasm of bacteria, disrupting the bioenzymes inside the bacteria cells, and as a result, killing the bacteria. Here are some infiltration-type bactericides.

$$CH_2O$$

(Formaldehyde)

$$OHC{-}\!\left(CH_2\right)_3\!{-}CHO$$

(Glutaraldehyde)
(Xu et al. 1993; Wrench 1991)

$$CH_3{-}CH{=}CH{-}CN$$

(2-butenenitrile) (Dria et al. 1985)

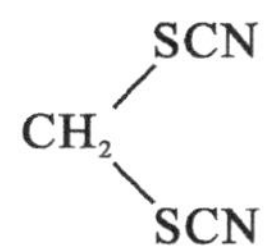

(Methane dithiocyanate) (Beck et al. 1985)

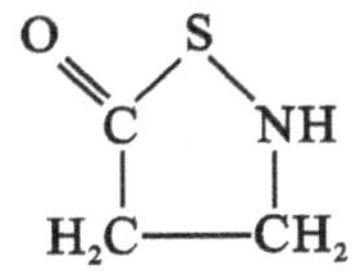

(Isothiazolidone)

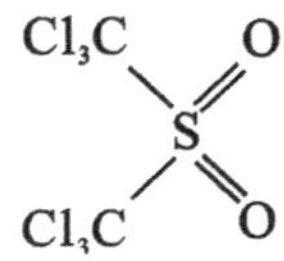

(Hexachloro-dimethyl sulfone)

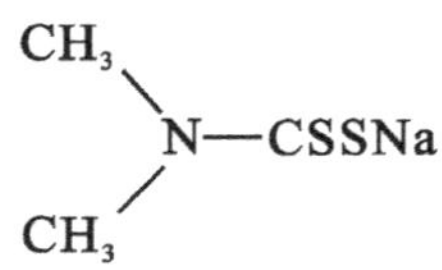

(Dimethylamino sodium dithiocarbamate)

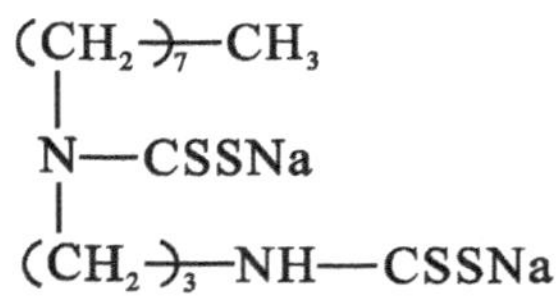

[*N*-octyl-1,3- propylidene-bis(sodium dithiocarbamate)]

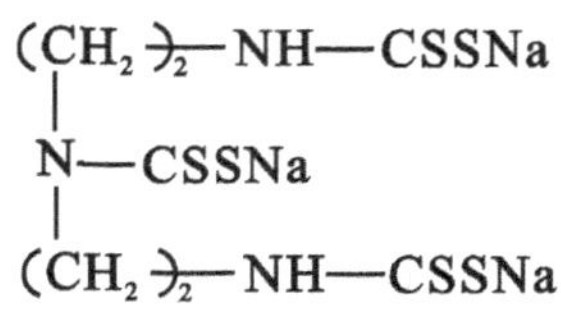

[Bis(1,2-ethylidene)-tris(sodium dithiocarbamate)]

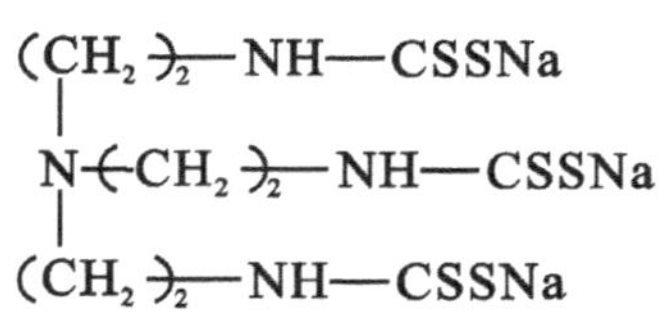

[1,7-heptyliene-4-methylene tris(sodium dithiocarbamate)]

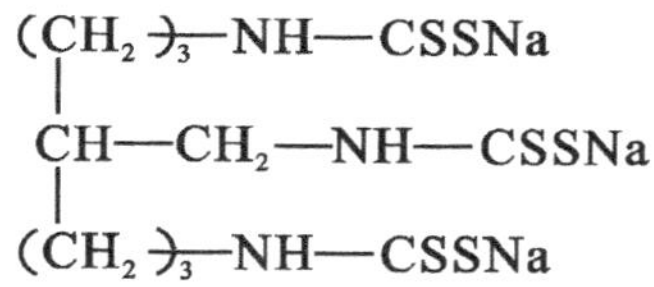

[Nitrilo-tris(1,2-ethylidene)-tis(sodium dithiocarbamate)]

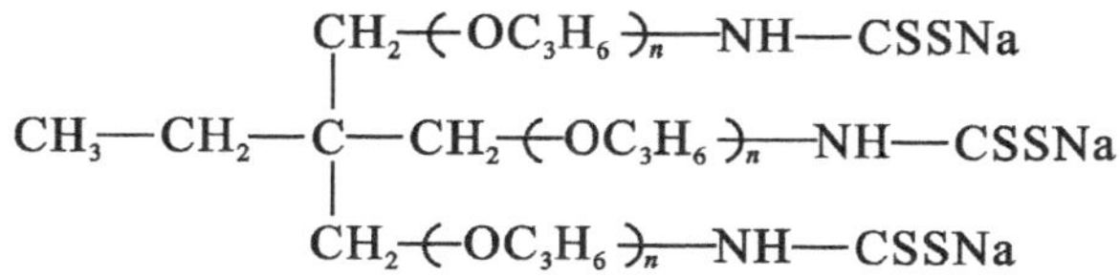

[Polyoxypropylene-2,2-bis(hydroxymethyl)-*n*-butyl ether tris (sodium dithiocarbamate)]

In most cases, bactericides are used as complexes. Complex bactericides are more effective than unitary bactericide.

The bactericides must be used alternately, as the application of single type of bactericide for a long time will make bacteria develop drug resistance, which will significantly reduce the bactericidal effect.

The starting concentration of bactericide must be high, while when the amount of bacteria is under control, the bactericide can be used in lower concentration. In this way, the reproduction of bacteria can be efficiently controlled.

Bactericides can be added continuously or intermittently. When added continuously, the mass concentration of bactericide is normally in the range of 10–50 mg/L, while the mass concentration of bactericide is normally in the range of 100 and 200 mg/L (Hudgins 1992) when added intermittently.

References

Beck D, Miller JW, Wernau WC et al (1985) Biopolymer composition having enhanced filterability. EP Patent 145,217, 19 June 1985

Boffardi BP, Cook MM, Ralston PH (1976) Scale inhibition and corrosion inhibition. US Patent 3,965,027, 22 June 1976

Cohen M (1979) Dissolution of iron. In: Brubaker GR, Phipps PBP (eds) Corrosion inhibitors, ACS symposium, ACS, pp 126–152

Doyle MJ, Ostovar P, Walker PAM (1992) Method for inhibiting scale formation. GB Patent 2,248,831, 22 Apr 1992

Dria DE, Bremer NJ, Dria MA et al (1985) Nitrile compounds as oil field biocides. US Patent 4,507,212, 26 Mar 1985

Durham DK, Conkle UG, Downs HH (1989) Additive for clarifying aqueous systems. GB Patent 2,219,291, 6 Dec 1989

Falk DO, Dormish FL, Beazley PM et al (1992) Polyvinyl sulfonate scale inhibitor. US Patent 5,092,404, 3 Mar 1992

Gill ST, Rey SP, Wiernik JH (1990) Methods for controlling silica/silicate deposition in aqueous systems using phosphonates and carboxylic/sulfonic polymers. US Patent 4,933,090, 12 June 1990
Greenfield GL (1975) Process of inhibiting scale using an acrylic composition. US Patent 3,904,522, 9 Sept 1975
Han Z, Sun A, Li Q et al (2009) Polyepoxysuccinic acid for scale inhibition in gathering system at Weizhou-12-1 offshore oil field. Oilfield Chem 26(2):158–160
Han L, Wang B, Chen Y et al (2008) Synthesis of potassium ferrate (VI) as sterilization agent for oilfield sewage treatment. Oil-Gasfield Surf Eng 27(10):33–34
Hann WM, Robertson ST, Weinstein B (1998) Method for metal sulfate scale control in harsh oilfield conditions. US Patent 5,755,972, 26 May 1998
He X, Jin H, Hu Z et al (2008) Applied research on electrolytic brine sterilization technique in oilfield sewage treatment. Petrol Eng Constr 34(6):4–6
Henson ER, Doty PA (1990) Corrosion inhibitors for aqueous brines. US Patent 4,980,074, 25 Dec 1990
Hudgins CM Jr (1992) Chemical treatments and usage in offshore oil and gas production systems. JPT 44(5):604–611
Jin H, Ding J, Yuan R et al (2008) Research on novel sewage sterilization techniques. Oil-Gas field Surf Eng 27(9):16–17
Johnson D, Mizuno WG (1973) Treatment of water used in heat transfer equipment. US Patent 3,715,307, 6 Feb 1973
Knight BL (1973) Reservoir stability of polymer solutions. JPT 25(5):618–625
Liu J, Huang Y (2008) Synthesis of green scale inhibitor polyepo-xysuccinicacid. Technol & Dev Chem Ind 37(11):17–19, 32
Lu J, Peng B, Li M et al (2009) Synthesis and performance evaluation of cationic flocculants modified from starch. J China Univ Petrol (Ed Nat Sci) 33(1):127–130, 135
Mazzolini EI, Bertero L, Truefitt CS (1992) Scale prediction and laboratory evaluation of $BaSO_4$ scale inhibitors for seawater flood in a high-barium environment. SPE Prod Eng 7(2):186–192
Miller JW, O'Neil JJ (1991) Structure/performance relationships for barium sulfate and strontium sulfate antiscalants. SPE 23608
Ning T (1998) Bactericides for injection waters: developments and uses in Shengli oil field. Oilfield Chem 15(3):285–288
Qu C, Wang X, Xie J et al (2006) Using cationic polymer for treating the produced water of oilfields. J Xi'an Shiyou Univ 21(20):23–25
Quinlan PM, Groves W (1980) Microbiocidal quaternaries of halogen derivatives of alkynoxymethyl amines. US Patent 4,206,233, 3 June 1980
Quinlan PM (1978) Use of polyquaternary ammonium methylene phosphonates as microbiocides. US Patent 4,084,950, 18 Apr 1978
Ralston PH (1973) Phosphonates—a chemical answer to oilfield water problems. SPE 4350
Rivers GT (1992) Water clarifier. US Patent 5,152,927, 6 Oct 1992
Schremp FW, Chittum JF, Arczynski TS (1961) Use of oxygen scavengers to control external corrosion of oil-string casing. JPT 13(7):703–711
Shao H, Leng Y (2001) A study on scale inhibition properties of polyaspartate. Oilfield Chem 18(2):181–183
Sivakumar A, Ramesh M (1996) Demulsification of oily waste waters using silicon containing polymer. US Patent 5,560,832, 1 Oct 1996
Son AJ, Kuzlik MS (1985) Corrosion inhibitor for heavy brines. US Patent 4,539,122, 3 Sept 1985
Song DS, Duffy RJ, Witschonke CR et al (1977) Low molecular weight hydrolyzed polyacrylamide and use thereof as scale inhibitor in water systems. US Patent 4,001,161, 4 Jan 1977
Tate JF, Shupe RD, Maddox J Jr (1976) Composition and method for treating scale. US Patent 3,975,282, 17 Aug 1976
Thompson NES, Asperger RG (1990) Methods for treating hydrocarbon recovery operations and industrial waters. EP Patent 349,681, 10 Jan 1990

Wang Y, Yang H (1997) Study of scale control by nonionic-anio-nicamprolytic surfactants. J Univ Petrol China (Ed Nat Sci) 21(4):109–111

Wrench E (1991) Anti-microbial agent. GB Patent 2,244,216, 27 Nov 1991

Wu R, Zhou L, Li S et al (2009) Preparation and performance evaluation of $BaSO_4$ scale-inhibitor. Fault-Block Oil & Gas Field 16(1):111–112, 120

Xia M, Wang F, Lei W (2008) Scale inhibiting effect and mechanism of carboxylic polymers on sulfate sediments. Acta Petrolei Sin (Petrol Process Sect) 24(4):460–464

Xu L, Liu J, Tang H et al (1993) Bactericidal power of aldehydic compounds for sulfate-reducing bacteria. Oilfield Chem 10(3):260–263

Yu M, Liu M, Huang J (2005) Preparation of the copolymer of MA-AA-MAS and study on its scale inhibiting performance. J Xi'an Shiyou Univ (Nat Sci Ed) 20(6):56–58

Zhao X, Chen X, Huo S et al (2008) Experiments and application of oilfield sewage sterilization techniques. Petrol Plan Eng 19(6):31–34

Printed by Printforce, the Netherlands